Financial Mathematics for Actuarial Science

The Theory of Interest

Financial Mathematics for Actuarial Science

The Theory of Interest

Richard James Wilders
North Central College

CRC Press
Taylor & Francis Group
Boca Raton London New York

CRC Press is an imprint of the
Taylor & Francis Group, an **informa** business

CRC Press
Taylor & Francis Group
6000 Broken Sound Parkway NW, Suite 300
Boca Raton, FL 33487-2742

Printed on acid-free paper

International Standard Book Number-13: 978-0-367-25308-0 (Hardback)

Visit the Taylor & Francis Web site at
http://www.taylorandfrancis.com

and the CRC Press Web site at
http://www.crcpress.com

To my life partner, Kathy Wilders

Contents

Preface   xi

Author   xiii

1 Overview and Mathematical Prerequisites   **1**
1.1 Introduction 1
1.2 Calculators and Computers 1
    1.2.1 Sequences and Series 6
1.3 Approximation Techniques 9
    1.3.1 Newton's Method (Also Called the Newton-Raphson
        Method) 10
    1.3.2 Approximations Using Taylor Series 11
1.4 Exponents and Logarithms 14
Exercises for Chapter 1 16

2 Measuring Interest   **19**
2.1 Introduction 19
2.2 The Accumulation and Amount Functions 19
2.3 The Effective Rate of Interest 22
2.4 Simple Interest 24
2.5 Compound Interest 27
2.6 Other Accumulation Functions 32
2.7 Present and Future Value: Equations of Value 36
2.8 Nominal and Effective Rates of Interest 40
2.9 Discount Rates 48
    2.9.1 Nominal Rate of Discount 51
2.10 Forces of Interest and Discount 54
    2.10.1 Varying Rates of Interest 57
Exercises for Chapter 2 61

3 Solving Problems in Interest   **67**
3.1 Introduction 67
3.2 Measuring Time Periods: Simple Interest
    and Fractional Time Periods 67
3.3 Fractional Time Periods 69
3.4 Equations of Value at any Time 70

3.5   Unknown Time   74
3.6   Doubling Time   79
      3.6.1   The Rule of 72   82
3.7   Finding the Rate of Interest   82
3.8   Deposits and Withdrawals: Cash Flow Problems   86
Exercises for Chapter 3   92

4   Annuities   **97**
4.1   Introduction   97
4.2   Fixed Term Annuities-Immediate with Constant Payments .   98
4.3   Fixed Term Annuities-Due 108
4.4   The Value of an Annuity at any Date 114
      4.4.1   The Value of an Annuity prior to its Inception 114
      4.4.2   The Value of an Annuity after the Final Payment
              Is Made 116
      4.4.3   The Value of an Annuity at any Time between the
              First and Last Payments 117
4.5   Perpetuities 119
4.6   Non Integer Time Periods 122
4.7   Unknown Time 123
4.8   Unknown Rate of Interest 124
4.9   Varying Rates of Interest 125
      4.9.1   Present Value of Annuities with Varying Interest
              Rates 126
      4.9.2   Accumulated Value of Annuities under Varying
              Interest Rates 126
4.10 Annuities Payable at Different Frequencies than Interest Is
      Convertible 129
4.11 Alternative Method: Annuities Payable Less
      Frequently than Interest Is Convertible 131
4.12 Annuities Paid More Frequently than Interest Is Converted . 135
4.13 Perpetuities Paid More Frequently than Interest Is Convertible 138
4.14 Annuities with Varying Payments 139
4.15 Varying Perpetuities 142
4.16 Varying Annuities Paid Less Frequently than Interest Is
      Convertible 146
4.17 Continuous Annuities 146
Exercises for Chapter 4 148

5   Amortization Schedules and Sinking Funds   **153**
5.1   Introduction 153
5.2   Amortization Method 153
5.3   Amortization of a Loan 156
5.4   Methods for Computing the Loan Balance 157

5.5   Allocation of Loan Payments between Principal and Interest   161

5.6   Formulas for the Balance, Amount to Interest, and Amount to Principal at any Time   165

5.7   Examples Using the TI BA II Plus to Create Amortization Tables   170

5.8   Creating Annualized Amortization Tables Using Excel   176

5.9   How to Construct the Sinking Fund Schedule in Excel   183

5.10 Amortization with Non-Standard Payments   186

5.11 Truth in Lending   189

5.12 Real Estate Loans – Home Loans   191

5.13 Balloon and Drop Payments   195

5.14 Problems Involving Loans   196

5.15 Important Concepts from this Chapter   198

Exercises for Chapter 5   198

6 Yield Rates   207

6.1   Discounted Cash Flow Analysis   207

     6.1.1   Using the TI BA II Plus Cash Flow Worksheet to Compute Internal Rate of Return   211

6.2   Uniqueness of the Yield Rate   213

6.3   Reinvestment   218

6.4   Interest Measurement of a Fund   220

     6.4.1   Dollar-Weighted Estimate for the IRR   223

     6.4.2   Time-Weighted Estimate for the IRR   226

6.5   Selling Loans   231

6.6   Investment Year and Portfolio Methods   233

Exercises for Chapter 6   236

7 Bonds   241

7.1   Introduction   241

7.2   Basic Bond Terminology   244

7.3   Pricing a Bond   247

7.4   Premium and Discount   252

7.5   Determination of Yield Rates   260

7.6   The Term Structure of Interest   262

7.7   Forward Rates   271

7.8   Converting from Forward Rates to Spot Rates   273

7.9   Price of a Bond between Coupon Payments   279

7.10 Callable Bonds   280

7.11 Bonds with Varying Payments   283

Exercises for Chapter 7   285

8 Exact Asset Matching and Swaps   289
   8.1   Exact Asset Matching   289
   8.2   Swap Rates   294
   8.3   Interest Rate Swaps   297
   8.4   Deferred Interest Rate Swaps   298
   8.5   Varying Notional Amounts   300
   8.6   The Market Value of an Interest Rate Swap 302
   Exercises for Chapter 8 303

9 Interest Rate Sensitivity   311
   9.1   Introduction   311
   9.2   The Price Curve: Approximations Using Tangent Lines and
         Taylor Polynomials   313
   9.3   Measuring Sensitivity to Interest Rate Fluctuation: Duration 318
   9.4   Macaulay Duration $(D(i,\infty)) = D_{\text{mac}}(i)$ 320
         9.4.1   Duration of a Coupon Bond for which $F = P = C$ . . 323
         9.4.2   Duration of an Amortized Loan 325
   9.5   Duration of a Portfolio 326
         9.5.1   Duration of a Portfolio of Instruments
                 of Known Duration 328
   9.6   Convexity   328
         9.6.1   Convexity of a Portfolio of Instruments with Known
                 Convexities 330
   9.7   Immunization   330
   9.8   Full Immunization   334
   Exercises for Chapter 9 335

10 Determinants of Interest Rates   339
   10.1 Introduction   339
   10.2 Equilibrium Interest Rates   339
   10.3 T-Bills 340
   10.4 Effective and Continously Compounded Rates 342
   10.5 Interest Rates Assuming No Inflation or Risk
         of Default   343
   10.6 Interest Rates Assuming a Default Risk but
         No Inflation 344
   10.7 Inflation 347
   Exercises for Chapter 10   348
   Appendix: Basic Setup for the TI BA II Plus   351

Answers To Odd-Numbered Exercises   353

Index   373

Preface

To the Student:

This book is designed to provide the mathematical background needed to prepare successfully for the SOA Financial Mathematics (FM) Exam. It is assumed that you have had at least a semester of calculus and have solid problem-solving ability.

Many problems require the use of a specialized calculator. You should purchase both the TI BA II Plus and the TI-30XS MultiView calculators. These are the most useful of the small number of calculators permitted in the exam room. Assistance in using these calculators is a key feature of this book.

Virtually every problem you will encounter in this book is a "word" problem – that is, a paragraph which you will need to decipher. In almost all cases, your answer will be a number: an interest rate, a price, a required payment, etc. It will be crucial that you pay attention to units when crafting your answer. If the problem asks for time in months, make sure your answer is stated in months.

There are numerous technical terms with specialized meanings which you will need to come to grips with. Make a list of these words and their meanings, as well as one or more specific problems in which they appear.

As is the case with any mathematics text, the example problems in the body of each section are a key part of the learning process. You should work through them as you read the text. You should also work a large portion of the exercises at the end of each section. On the other hand, no book can contain all the variations on a particular scenario. If you intend to sit for the FM Exam, you should plan on spending up to 300 hours reviewing and working problems after completing this book. Several companies provide review materials as well as sample exams. Consult with your professor as to what might be best for you.

There are a large number of formulas in this book. You should memorize most of them. While you can solve many problems using the calculator (thus avoiding the formula), there are always a few problems on the FM Exam which require that you know the formula involved.

As with any text, there are undoubtedly errors, for which I apologize. Your assistance in correcting them for future students would be appreciated. You can contact me with questions or corrections at rjwilders@noctrl.edu

Best wishes on your journey toward a career as an actuary!

Richard James Wilders, Ph.D.

North Central College

Author

Richard James Wilders is Marie and Bernice Gantzert Professor in the Liberal Arts and Sciences and Professor of Mathematics at North Central College in Naperville, IL. He earned his B.S. from Carnegie Mellon University and his M.S. and Ph.D. from The Ohio State University.

He has taught the Financial Mathematics course at North Central since creating it some 20 years ago and wrote this book based on his experiences with North Central students. Draft copies of the present work have been used by 100 or so students over the past few years, many of whom have made useful suggestions. In addition to financial mathematics, Dr. Wilders teaches calculus, statistics, finite mathematics, abstract algebra, and the history of mathematics and of science. He is a member of the Mathematical Association of America (MAA) and of the National Council of Teachers of Mathematics (NCTM).

He and his wife enjoy the Chicago theater and dance scene and are avid Ohio State football fans.

Chapter 1

Overview and Mathematical Prerequisites

1.1 Introduction

This book is concerned with the **measurement of interest** and the various ways interest affects what is often called the **time value of money (TVM)**. Interest is usually defined as the compensation that a borrower pays to a lender for the use of capital. It can be thought of as a form of rent. While interest can be paid in any units we will assume in all cases that interest is expressed in terms of money.

The goal of this book is simple – to provide the mathematical understandings of interest and the time value of money needed to succeed on the actuarial examination covering interest theory (FM/1). There is other material on this exam – visit the SOA or CAS websites for further information.

The syllabus for this exam along with past exams and solutions can be found at http://www.soa.org/education/exam-req/edu-exam-fm-detail.aspx

While the calculations in this course can become quite involved, the basic principles are quite simple. We begin with a few basic mathematical techniques which will be used throughout this text. Consult any undergraduate calculus text for more details.

1.2 Calculators and Computers

Because the calculations are often tedious if done on a standard calculator, you should purchase a financial calculator, especially if you intend to take the SOA/CAS common exam on Financial Mathematics (FM). Here is a list of the acceptable calculators as provided by the SOA website:

Calculators – For all exams (except EA exams): only the following Texas Instrument calculator models may be used:

- BA-35 TI-30Xa

- BA II Plus* TI-30XIIS*

- BA II Plus Professional Edition* TI-30XIIB*

- TI-30XS MultiView* TI-30XB MultiView*

* Upon entrance to the exam room, candidates must show the supervisor that the memory has been cleared. For the BA II Plus and the BA II Plus Professional Edition, clearing will reset the calculator to the factory default settings.

You may bring 2 calculators into the exam room. They should be these two:

TI BA II Plus Professional Edition
TI-30XS MultiView

Keystroke instruction will be provided within this text for the TI BA II Plus and TI BA II Plus Professional as well as the TI-30XS MultiView. **Prior to using your TI BA II Plus you should go to the Appendix: Basic Setup and change some of the default settings.** Most of the examples in the text will be worked out using the **TI BA II Plus**, and the keystrokes will be presented as part of the solution.

Please be aware that not all of the problems on the FM exam can be solved using only the financial functions on the calculator. You will need to know the formulas and how to use them! Indeed, some of the questions asked require that you recognize various forms of the important formulas.

Using the TI-30XS Calculator

The calculator is pictured below (Figure 1.1):

FIGURE 1.1

This is a brief introduction to the functions of the **TI-30XS MultiView** we will use most often. Refer to the reference manual for more information. Unless illustrated with an image, keys will be displayed within parentheses. For example (+) would indicate you are to press the key with a + sign on it.

As with the TI BA II Plus you need to set the mode so at to display things in the most useful fashion. Click on (mode) and then use the arrow button to scroll through the choices. Click on the (enter) key to confirm each choice. The most important choice is mathprint which will display math in the way it is usually entered in texts: 3^2 instead of $3 \wedge 2$, for example. When you are done chose (2nd)(quit) to exit the mode choice menu.

There is one button which is very handy:

Use this button to toggle between displaying answers as fractions or as decimals. If your answer displays as a fraction click on the button to convert to the decimal representation.

There is no = button. Instead, you complete a calculation by using the (enter) button. Here is the keystroke sequence to compute $(3 + 2) * 5$:

Exponents are computed as follows:
The key sequence to compute 3^5 is

The key sequence to compute 3^{-5} is

For more complicated calculations you need to exit exponent mode using the arrow button displayed as Figure 1.5.

The key sequence to compute $3^5 * (4 + 11)$ is

You can reduce errors in complicated calculations by storing numbers. There are seven registers labeled x, y, z, t, a, b, c. The key sequence below stores the value 4 in the x register

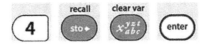

Pressing  multiple times will cycle through the available variables. The calculator will display the variable being stored or retrieved. The key sequence below stores 4 in the z register:

Here is the screen display for this action.

Example 1.1: Compute the value of $\left(1 - \frac{1}{x^2}\right) * \left(1 + x^3 y\right)$ for $x = 1.332$, $y = 2.446$

Step 1: Store the two variables

Note: the subscript (2) on the second use of the indicates that it was pressed 2 times, thus storing the second number in the y register.

Step Two Calculate the value of

$$\left(1 - \frac{1}{x^2}\right) * (1 + x^3 y) \text{ for } x = 1.332, y = 2.446$$

To see the current value of all stored variables use the key sequence below

Stored variables are retained until you clear them or change their value.

We will often need to evaluate the same expression at multiple values of the input variable. To do that we can use the table feature which is accessed by pressing the (table) key. This will display a function of x which you can edit as shown below

$$y=450*(.02*x^2+1)$$

Press the enter key when you are done to display the menu shown below

```
DEG
Start=■
Step=1
Auto    Ask-x
                OK
```

You can either set a start value and the step or increment or use the Ask-x option and enter values of x. This option is the most useful in our case. Use the arrow button to move around in this screen. Click on OK and then Enter when you are done. The calculator will display this screen pictured below.

```
DEG
x       y


x=
```

Type in a value of x and then press the (enter) key to see the computed value of the function. Up to three values can be displayed, but you can enter as many as you like. The screen will show the last three values entered as well as the computed value of the function. Here is the result of entering $x = 5.6$, 8.2, −25 into the formula entered above.

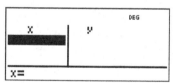

The table feature can be used to solve multiple choice questions by simply entering all the proposed answers and seeing which one works. This can sometimes be easier than solving the equation.

Excel

Microsoft Excel is a crucial tool for actuaries. Each chapter has one or more Projects involving Excel. Excel is the standard software package for business applications. All of the problems involving the creation of a substantial table of values will be solved using Excel. You will learn how to write basic Excel macros (which automate data entry and analysis). Past graduates have reported that knowledge of Excel was a big help in getting their first job and also key to success on the job. You must become familiar with Excel if you intend to be an actuary, accountant, financial analyst, or work in any other field in which financial calculations are required.

1.2.1 Sequences and Series

A sequence of payments over time is known as an annuity. We will often need to compute the value of an annuity at a particular point in time. To do so we compute the value of each payment in the sequence (which will depend on the time that payment will be made) and then add those values to obtain the total value. The sequence of sums obtained by adding the terms of a sequence is called a series. For example, if our sequence of terms (payments, usually) is $100, 200, 300, 400$ the series of sums is $100, 100 + 200, 100 + 200 + 300, 100 + 200 + 300 + 400$.

Geometric Series

If we compute the sum of the values of the payments at the current time, the result is called the present value (PV) of the annuity. If we compute the accumulated values of the payments at some time in the future, the result is called the future value (FV) of the annuity. In either case, we will usually end up with a geometric series (the sum of a sequence where each term is a constant multiple of the preceding term) and so need the formula for the sum of such a series:

$$\sum_{i=0}^{n-1} av^i = a + av + av^2 + \cdots + av^{n-1} = a\frac{1 - v^n}{1 - v} \tag{1.1}$$

Here a is the initial term and v is the common multiple[1].

If $|v| < 1$ then $\lim_{n\to\infty} v^n = 0$ and we can compute the sum of an infinite series of payments (called a perpetuity) as well:

$$\sum_{i=0}^{\infty} av^i = \lim_{n\to\infty} a\frac{1 - v^n}{1 - v} = \frac{a}{1 - v} \tag{1.2}$$

---

[1]The letter v usually represents the quantity $\frac{1}{1+i}$ where i is the interest rate, so $0 < v < 1$ for most cases we will deal with.

Using Equations 1.1 and 1.2 can be a bit tricky as not all series start at $i = 0$. The most direct way to deal with this is to write down a few terms of the series you are dealing with and match them up with Equation 1.1 or Equation 1.2. Note that you don't need to figure out the last term since

$$a = \text{first term}$$
$$v = \text{common multiple}$$
$$n = \text{number of terms.}$$

Example 1.2 Find the value of the expression: $\sum_{i=1}^{20} \frac{3 \cdot 4^{2i}}{5^i}$

Solution: We write out a few terms and compare to Equation 1.1

$$\frac{3 \cdot 4^4}{5^2} + \frac{3 \cdot 4^6}{5^3} + \cdots + \frac{3 \cdot 4^{40}}{5^{20}}$$
$$a + av + \cdots + av^{n-1}$$

We see that $a = \frac{3 \cdot 4^4}{5^2} = \frac{768}{25}$ and that $v = \frac{16}{5}$. Since we are summing from $i = 2$ to $i = 20$, there are 19 terms[2].
   We then have

$$a\frac{1 - v^n}{1 - v} = \frac{768}{25}\frac{\left(1 - \left(\frac{16}{5}\right)^{19}\right)}{\left(1 - \frac{16}{5}\right)} = 55,315,662,541$$

Note that we can use the **TI-30XS** to evaluate this expression. First compute $\frac{16}{5}$ and store it as x. Then enter the following:

$$(768 \div 25) \times (1 - x^{19}) \div (1 - x) \text{ enter}$$

The value returned should be $5.531566254 \times 10^{10}$.

Example 1.3 Find the indicated sum $\sum_{i=6}^{\infty} \frac{4*(3)^{2i}}{5^{2i}}$.

Solution: We write out the first few terms and use Equation 1.2

$$\frac{4 * (3)^{12}}{5^{12}} + \frac{4 * (3)^{14}}{5^{14}} + \frac{4 * (3)^{16}}{5^{16}}$$

We deduce that $a = \frac{4*(3)^{12}}{5^{12}}$ and $v = \frac{(3)^2}{5^2} = \frac{9}{25}$. Using Equation 1.2, we obtain

$$\frac{\frac{4*(3)^{12}}{5^{12}}}{1 - \frac{9}{25}} = .01360489$$

---

[2]In general if we have the sum $\sum_{i=a}^{b} f(i)$ there are $b - a + 1$ terms. This is the n we use in the formula.

Arithmetic Series

An arithmetic series is created by adding the terms of a sequence where a constant (denoted by d in the formula below) is added to each term to get the next term. In the case of an arithmetic series we have

$$a + (a + d) + (a + 2d) + \cdots + (a + (n-1)d) = \frac{n(2a + (n-1)d)}{2} \qquad (1.3)$$

Example 1.4: In the simplest case $a = d = 1$ and we have the formula Carl Friederich Gauss supposedly proved at age six.

$$\sum_{i=1}^{n} i = 1 + 2 + 3 + \cdots + n = \frac{n(n+1)}{2} \qquad (1.4)$$

In some cases, we will need to deal with a combination of an arithmetic and a geometric series:

$$A = Pv + (P+Q)v^2 + (P+2Q)v^3 + \cdots + (P + (n-1)Q)v^n \qquad (1.5)$$

This situation (which we will refer to as a *P-Q Series*) arises when we have an annuity[3] which starts with an initial payment which is then incremented by Q at the end of each subsequent period (Q can be positive or negative). In many cases Q is added to account for inflation. To simplify this expression we first divide both sides by v, obtaining:

$$\frac{A}{v} = P + (P+Q)v + (P+2Q)v^2 + (P+3Q)v^3 + \cdots + (P + (n-1)Q)v^{n-1} \qquad (1.6)$$

Subtracting Equation 1.5 from Equation 1.6 gives us:

$$A\left(\frac{1}{v} - 1\right) = Ai = P(1 - v^n) + Q(v + v^2 + v^3 + \cdots + v^n) - Qnv^n \qquad (1.7)$$

In Equation 1.7, we observe that Q is multiplied by a geometric series with initial term v and common ratio v. Using Equation 1.1, we obtain the following simplification of Equation 1.

$$A = \frac{1}{i}P(1 - v^n) + \frac{1}{i}Q\left(\frac{1 - v^n}{i} - nv^n\right)$$

$$= Pa_{\overline{n}|,i} + Q\left(\frac{a_{\overline{n}|,i} - nv^n}{i}\right) \qquad (1.8)$$

In the second line we have used the term $a_{\overline{n}|,i}$ which we will encounter in Chapter 4 (Annuities). The symbol $a_{\overline{n}|,i}$ is used to represent the present value of an annuity of \$1 at an interest rate of i.

---

[3] An annuity is a sequence of payments. If the payments are constant the annuity is called a level annuity.

Using Equation 1.1 and $v = \frac{1}{1+i}$ (as we will see in Chapter 4) we can see that

$$a_{\overline{n}|,i} = \frac{1 - v^n}{i}.$$

We used: $v = \frac{1}{1+i}$, $i = \frac{1}{v} - 1$ to obtain this result.

To calculate the value of Equation 1.8 it is easiest to use the **TI-30XS**. Compute v and $a_{\overline{n}|,i}$ and store them as x and y. With this done, Equation 1.8 becomes

$$P \times y + Q \times (y - n \times x^n) \div i \tag{1.9}$$

Example 1.5 Find the value of an annuity with $n = 10$ payments. The first payment is $P = 100$ and subsequent payments are increased by $Q = 10$. The interest rate is $i = .05$.

Solution: We have $v = \frac{1}{1+.05} = .952380952$ which we store as x in the **TI-30XS**. We compute the value of $a_{\overline{n}|,i} = \frac{1-x^{10}}{.05} = 7.721734929$ which we store as y. Now we compute our answer using Equation 1.9

$$100 \times y + 10 \times (y - 10 \times x^{10}) \div .05 = 1088.69$$

1.3 Approximation Techniques

In many cases, we will need to solve equations for which no direct method applies. You are probably familiar with the quadratic formula: The solutions to $ax^2 + bx + c = 0$ are

$$x = \frac{-b \pm \sqrt{b^2 - 4ac}}{2a} \tag{1.10}$$

There are similar equations for polynomials of degrees 3 and 4, but no such formula exists for polynomials of degree 5 or higher. In some cases, we can reduce a higher degree polynomial to a quadratic, but these techniques won't always work. As a result, we will utilize approximating techniques to solve such equations. We will use four methods.

a) Excel's financial functions.

b) Newton's Method (not used much anymore, provided as an historical note).

c) MAPLE (very powerful tool, but requires interpretation of results); MAPLE seems little used by financial folk.

d) TI Calculator internal calculation. Along with Excel, this will be the tool you will use most often in "the real world."

1.3.1   Newton's Method (Also Called the Newton-Raphson Method)

Isaac Newton (1643–1727), an English philosopher and mathematician, did important work in both physics and calculus. His method for approximating roots to polynomials is a very nice application of the tangent line. Joseph Raphson (1648–1715), also English, was made a member of the Royal Society prior to his graduation from Cambridge. See more about these two at the MacTutor History of Mathematics site: http://www-history.mcs.st-andrews.ac.uk/index.html

Newton's Method solves the equation $f(x) = 0$ using an iteration technique. An iteration technique involves three stages:

1) Determining an initial guess (or approximation) called x_0,

2) Constructing an algorithm to compute x_{i+1} in terms of x_i,

3) A proof that the sequence x_n converges to the required value, in our case a solution of the equation $f(x) = 0$.

The process starts with the initial approximation x_0 and then computes x_1, x_2, etc., until a desired degree of accuracy is attained. We will discuss how to make an educated guess (the x_0) in the context of specific problems[4]. At this point, we are interested only in describing how Newton's Method generates the iteration sequence in 2). A proof that the method works is beyond the scope of this text – consult an Advanced Calculus text, if you would like to see a proof.

To create the sequence of approximations using the Newton-Raphson Method, we start with a reasonable first approximation, x_0. Often this is done by using a graphing calculator to graph the function and then reading off an estimate from the graph. To find x_1, we first construct the tangent line to the graph of f at the point $(x_0, f(x_0))$. The second estimate, x_1, is the x-intercept of this tangent line.

In general, x_{i+1} is the x-intercept of the tangent line to the graph of f at the point $(x_i, f(x_i))$.

You can verify that this technique yields the following formula for x_{i+1}

$$x_{i+1} = x_i - \frac{f(x_i)}{f'(x_i)} \qquad (1.11)$$

Note that if $f(x_i) = 0$, the procedure terminates (as it should), since we are trying to solve $f(x) = 0$). If $f'(x_i) = 0$ the procedure yields an undefined value. Newton's method can be shown to converge so long as the initial guess is sufficiently close to the desired root. In particular there can't be a maximum

---

[4]With a graphing calculator, or Maple, you can use the graph to get a very good first approximation.

or minimum of the function between the initial guess and the desired root because $f'(x) = 0$ at maximums and minimums.

We will use the Newton-Raphson Method to find several approximations to a solution of $x^3 - 3x^2 + x - 4 = 0$ starting with an initial guess of $x_0 = 4$.

Solution: The curve and its tangent line at $x_0 = 4$ is sketched below (Figure 1.2).

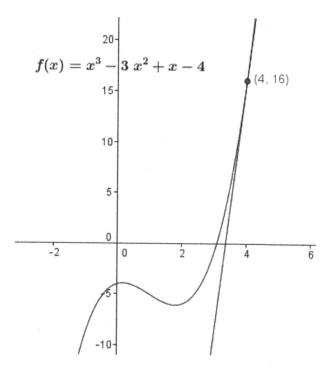

FIGURE 1.2

Since $f(4) = 16$, we will compute the equation of the tangent line at the point $(4, 16)$. We have $f'(x) = 3x^2 - 6x + 1$ and so $f'(4) = 25$. The point-slope form of the equation of the tangent line is thus $y - 16 = 25(x - 4)$. Setting $y = 0$ and solving for x produces an x-intercept of this line of $x_1 = \frac{84}{25}$. Substituting into our original function yields $y = f\left(\frac{84}{25}\right) = 3.42$, much closer to 0 than 16. To continue the process, we compute the slope at $x = \frac{84}{25}$ and use Equation 1.10 to find $x_2 = \frac{718708}{229825}$. Calculating, we find that $f(x_2) = .3711$ which is even closer.

1.3.2 Approximations Using Taylor Series

If $f(x)$ is a function which has derivatives of all orders (f', f'', f''' etc., all exist) it can be shown that (under certain restrictions) $f(x)$ can be computed as an infinite sum of terms involving its derivatives.

$$f(x) = f(a) + f'(a)(x-a)^2 + \frac{f^{(2)}(a)}{2!}(x-a)^2 + \cdots + \frac{f^{(n)}(a)}{n!}(x-a)^n + \cdots$$

$$(1.12)$$

In the expression above $f^{(n)}(a)$ refers to the nth derivative of f evaluated at a. We can compute approximate values of $f(x)$ near a known value $f(a)$ by using the first few terms in 1.12.

Example 1.6: Use the first four terms of Equation 1.12 and $a = 0$ to approximate $\sin(x)$

Solution:

$$\sin(x) = \sin(0) + \sin'(0)(x-0) + \frac{\sin''(0)}{2!}(x-0)^2 + \cdots + \frac{\sin'''(0)}{3!}(x-0)^3$$

$$= 0 + \cos(0)(x-0) + \frac{-\sin(0)}{2!}(x-0)^2 + \frac{-\cos(0)}{3!}(x-0)^2$$

$$= x - \frac{x^3}{6}$$

As the sketch below illustrates this approximation is quite good for values of x close to 0 (Figure 1.3).

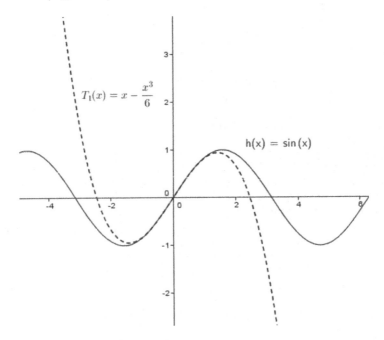

FIGURE 1.3

If we use just the first two terms of 1.12, we obtain the first-order Taylor Series Approximation – also known as the tangent line!

First Order Taylor Approximation

$$f(x) \approx f(a) + f'(a)(x - a) = T_1(x) \tag{1.13}$$

If we use the first three terms we obtain the Second Order Taylor Approximation.

Second Order Taylor Approximation

$$f(x) \approx f(a) + f'(a)(x - a) + \frac{f''(a)}{2!}(x - a)^2 = T_2(x) \tag{1.14}$$

We will use these formulas in Chapter 9 to approximate the change in value of an investment portfolio when interest rates change.

Example 1.7: Find the first and second order Taylor approximations to e^x at $a = 0$.

Solution:

$$T_1(x) = 1 + x$$

$$T_2(x) = 1 + x + \frac{x^2}{2}$$

Figure 1.4 is a sketch of the three functions in Example 1.7:

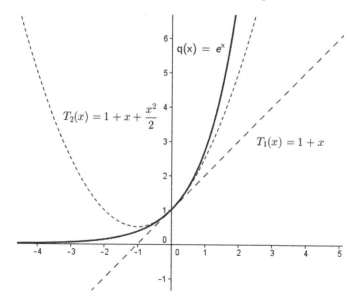

FIGURE 1.4

It is possible (see any calculus text for the details) to compute the entire Taylor Series for the common transcendental functions (sin, cos, ln, e^x, etc.) The one we will use most often is the Taylor Series for e^x with $a = 0$:

$$e^x = 1 + x + \frac{x^2}{2!} + \frac{x^3}{3!} + \cdots + \frac{x^n}{n!} + \cdots = T_\infty(x) \qquad (1.15)$$

This formula can be used to compute the values of other functions which are based on e^x:

Example 1.8. Find the Taylor Series for the function e^{-x^2} at $a = 0$.

Solution: We substitute $-x^2$ into $T_\infty(x)$:

$$e^{-x^2} = 1 + (-x^2) + \frac{(-x^2)^2}{2!} + \frac{(-x^2)^3}{3!} + \cdots$$

$$= 1 - x^2 + \frac{x^4}{2!} - \frac{x^6}{3!} + \cdots$$

Figure 1.5 is a graph of the function e^{-x^2} and the first four terms of the Taylor Series:

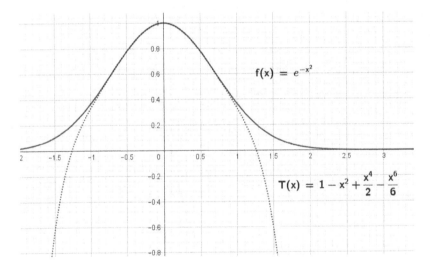

FIGURE 1.5

1.4   Exponents and Logarithms

For convenience we state some of the basic properties of the exponential and logarithmic functions. Unless otherwise stated, we will use the logarithm to base e, indicated as $\ln(x)$ and often referred to as the natural logarithm The **TI BA II Plus** and **TI-30XS** both have a button devoted to ln - the 2ND function for this button is e^x.

BASIC IDENTITIES

$$\ln(ab) = \ln(a) + \ln(b)$$
$$\ln(a^r) = r\ln(a)$$
$$\ln(\frac{a}{b}) = \ln(a) - \ln(b)$$
$$\frac{d\ln(x)}{dx} = \frac{1}{x}, \quad \frac{d\ln(1+i)}{di} = \frac{1}{1+i}$$
$$\ln(e^x) = e^{\ln(x)} = x$$
$$\frac{de^x}{dx} = e^x$$
$$\int e^u du = e^u + c$$
$$\int \frac{1}{u} du = \ln(|u|) + c$$

Example 1.9: Solve $(1.05)^n = 2$.

Solution: We take ln of both sides to obtain $n \cdot \ln(1.05) = \ln(2)$. Thus, $n = \frac{\ln(2)}{\ln(1.05)} \approx 14.21$.

Example 1.10: Solve for i:

$$(1+i)^3 = 1 + 3 \cdot (.05) = 1.15.$$

Solution: We take ln of both sides to obtain $3\ln(1+i) = \ln(1.15)$. This gives us $\ln(1+i) = \frac{\ln(1.15)}{3} = 0.04658731412$. As a result, $1 + i = e^{0.04658731412} = 1.047689553$. Hence $i = .047686553$. We could also solve this problem by taking the cube root of both sides of the equation. $(1 + i) = \sqrt[3]{1.15} = (1.15)^{\frac{1}{3}} = 1.047689553$.

Note: Don't round intermediate values in your calculations. It is appropriate to round the final answer. For example $145.8802 would be reported as $145.88.

The **TI BA II Plus** does not have an nth root button, so you need to use the y^x button with $x = \frac{1}{3}$. If you don't know the decimal value of $\frac{1}{x}$ use the $\frac{1}{x}$ key. Here is how the calculation looks on the **TI BA II Plus** for $1.08^{\frac{1}{7}}$, 1.08, $y^x, 7, \frac{1}{x}$ = Result: **1.011055**.

TI 30XS $1.08(\wedge)7(x^{-1})$ Enter

Integration

We will need to compute a few integrals. The most common one will be a standard substitution of the form

$$\int \frac{1}{u} du = \ln(|u|) + C$$

Example 1.11: Compute $\int \frac{2x}{x^2+3} dx$.

Solution: We make the substitution $u = x^2 + 3$ and obtain

$$\int \frac{2x}{u} dx = \int \frac{1}{u} du = \ln(|u|) + C = \ln(x^2 + 3) + C$$

In Chapter 2 we will need to compute the value of expressions of the form

$$a(t) = e^{\int_0^t \delta(x)dx} \qquad (1.16)$$

Example 1.12: Find $a(t)$ if $\delta(x) = \frac{x}{x^2+3}$

Solution: We begin with the integral:

$$\int_0^t \frac{x}{x^2+3} dx = \int_3^{t^2+3} \frac{.5du}{u} du$$

$$= .5(\ln(t^2+3) - \ln(3)) = .5\ln\left(\frac{t^2+3}{3}\right)$$

We now compute

$$e^{.5\ln\left(\frac{t^2+3}{3}\right)} = \sqrt{\frac{t^2+3}{3}}$$

Exercises for Chapter 1

Note: If you are rusty on logs and exponents, it might be a good idea to review them using a pre-calculus text.

1) Find the sum of each series by writing out the first few terms and comparing the result to Equation 1.1 or 1.2 or 1.3. Use the **TI-30XS** to do the calculations.

   a) $\sum_{i=1}^{30} 3\left(\frac{2}{3}\right)^i$

   b) $\sum_{i=0}^{\infty} 2\left(\frac{3}{4}\right)^i$

   c) $\sum_{i=1}^{10} 4\left(\frac{5}{3}\right)^{3i+1}$

   d) $(15)v + (20)v^2 + (25)v^3 + \cdots (60)v^{10}$ where $v = \frac{1}{1.05}$

2) Solve.

   a) $1.4^n = 5.6$

   b) $\ln(3a) = 6$

   c) $(1+i)^{10} = 2$

3) Compute the value of each integral.

   a) $\int_1^3 \frac{x}{x^2+3} dx$

   b) $\int_1^t \frac{x+1}{x^2+2x+1} dx$

4) Find $a(t)$ using Equation 1.16.

   a) $\delta(x) = \frac{2x+1}{\sqrt[3]{x^2+x+1}}$

   b) $\delta(x) = x^2 + 2$

5) Find the first and second order Taylor Series approximations for the functions at the indicated value of a.

   a) $\ln(1+x)$, $a = 0$

   b) $\sin(x)$, $x = \frac{\pi}{2}$

6) Find the Taylor Series for e^{x^3} using Equation 1.15.

Chapter 2

Measuring Interest

2.1 Introduction

Perhaps the most common financial transaction is the investment of a certain amount of money at a specified rate of interest. A person might deposit money in a savings account with the expectation of earning interest on the amount deposited. Conversely, a bank makes loans with the expectation of being paid interest in addition to repayment of the principal. The term principal refers to the value of the loan/deposit at the time the transaction is made. The term present value is also used in this context.

In each case the interest is paid in compensation for the use of funds during the period of the transaction. The initial deposit or loan is called the **principal** and the total amount paid back after a period of time is called the **accumulated value**. The difference between the accumulated value and the initial deposit is the amount of **interest**.

Example 2.1 A loan of $1,000 is paid off with ten equal payments of $120 each. What is the principal for this transaction? What is the amount of interest paid? Can we compute the interest rate for this transaction?

Solution: The principal is the amount of the loan: $1,000. The payments total $1,200 on a principal of $1,000 so the interest paid is $200. Although $200 is 20% of $1,000, the interest rate on this loan is almost certainly not 20%. To compute the interest rate we need to know when the payments took place. We will discuss the computation of interest rates for loans in Chapter 5.

We begin this chapter by discussing the various ways interest is computed when a single deposit is made and later withdrawn. Later, we will discuss those (much more common) cases where deposits and withdrawals occur throughout the period of an investment.

2.2 The Accumulation and Amount Functions

We assume an amount P_o is deposited and wish to compute the value of this deposit at any time in the future. The initial amount is also known as the principal or present value (PV) while the value at a later date is known as the accumulated amount or future value (FV). The difference between the accumulated amount and the principal is the interest earned (or paid - it all depends on whose point of view you take!) on the transaction.

$$\text{Interest Paid(earned)} = I = FV - PV \qquad (2.1)$$

Example 2.2: $500 is deposited in a savings account at Big Olde Bank. Three years later, the account has an accumulated value of $546.36. Discuss the situation from the point of view of

a) the customer

b) the bank.

Solution:

a) From the point of view of the customer, the interest **earned** is $46.36.

b) From Big Olde Bank's point of view, the interest paid is $46.36. Big Olde Bank will need to find a way to put the $500 to work so as to earn at least $46.36 in order to turn a profit on this transaction.

In almost all cases, the accumulated amount will depend on the length of time between deposit and withdrawal. We will use the letters t and n for the length of time of a given transaction. In general, n will indicate an integral number of some time measurement, e.g., 3 years or 7 months. By contrast, t will be used in cases where the time period is not assumed to be an integer, e.g., 3.7 years or 14.56 days.

The unit in which time is measured is known as the **period** or the **interest conversion period**. We can measure time using any time period provided we make it clear what that period is. Time is measured in years unless stated otherwise. When working problems it is important to determine the interest conversion period first. Finally, while we will often consider the case when t is an integer $(0, 1, 2, 3..)$, all of the analysis applies if t is any real number.

NOTE: In some cases interest can only be withdrawn at integer time periods. You will encounter this on some FM problems. In some cases a different formula is used to calculate the interest earned during a fractional period.

The Accumulation Function

Suppose we invest one unit ($1 typically) of principal. We define the accumulation function $a(t)$ as the accumulated value (FV) of that $1 after t years.

The Accumulation Function
$$\alpha(t) = \text{accumulated value of 1 unit after } t \text{ years where } t \geq 0 \qquad (2.2)$$

The accumulation function is almost always assumed to have three properties:

Properties of an Accumulation Function

1) $a(0) = 1$.

2) $a(t)$ is an increasing function.

3) $a(t)$ is a continuous and differentiable function.

Example 2.3:

a) Verify that $a(t) = 3t^2 + 1$ satisfies the three properties of an accumulation function.

b) If \$1 is deposited at time t how much will be on hand at time $t = 4$? Note that we can also state this problem as follows: If $PV = 1$, find FV when $t = 4$.

Solution:

a) $a(0) = 3 \times 0^2 + 1 = 1$

$a'(t) = 6t$ which is positive for $t > 0$ so our function is differentiable, hence continous and is also increasing.

b) Amount on hand is $a(4) = 49$.

If we deposit $PV = P_0$ units instead of just 1, the accumulated amount is given by the amount function:

The Amount Function

$$A(t) = P_0 \cdot a(t) \tag{2.3}$$

If we denote the initial deposit (principal) by PV as is common and the accumulated amount by FV, we can write this as

The Amount Function Using FV and PV Notation

$$FV = A(t) = PV \cdot a(t) \tag{2.4}$$

In general, the dollar amount of interest earned varies from period to period[1]. We denote the amount of interest earned during the nth period by I_n. Since the first period is from $t = 0$ to $t = 1$, the nth period is the time from $t = n - 1$ to $t = n$. We can compute I_n using our basic definition of interest as the difference between the initial and final principal amounts. In this case we have

Interest Earned in the nth Period

$$I_n = A(n) - A(n-1) = P_0 \cdot (a(n) - a(n-1)) \tag{2.5}$$

---

[1] The rate of interest may also vary from period to period! We will assume for now that the rate of interest is constant.

2.3   The Effective Rate of Interest

Definition: The **effective rate of interest for a given period** is the amount of money that one unit of principal invested at the start of a particular period will earn during that one period. It is assumed that interest is paid at the end of the period in question[2].

If the period is one year, the effective rate of interest is also referred to as the **effective annual rate of interest**. This is sometimes called the APR (annual percentage rate) when expressed as a percentage. An annual effective rate of .05 would correspond to an APR of 5%. As we shall see, the APR reported by various consumer credit companies is not always the same as the effective annual rate. As a result, we will use the term **effective annual interest rate** (or effective monthly interest rate or ...) except during our discussion of the Truth in Lending Laws at the end of Chapter 6.

We will use the symbol i for the effective rate of interest. Again, unless stated otherwise, the period is assumed to be one year. In terms of $a(t)$ we have $i = a(1) - a(0) = a(1) - 1$ so we have:

Effective Rate of Interest, i_1 in period 1

$$i_1 = a(1) - 1 \tag{2.6}$$

We can express i in terms of $A(t)$ as well. Since $A(t) = P_0 a(t)$ we have:

$$
\begin{aligned}
i_1 &= \frac{a(1) - a(0)}{a(0)} \\
&= \frac{\frac{A(1)}{P_0} - \frac{A(0)}{P_0}}{\frac{A(0)}{P_0}} \\
&= \frac{A(1) - A(0)}{A(0)}
\end{aligned}
\tag{2.7}
$$

Example 2.4: A deposit of $550 earns $45 in interest at the end of one year. What is the effective annual interest rate?

Solution: We use Equation 2.7

$$i_1 = \frac{A(1) - A(0)}{A(0)} = \frac{595 - 550}{550} = .08182$$

This would typically be expressed as a percentage: 8.182%.

---

[2]If interest is paid at the start of the period, it is called the rate of discount.

Calculator Cautions

1) The **TI BA II Plus** has several worksheets and special keys. When using any of these financial keys or worksheets we enter interest as a percentage. The **TI BA II Plus** will return the answer as a percentage. When using the standard keys, we enter interest as a decimal. In the case of the **TI-30XS** interest rates are always entered and returned as decimals.

2) Excel allows you the option to format a cell in percentage format. Once you do so, you can enter interest rates as percentages rather than decimals. If the numerical rate is .05, the percentage rate is 5.0%, so it can make a big difference!

3) It is dangerous to round numbers until the very end. Use your calculator's memory feature. Learn how to use the memory feature of the **TI BA II Plus** and the **TI-30XS**.

We now define the effective rate of interest in a more precise fashion.

Definition: The effective rate of interest for any time period is the ratio:

$$\frac{\text{Ending Balance} - \text{Starting Balance}}{\text{Starting Balance}}$$

We can compute the effective rate of interest over any period using previously defined terms. Let i_n be the effective rate of interest during the nth period (note that this is the period from $t = n - 1$ to $t = n$).

Formula for the Effective Rate of Interest

$$i_n = \frac{A(n) - A(n-1)}{A(n-1)} = \frac{I_n}{A(n-1)} \qquad (2.8)$$

The Two Main Accumulation Functions

What sort of functions could qualify as accumulation functions? Recall that we have defined i by the formula $a(1) = 1 + i$. This gives us two points on the accumulation function: $(0, 1)$ and $(1, 1 + i)$. Any increasing differentiable function which goes through these two points is a potential accumulation function.

There are an infinite number of continuous, increasing, differentiable functions which pass through these two points. Each such function represents a possible accumulation function and hence a potential method for computing the accumulated interest. From this infinite collection, only two are used in practice - they are those which result in what we refer to as **simple** and **compound** interest.

2.4   Simple Interest

The first method of computing interest we will consider assumes that $a(t)$ is a linear function of t. This method is known as **simple interest**. It was in common use prior to the advent of calculators and computers since it is very easy to compute. It is still used in a few cases, most notably in the case of fractional time periods.

In the case of simple interest we know that $a(t)$ is a linear function and so its graph is a line. We are given two points on this line so it is easy to obtain an expression for the value of $a(t)$ at any time t. The slope of our line is given by

$$\frac{(1+i)-1}{1-0} = i$$

Since the line contains the point $(0, 1)$ it's equation is given by:

Accumulation Function for Simple Interest

$$a(t) = 1 + it \qquad (2.9)$$

In the case of simple interest, the accumulation function is a linear function with slope i. Figure 2.1 is a plot of the accumulation function for the case that $i = .05$:

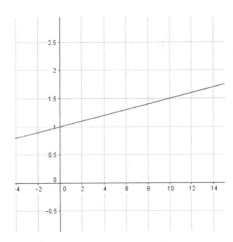

FIGURE 2.1

The amount function is just as easy to compute:

Amount Function for Simple Interest

$$A(t) = FV = PV \cdot (1 + it) \tag{2.10}$$

This is a line with y-intercept $= PV$ and slope $= PV_i$.

Example 2.5 \$550 is deposited at 4.6% simple interest for five years. What is the accumulated amount at the end of this period?

Solution: We have $PV = 550$, $i = .046$ and $t = 5$.

We use Equation 2.10 to compute the requested number:

$$A(5) = PVa(5) = 550(1 + 0.46 \cdot 5) = \$676.50$$

Example 2.6 At what rate of simple interest will \$550 accumulate to \$645 in three years?

Solution: We have $550(1+3i) = 645$ Solving for i yields $i = .057575 = 5.75\%$

The Effective Rate Interest in the Case of Simple Interest

A constant rate of simple interest results in a declining effective rate of interest. Using Equation 2.8, we can compute the effective rate of interest in period in the case of simple interest:

Effective Rate of Interest: Simple Interest

$$i_n = \frac{A(n) - A(n-1)}{A(n-1)} = \frac{(1+in) - (1+i(n-1))}{1+i(n-1)} = \frac{i}{1+i(n-1)} \tag{2.11}$$

NOTE: Equation 2.11 was simplified by canceling the term PV which was common to all three terms.

Using Equation 2.11 we can see that

$$\lim_{n \to \infty} i_n = \lim_{n \to \infty} \frac{i}{1+i(n-1)} = 0 \tag{2.12}$$

This means that the effective rate of interest declines to 0 in the case of simple interest. Figure 2.2 is a plot of the effective rate of interest as a function of time in the case that $i = .05$.

Example 2.7 What is the effective rate of interest in the sixth period if the rate of simple interest is 5.6%?

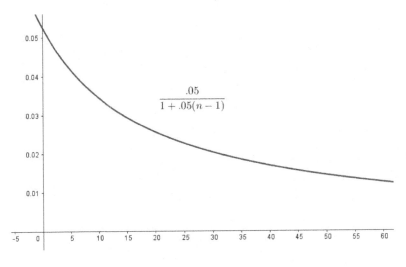

FIGURE 2.2

Solution: We use Equation 2.11 to obtain

$$i_6 = \frac{.056}{1 + (.056) \cdot 5} = .04375 = 4.375\%$$

Example 2.8 An account earns simple interest of 8% annually. After how many years will the effective rate of interest decline to 4%?

Solution: We are solving for n given the values of i_n and i

$$\frac{.08}{1 + .08 \cdot (n - 1)} = .04$$

Cross multiplying gives us

$$.08 = .04 + .0032(n - 1)$$
$$.04 = .0032(n - 1)$$
$$n = 13.5 \text{years}$$

If we are asked to express this as a whole number of years, we would choose fourteen since the effective rate of interest is higher than 4% after thirteen years.

Example 2.9: In the eighth year an account earning simple interest has an effective rate of 5%. What is the rate of simple interest?

Solution: We are given n and i_n and need to find i. Using Equation 2.11 we have:

$$\frac{i}{1 + i \times 7} = .05$$
$$i = .05 + .35i$$
$$.65i = .05 \Rightarrow i = 7.69\%$$

2.5   Compound Interest

Simple interest computes the interest in each period based solely on the amount of the initial deposit. If there are no withdrawals during the life of the transaction, the amount on deposit increases over time while the amount of interest paid at the end of each period remains constant. As we just saw this results in a declining effective rate of interest when we use simple interest. If $400 is deposited at 5% simple interest the amount on deposit after three years is $460. However (using simple interest) the interest paid at the end of year 4 is still only $20 (5% of $400). If interest was paid based on the amount on deposit at the end of year 3, the interest at the end of year 4 would be $(.05) \cdot \$460 = \$23$.

When the interest paid at the end of a given period is based on the accumulated value of the principal at the start of that period rather than on the amount of the original deposit we obtain what is known as ***compound interest***.

To find the formula for the accumulation function in the case of compound interest, we compute the earned interest at the end of each period, add this to the previous balance, and use that number as the principal in computing the interest for the next period. See Table 2.1.

TABLE 2.1

Period	Starting Principal	Interest Earned	New Principal
0	1		
1	1	i	$1 + i$
2	$1 + i$	$i(1 + i)$	$(1 + i) + i(1 + i) = (1 + i)^2$
3	$(1 + i)^2$	$i(1 + i)^2$	$(1 + i)^2 + i(1 + i)^2 = (1 + i)^3$
n	$(1 + i)^{n-1}$	$i(1 + i)^{n-1}$	$(1 + i)^{n-1} + i(1 + i)^{n-1}$ $= (1 + i)^n$

This leads us to believe that the following formula holds for the accumulation function in the case of compound interest

Accumulation Function: Compound Interest

$$a(n) = (1 + i)^n \tag{2.13}$$

You can prove Equation 2.13 for whole numbers using mathematical induction. We have calculated this result using only integral values for the time on deposit. If interest is only paid at integral multiples of the period, $a(n)$ is a step function:

$$a(n) = \begin{cases} 1 & 0 \le n < 1 \\ 1+i & 1 \le n < 2 \\ (1+i)^2 & 2 \le n < 3 \\ (1+i)^3 & 3 \le n < 4 \\ etc.. \end{cases}$$

If we assume that interest can be collected at any time in the life of the investment, it's natural to extend this step function to include real $t \ge 0$. We thus obtain the function $a(t) = (1+i)^t$ defined for all real numbers t which is the continuous, differentiable extension of Equation 2.3

Accmulation Function for Compound Interest

$$a(t) = (1+i)^t \tag{2.14}$$

This is an exponential function whose graph is very different from the linear function we obtained for simple interest. See Figure 2.3.

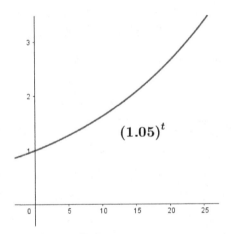

FIGURE 2.3

The amount function for compound interest is

Amount Function for Compound Interest

$$A(t) = P_0 a(t) = P_0(1+i)^t$$
$$FV = PV(1+i)^t \tag{2.15}$$

Compound interest is now much more commonly used than simple interest. One reason for that is the following result:

Theorem: In the case of compound interest at a rate of i, the effective rate of interest is constant: That is: $i_n = i$ for all n.

Proof: we compute and simplify the expression for i_n using Equation 2.8

$$i_n = \frac{A(n) - A(n-1)}{A(n-1)} = \frac{(1+i)^n - (1+i)^{n-1}}{(1+i)^{n-1}} = \frac{(1+i) - 1}{1} = i \quad (2.16)$$

Example 2.10 \$4,000 is deposited at 7.6% effective annual compound interest. Compute the balance in the account at the end of seven years.

Solution: We have $A(t) = 4000a(t) = 4000(1 + .076)^t = 4000(1.076)^7 = \$6,679.53$

TI BA II Plus Solution: (Table 2.2) This is our first example of a problem which can be solved easily using the **TI BA II Plus** calculator. We will use the TVM keys - the row of keys which reads N, I/Y, PV, PMT, FV.

TABLE 2.2

Key	Meaning
N	Number of Periods (years, months, etc.)
I/Y	Interest per period as a percentage
PV	Initial Deposit (present value)
PMT	Payments made, if any after the initial deposit
FV	Accumulated value

Entries into the TVM keys are stored until you re-enter or clear them. They are not cleared when the calculator is turned off. This can be an advantage, but you need to be aware that you need to either clear all entries when you start a new problem (Using 2ND CLR TVM) or enter the new number for each key.

NOTE: When using the TVM keys interest rates are entered and reported as percentages, not decimals.

To enter a number, key in the number and then press the appropriate key. The calculator will display the key symbol and the value (Example: $N = 12$). You can enter any four of the FV numbers in the TVM set and the calculator will compute the fifth number. To compute a value first use the CPT (for compute) key and then the key representing the value you need.

With all that out of the way, here is how to solve Example 2.9 on the **TI BA II Plus**. We know the values of N, I/Y, PV, and PMT and want to compute the value of FV. In this and all subsequent examples using the TVM keys, the number computed will be displayed as $CPT =$

Note that the value for FV is displayed as a negative number. The **TI BA II Plus** always displays (and requires you to enter) the values of PV and FV with opposite signs. If we enter PV as -4000, FV is displayed as a positive number. This makes sense if we think of a simple bank account in which we deposit \$4,000 and then withdraw \$6,679.53 in seven years:

From the bank's point of view: The bank receives \$4,000 and then pays out \$6,679.53 in seven years. From our point of view: We pay \$4,000 and then receive \$,6679.53 in seven years.

The **TI BA II Plus** also displays the commas which group numbers by hundreds - this makes your answer easier to read. Note that we entered the interest rate as 7.6, not .076.

TI-30XS Solution: $4000 \times 1.076 \wedge 7$ (enter) 6769.53

Example 2.11 Find the rate of compound interest at which \$1 will accumulate to the same amount in five years as it will at 8% simple interest.

Solution: The simple interest account will accumulate to $1 \cdot (1 + 5 \cdot .08) =$ \$1.40. We must find i so that $(1 + i)^5 = 1.4$. We could take the fifth root of both sides of the equation $(1 + i)^5 = 1.4$ giving us $1 + i = \sqrt[5]{1.4} = 1.06961$ or $i = .06961$.

The **TI BA II Plus** solution is much easier! We enter the values we are given and compute the unknown interest rate (Table 2.3).

TABLE 2.3

N	I/Y	PV	PMT	FV
5	$CPT = 6.96$	1	0	-1.4

Note that the FV was entered as a negative and that interest is reported as a percentage. If you enter $PV = 1$ and $FV = 1.4$ the calculator will return an error message. You can enter $PV = -1$ and $FV = +1.4$ if you prefer.

The **TI-30XS** would compute the decimal value of the interest rate as $\sqrt[5]{1.4} - 1$.

Example 2.12 You have $500 on deposit earning 8.2% annual compound interest. How long will it be before your account balance is $856?

Solution: In this case we enter I/Y, PV and FV and compute N (Table 2.4).

TABLE 2.4

N	I/Y	PV	PMT	FV
$CPT = 6.8221574$	8.2	500	0	-856

Example 2.13 After eleven years an account earning 4.5% annual compound interest has accumulated to $4,500. What was the value of the original deposit?

Solution: We enter N, I/Y and FV and compute PV (Table 2.5).

TABLE 2.5

N	I/Y	PV	PMT	FV
11	4.5	$CPT = 2,772.89$	0	-4500

Example 2.14 $500 is deposited in an account earning annual compound interest of 5.7%. At the end of two years the accumulated amount is transferred to an account which pays an unknown annual compound interest. At the end of three additional years the account shows a balance of $600. What was the rate of annual compound interest during the final three years?

Solution: With problems like this it is useful to draw number lines with the known and unknown quantities indicated. These are called time lines (Figure 2.4).

500				600
0	5%	2	?%	5

FIGURE 2.4

We first compute the amount on deposit at the end of two years (Table 2.6):

TABLE 2.6

N	I/Y	PV	PMT	FV
2	5.7	500	0	$CPT = -558.62$

Now we need to find the interest rate which will increase $558.62 to $600 in three years. We use the **TI BA II Plus** TVM keys again. Simply press PV to convert the FV result into our new present value. We then enter the new $N = 3$ and the $FV = 600 and compute I/Y (Table 2.7).

TABLE 2.7

N	I/Y	PV	PMT	FV
3	$CPT = 2.41032801$	-558.62	0	600

Note that we entered a positive value for FV since the PV was a negative number.

2.6 Other Accumulation Functions

In theory any function $a(t)$ which is continuous, increasing, and satisfies $a(0) = 1$ can serve as an accumulation function. We now consider a few problems involving non-standard accumulation functions. While these don't show up in "real life," they do appear on the actuarial exams.

Example 2.15 Suppose that $a(t) = .1t^2 + b$. The only investment is $300 made at time $t = 1$. What is the accumulated value of the investment at time $t = 10$?

Solution: Since it is always required that $a(0) = 1$ we must have $b = 1$, so $a(t) = .1t^2 + 1$. We are interested in computing $A(t)$ at $t = 10$. We know that $A(t) = P_0 a(t)$ and so need to find a value for P_0. The trick is to pretend that the $300 is not the only investment but rather the value of the investment at time $t = 1$. We then have

$$300 = A(1) = P_0 \cdot a(1)$$

$$P_0 = \frac{300}{a(1)} = \frac{300}{1.1}$$

We then have:

$$A(10) = P_0 \cdot a(10) = \frac{300}{a(1)} \cdot a(10) = \frac{300}{1.1} \cdot 11 = \$3,000$$

The technique used in this example can be generalized to any problem involving an accumulation function. Suppose we know $A(t_1)$ and want to find $A(t_2)$. Since we do not know the value of P_0, we can't compute $A(t_2)$ directly. However, we do know that $A(t_1) = P_0 \cdot a(t_1)$ and $A(t_2) = P_0 \cdot a(t_1)$ hence $P_0 = \frac{A(t_1)}{a(t_1)} = \frac{A(t_2)}{a(t_2)}$. We solve this equation for $A(t_2)$ to obtain a very useful formula which computes the amount function at any time based on the amount at any other time.

Amount Function Calculator

$$A(t_2) = \frac{a(t_2)}{a(t_1)} A(t_1) \tag{2.17}$$

This is valid for any accumulation function and does not require that $t_1 > t_2$.

Example 2.16 Suppose that the accumulation function is given by $a(t) = at^2 + 1$. Further assume that $100 invested at time 0 accumulates to $280 by time 6. Find the accumulated value at time $t = 20$ of $500 invested at time 5.

Solution: We can use Equation 2.17 to obtain an expression for $A(20)$ in terms of $a(20)$ and $a(5)$

$$A(20) = \frac{a(20)}{a(5)} A(5) = \frac{a(20)}{a(5)} 500$$

To complete the problem, we need a formula for $a(t)$. Since $100 invested at 0 increased to $280 when $t = 6$, we have

$$A(6) = 280 = \frac{a(6)}{a(0)} 100 = \frac{36a + 1}{1} 100$$

Hence $36a + 1 = 2.8$ so $36a = 1.8$ and $a = .05$. Hence $a(t) = .05t^2 + 1$. Therefore

$$A(20) = \frac{a(20)}{a(5)} 500 = \frac{.05 \cdot 20^2 + 1}{.05 \cdot 5^2 + 1} 500 = \frac{21}{2.25} 500 = \$4,666.67$$

Example 2.17 Suppose $a(t) = a^2 + 1$ and that an investment of $500 at $t = 1$ accumulates to $1,500 at $t = 5$. What will be the accumulated value at $t = 9$ of an investment of $200 at $t = 3$?

Solution: We use Equation 2.17

$$A(5) = 1500 = \frac{a(5)}{a(1)} 500$$

$$\frac{a(5)}{a(1)} = \frac{25a + 1}{a + 1} = \frac{1500}{500} = 3$$

Thus, $25a + 1 = 3a + 3$ and so $a = \frac{1}{11}$. This means that $a(t) = \frac{1}{11}t^2 + 1$. Now we use Equation 2.17 yet again to complete the solution

$$A(9) = \frac{a(9)}{a(3)} 200 = \frac{\frac{81}{11} + 1}{\frac{9}{11} + 1} 200 = \frac{92}{20} 200 = \$920$$

You can use Equation 2.17 in many situations, since it is valid regardless of the formula for $a(t)$. Here is an example involving compound interest.

Example 2.18 An investment is earning compound interest. If $100 invested in year 2 accumulates to $105 by year 4, how much will $500 invested in year 5 be worth in year 10?

Solution: The initial information gives us

$$105 = \frac{(1+i)^4}{(1+i)^2}100 = 100(1+i)^2$$

$$(1+i)^2 = \frac{105}{100} = \frac{21}{20}$$

$$1+i = \sqrt{\frac{21}{20}}$$

$$i = \sqrt{\frac{21}{20}} - 1 \approx 0.0247$$

We can now compute the value of $500 invested at $t = 5$ in year $t = 10$

$$A(10) = \frac{(1.02469507660)^{10}}{(1.02469507660)^5}500$$

$$= (1.024695077)^{10-5} \cdot 500 \approx \$564.86$$

TI BA II Plus Solution. Since we are dealing with compound interest, we can solve this problem using the TVM keys. We first compute the compute the effective annual interest rate and then compute the required accumulated value. We know that $100 accumulates to $105 in two years (from year 2 to year 4). We compute the interest rate as follows (Table 2.8):

TABLE 2.8

N	1/Y	PV	PMT	FV
2	$CPT = 2.46951$	100	0	-105

Now we need to accumulate $500 for five years at this same interest rate. Without clearing the TVM keys, we just enter the $PV = 500$ and $N = 5$ and compute FV (Table 2.9)

TABLE 2.9

N	1/Y	PV	PMT	FV
5		500		$CPT = 564.863$

For the case of **compound interest**, if we have an amount A_{t_0} at time t_0 earning compound interest with effective annual interest of i, we can compute the amount A_t at any other time t by the equation

$$A_t = A_{t_0}(1 + i)^{t-t_0} \tag{2.18}$$

Equation 2.18 is the compound interest version of Equation 2.17. As the next example illustrates, Equation 2.18 produces the correct result even if $t < t_0$.

Example 2.19 An account is earning interest at an effective annual interest rate of 6.5%. The account balance is $450 at $t = 4$.

a) What will the balance be at time $t = 7$?

b) What was the original balance in the account?

Solution:

a) We use Equation 2.18 with $A_{t_0} = 450$ and $t_0 = 4$ and $t = 7$

$$A_7 = 450(1.065)^{7-4} = \$543.5773313.$$

b) In this case, $t = 0$ and we have $A_0 = 450(1.065)^{0-4} = \$349.7953909$

Solution Using the TI BA II Plus: We enter $450 as the PV, 6.5 as I/Y and a payment of 0.

a) We are interested in the balance three years later (Table 2.10):

TABLE 2.10

N	I/Y	PV	PMT	FV
3	6.5	450	0	$CPT = -543.58$

b) We are interested in the balance four years prior to the time at which it is $450, so we enter $N = -4$ and re compute FV (Table 2.11).

TABLE 2.11

N	I/Y	PV	PMT	FV
-4				$CPT = -349.79$

We can check this result by setting the PV to -349.795 and accumulating it for four years. Press the PV key to transfer the -349.79 to PV and then press 4 and then N (Table 2.12).

Using the **TI-30XS** we can compute these directly as well. TI-30XS keys are placed in parentheses

a) 450(×)1.065(∧)3(ENTER) 543.58

b) 450(×)1.065(∧)(−)4(ENTER) 349.795

TABLE 2.12

N	I/Y	PV	PMT	FV
4		$= -349.795$		$CPT = 450$

2.7   Present and Future Value: Equations of Value

We now have formulas which compute the accumulated amount (also called the future value or FV) of P_0 at an interest rate of i over t periods. The amount P_0 is usually called the present value (PV). In many cases, we must answer the converse question: how much must be deposited at an interest rate of i so that it will accumulate to a given future amount FV? As an example, we might be putting money away toward the purchase of a car or a down payment on a house.

In some cases, money is being saved toward the payment of an annual tax or other obligation. In Chapter 9 we will discuss ways in which investors seek to ensure that the funds saved will suffice to cover a set of future obligations. Banks often require mortgage holders to contribute to an account to provide funds for the payment of property taxes. These sorts of accounts are known as escrow accounts.

We begin with the calculation of the relation between the present and future value of a single deposit for one year. We want to find the amount PV which is sufficient to accumulate to an amount FV at the end of one year. In one period a deposit of PV will accumulate to $PV(1 + i)$. We want $PV(1 + i) = FV$ and so $PV = FV \cdot \frac{1}{1+i}$. The term $\frac{1}{1+i}$ is called the ***discount factor*** and is referenced by the symbol v. Thus:

$$v = \frac{1}{1 + i}$$
$$i = \frac{1 - v}{v} = \frac{1}{v} - 1 \tag{2.19}$$

Over t periods we have one set of formulas for simple interest and a second for compound interest. In each case, we merely solved the first equation (relating FV to PV) for the present value in order to get an equation with computes PV in terms of FV.

Present and Future Value: Simple Interest

$$FV = PV(1 + it)$$
$$PV = FV \frac{1}{1 + it} \tag{2.20}$$

Present and Future Value: Compound Interest

$$FV = PV(1+i)^t$$
$$PV = FV(1+i)^{-t} = FVv^t \tag{2.21}$$

Note that in each case there are four variables (PV, FV, i and t). Knowing any three of the four enables us to solve for the fourth variable. As we have seen the **TI BA II Plus** will perform these calculations. The examples below show how this can be done.

Example 2.20 An investor is saving to pay off an obligation of \$50,000 which will be due in seven years. If the investor is earning 7.5% effective annual interest rate, how much must be deposited as a single deposit now to meet this obligation?

a) If we use simple interest.

b) If we use compound interest.

Solution We have $t = 7$, $i = .075$, and $FV = 50000$ and want to compute PV

a) For simple interest we use Equation 2.20. This calculation is most easily completed using the **TI-30XS**.

$$PV = \frac{50000}{(1+.075 \cdot 7)} = \$32,786.89$$

b)

$$PV = 50000(1.075)^{-7} = \$30,137.75$$

Note that both of these can be entered into the **TI-30XS** very easily. For part b) the **TI BA II Plus** is a bit quicker.

TI BA II Plus Solution in the case of compound interest (Table 2.13):

TABLE 2.13

N	1/Y	PV	PMT	FV
7	7.5	$CPT = 30,137.75$	0	$-50,000$

NOTE: If we had entered the FV as $+50,000$ the PV would have been returned as a negative. The **TI BA II Plus** always reverses the signs. So long as you are aware of this little quirk, it presents no difficulties. A positive quantity represents a deposit by you, while a negative quantity represents a withdrawal (or potential withdrawal) by you. Note also that we keyed in \$0 for the PMT. We are assuming (for now) a single deposit which is entered into the **TI BA II Plus** as the PV entry.

Example 2.21 $4,000 is deposited at 7.6% annual interest. What is the accumulated value of the account at the end of seven years? Make the calculation assuming:

    a) simple interest.

    b) compound interest.

Solution: We are given $PV = 4000$, $i = .076$, $t = 7$ and need to find FV

    a) Simple interest: $FV = PV(1 + it) = 4000(1 + .076 \cdot 7) = 4000 \cdot 1.53200000 = \$6,128.00$.

    b) Compound interest: $FV = PV(1 + i)^t = 4000(1.076)^7 = 4000 \cdot 1.66988245 \approx \$6,679.53$ (Table 2.14).

TABLE 2.14

N	1/Y	PV	PMT	FV
7	7.6	4,000	0	$CPT = -6,679.53$

NOTE: The **TI BA II Plus** has no special set of keys for simple interest so these problems are best solved using the **TI-30XS**.

Example 2.22 What is the accumulated amount in Example 2.21 if the 4,000 is left on deposit at compound interest for ten years instead of seven?

Solution: $FV = 4000(1.076)^{10} = \$, 8321.137619$

This is really easy to do on the **TI BA II Plus**. So long as you haven't cleared anything, just enter 10 for N and compute the FV. The calculator keeps all the TVM variables in memory (even if it shuts down or you turn it off). While that is convenient, it also means you need to remember to clear the TVM registers prior to starting a new problem (Table 2.15).

TABLE 2.15

N	I/Y	PV	PMT	FV
10			$CPT = -8,321.14$	

Example 2.23 A deposit of $456 accumulates to $600 in five years what is the rate of

    a) Simple interest.

    b) Compound interest.

...

Solutions

a) We use the first of equations 2.20. In this problem, $FV = 600$ and $PV = 456$. We thus have $600 = 456(1 + 5i)$. Thus $1 + 5i = \frac{600}{456}$ and so $i = \frac{1}{5}\left(\frac{600}{456} - 1\right) = .06315789480 \approx 6.316\%$.

b) Using Equation 2.21 we have $600 = 456(1 + i)^5$ We solve this problem by noting that $(1 + i)^5 = \frac{600}{456}$. We take the 5th root of both sides and then solve for i: $1 + i = \sqrt[5]{\frac{600}{456}}$

so that $i = \sqrt[5]{\frac{600}{456}} - 1 = .05642162235 = 5.64\%$.

Using the **TI BA II Plus** (Table 2.16)

TABLE 2.16

N	I/Y	PV	PMT	FV
5	$CPT = 5.64216223$	456	0	-600

Example 2.24 What deposit is required to accumulate $5,600 in six years at 5.6% simple interest? How much is required in the case of compound interest?

Solution: We are given $FV = \$5,600$ and $i = .056$ and must determine PV. In the case of simple interest we have

$$PV = FV \cdot \frac{1}{1 + it} = 5600\frac{1}{1 + .056 \cdot 6} = \$4,191.62$$

In the case of compound interest we have

$$PV = FV \cdot v^t = 5600\left(\frac{1}{1.056}\right)^6 \approx \$4,038.36$$

Using the TI BA II Plus (Table 2.17)

TABLE 2.17

N	I/Y	PV	PMT	FV
6	5.6	$CPT = 4,038.36$	0	$-5,600$

Example 2.25 At what rate of interest per quarter will $4,500 invested at an effective interest rate of i **per quarter compounded quarterly** accumulate to $10,000 in ten years?

Solution: This is our first example using a period other than years. Since the interest is expressed using a quarter year as the time period we convert ten years to forty quarters. Thus, $t = 40$ and we have

$$PV = FVv^t$$

$$4500 = 1/0000 \left(\frac{1}{1+i}\right)^{40}$$

This give us $\left(\frac{1}{1+i}\right)^{40} = .45$. We take the fortieth root of both sides and solve for $i = .02163279$ or about 2.016% per quarter.

Using the **TI BA II Plus** (Table 2.18)

TABLE 2.18

N	I/Y	PV	PMT	FV
40	$CPT = 2.016327948$	4500	0	$-10,000$

Note: The symbol I/Y stands for Interest per Year, but we can use any interest period as long as we enter N in the same units as the I/Y units. If we have interest per month, we enter N in months, not years.

2.8   Nominal and Effective Rates of Interest

We now consider cases where interest is paid either more or less often than the period used for measuring time. In many cases an annual interest rate is quoted even though interest may be compounded (or converted) more (or less) often than once per year. In order to compare such rates we must convert to a common time unit. This is typically a year, but in some problems another unit will be most useful. If the unit of measurement is one year we are computing what is called the **effective annual rate of interest**.

Example 2.26 Mammoth Credit offers a credit card with a monthly interest rate of 1.9%. Cards R Us offers a card with an annual rate of 23%. Which card is the better buy?

Solution: This may seem simple – multiply 1.9 times 12 to obtain 22.8. Hence the Mammoth card is the better deal. This is exactly what Mammoth wants us to do. In fact, Mammoth is allowed to describe their card as carrying an "APR" (Annual Percentage Rate) of 22.8%. However, this simple calculation only works if interest is computed using simple interest! Since all credit card interest is computed as compound interest, we need to convert the Mammoth rate to its effective annual interest rate. To do that we need an annual rate of interest, i, which is equivalent to our monthly rate of 1.9%. If we invest $1 at an annual rate of interest of i we will have $1 + i$ after one year. On the other hand, $1 invested at 1.9% per month for twelve months will accumulate to $1 \cdot (1.019)^{12} = 1.25340$. We thus we must have $1 + i = 1.25340$ so that $i = .25340 = 25.34\%$. The Mammoth Card has an effective annual interest rate of 25.34% making the Cards R Us card at 23% the better buy.

TI BA II Plus solution: We first compute the final amount based on the given monthly rate and then compute the interest rate required to achieve that in a single period (in this case one year) (Table 2.19).

TABLE 2.19

N	I/Y	PV	PMT	FV
12	1.9	1	0	$CPT = -1.253$

N	I/Y		PV	PMT	FV
1	$CPT = 25.340$				

The 22.8% APR calculated for Mammoth in the example is called the **nominal rate of interest**.

The 25.34% is the **effective annual interest rate** for the given **nominal rate** of interest of 22.8% compounded monthly. The effective annual interest rate is thus dependent on both the nominal rate of interest and the number of compounding periods per year.

Alternate **TI BA II Plus** solution: The ICONV worksheet.

The TI BA II Plus has several worksheets. These differ from the TVM keys in that they must be invoked to be used. After use you must exit the worksheet to return to normal editing. The ICONV worksheet is invoked by 2 ND $ICONV$ (above the 2). There are three entries which are accessed by using the ↑ or ↓ keys. Prior to using this (or any other worksheet) key in 2ND, $CLRWORK$ to erase previous entries. To enter a number, key in the number and press the ENTER key. The three keys in this worksheet are (Table 2.20):

TABLE 2.20

Key	Meaning
C/Y	Number of compounding periods per year (or other unit)
EFF	Effective rate of interest per year (or other unit)
NOM	Nominal rate of interest per year (APR)

We can enter C/Y and either EFF or NOM and then compute the other interest rate. As with all worksheets we enter interest rates as percentages.

We can use $ICONV$ to solve the previous problem as follows. We have $C/Y = 12$ and an interest rate of 1.9% per month. The nominal rate is thus $1.9 \cdot 12 = 22.8$ (Table 2.21).

2ND ICONV

TABLE 2.21

Key	
C/Y	12 ENTER
EFF	$CPT = 25.34$
NOM	22.8 ENTER

2ND QUIT

NOTES: To input a value for any of the three variables you must press the ENTER key after keying in the value. After doing so the calculator will display $NOM = 12.3$, for example. You must always enter the 2ND QUIT command to exit a worksheet. Failure to do so results in errors.

Notation

We will use the symbol $i^{(m)}$ for a nominal annual rate of interest for a period when interest is payable m times per period. The interest per period is then calculated as $\frac{i^{(m)}}{m}$. If the period is one year, $i^{(m)}$ is the nominal rate of annual interest. This is only equal to the annual effective rate of interest when $m = 1$.

In Example 2.25 We had $i^{(12)} = .228$ so that the interest per month was $\frac{i^{(12)}}{12} = \frac{.228}{12} = .019 = 1.9\%$. The nominal rate of annual interest is 22.8%, the monthly rate of interest is 1.9% and the effective annual rate of interest per year is 25.34%.

If i represents the effective rate for the period in question we have the fundamental equation

$$1 + i = \left(1 + \frac{i^{(m)}}{m}\right)^m \qquad (2.22)$$

This is true since both sides compute the accumulated value of \$1 at the end of a single period.

Computing Effective Rates of Interest

We can solve Equation 2.22 for $i^{(m)}$ or for i obtaining the following conversion formulas. If you know $i^{(m)}$, use the first to compute i. If you know i, use the second to compute $i^{(m)}$.

$$i = \left(1 + \frac{i^{(m)}}{m}\right) - 1$$

$$i^{(m)} = m\left((1 + i)^{\frac{1}{m}} - 1\right) \qquad (2.23)$$

Remember: the symbol $i^{(m)}$ represents the nominal annualized rate of interest. If $i^{(4)} = .04$, the interest per quarter is .01, not .04.

Example 2.27 What quarterly rate of interest is equal to an effective annual rate of 1.45%?

Solution: We have $m = 4$, $i = .0145$. We use the second of Equations 2.23

$$i^{(m)} = m\left((1+i)^{\frac{1}{m}} - 1\right) = 4(1.0145^{\frac{1}{4}} - 1) = .014421816$$

The quartely rate is $\frac{.014421816}{4} = .003605454 = .3605\%$

TI BA II Plus *ICONV* solution (Table 2.22):

TABLE 2.22

Key	
C/Y	4
EFF	1.45
NOM	$CPT = 1.442$

We divide by 4 to obtain the quarterly rate of .3605% Remember, NOM $= i^{(m)}$, the interest rate per period is $\frac{i^{(m)}}{m}$

Example 2.28 What nominal **monthly** rate of interest is equivalent to a **quarterly** rate of 3.4%?

Solution: Note that we are seeking an effective quarterly rate in this problem, not an effective annual interest rate. That means our base unit for time is the quarter. Since there are three months in a quarter, we are seeking $i^{(3)}$ given $i = .034$. We use *ICONV* (Table 2.23)

TABLE 2.23

Key	
C/Y	3
EFF	3.4
NOM	$cpt = 3.362$

Note that the actual **monthly** interest rate is not 3.3621784%, it's $\frac{3.3621784\%}{3} = 1.1207261\%$. The 3.3621784% figure is the nominal rate which is stated in terms of quarters. To obtain the monthly rate, we divide by 3 since there are three months in each quarter. What we have shown is that a monthly rate of 1.12107261% is exactly the same as a quarterly rate of 3.400000%. We can compute in quarters or in months, we just need to use the appropriate interest rate. As usual, rounding these figures can result in errors.

We can also use the TVM Keys:

We enter the future value of $1 using the given interest rate and period of 1 (quarter!). This is just 1.034 which is entered as -1.034 because of the **TI BA II Plus** convention. We then Compute the I/Y by entering $N = 3$ (since there are three months in a quarter) (Table 2.24).

TABLE 2.24

N	I/Y	PV	PMT	FV
3	$CPT = 1.1207614$	1	0	-1.034

Note that the TVM keys return the **actual monthly interest rate** as a percentage. The TVM keys are clearly the most straightforward method of solution.

Example 2.29 Given that $i^{(3)} = .0013$, find $i^{(7)}$ so that the two interest rates are equivalent.

TI BA II Plus: we first accumulate $1 at a rate of $i^{(3)}$ with $N = 3$. This gives us the accumulated amount after 1 year (Table 2.25).

TABLE 2.25

N	I/Y	PV	PMT	FV
3	$\frac{.13}{3} = .043333$	1	0	$CPT = -1.001300563$

We want a rate which accumulates to this same amount after seven periods so we change N to 7 and CPT I/Y (Table 2.26)

TABLE 2.26

N	I/Y	PV	PMT	FV
7	$CPT = .01856913$			

We have computed $\frac{i^{(7)}}{7}$ so we need to multiply by 7 to obtain $i^{(7)} = .129983908\% = 13\%$

Example 2.30 Suppose we have $100,000 on account at the monthly interest rate of 1.1207261%

Compare the accumulated amounts after ten years using 1.121% and the actual monthly rate of 1.1207261%.

Solution. Using Equation 1.12 and a rate of 1.121% we obtain

$100000(1.01121)^{120} = 381,033.0887$. Using the exact monthly interest rate we obtain

$100000(1.011207621)^{120} = 380,925.52$. a difference of $107.5563936. Not so much, you say?

It's more than enough to lead you to choose the wrong answer on a multiple choice exam.

Moral of this example: do not round any results except the final answer to a problem.

SOA Exam Hint: The SOA exams are all multiple choice and the answers are sometimes rounded quite a bit. You may need to round your **answer** appropriately to obtain the correct choice. There should be only one answer which can be obtained by rounding the complete correct answer.

Example 2.31 In an earlier example we saw that 1.1207261% per month is exactly the same as 3.4% per quarter. Use these two rates to compute the accumulated value of a deposit of $556 after seventeen months using

a) the monthly interest rate

b) the quarterly interest rate.

Solution

a) See Table 2.27. We have $t = 17$, $i = .01207261$, and $PV = 556$ thus

$$FV = PV(1+i)^t = 556(1.01207261)^{71} = \$671.98$$

TABLE 2.27

N	I/Y	PV	PMT	FV
17	1.1207261	556	0	$CPT = -671.98$

b) See Table 2.28. Measuring in quarters we have $t = \frac{17}{3} = 5\frac{2}{3}$, $i = .03362178$ and $PV = 556$

$$FV = 556(1.034)^{5\frac{2}{3}} = \$671.98$$

TABLE 2.28

N	I/Y	PV	PMT	FV
5.666667	3.4	556	0	$CPT = -671.98$

As expected, the two future values are equal.

Note: One more reminder: the negative sign is due to the TI's convention that the PV and FV must always have opposite signs.

Example 2.32 What is the effective annual interest for a car loan for which interest is compounded monthly at a nominal annual rate (APR) of 12.6%?

Solution: We are given $i^{(12)} = .126$ and need to find i. We use the first of equations 2.21

$$i = \left(1 + \frac{i^{(m)}}{m}\right)^m - 1 = \left(1 + \frac{.126}{12}\right)^{12} - 1 = .13354 = 13.354\%$$

TI BA II Plus using the ICONV worksheet (Table 2.29).

TABLE 2.29

Key	
C/Y	12
EFF	$CPT = 13.35$
NOM	12.6

TI BA II Plus using the TVM keys. We compute the FV of $1 over twelve months using a monthly rate of $\frac{12.6}{12}$. We then change N from 12 to 1 and CPT I/Y (Table 2.30)

TABLE 2.30

N	I/Y	PV	PMT	FV
12	$\frac{12.6}{12} = 1.05$	1	0	$CPT = -1.133537297$

Now enter $N = 1$ and CPT I/Y (Table 2.31).

TABLE 2.31

N	I/Y	PV	PMT	FV
1	$CPT = 13.35372966$			

NOTE: You can also read the interest rate directly from the first calculation. In one year $1 increased to 1.133537297 so the interest rate per year (the amount $1 earns) is .133537297 or 13.3537297%

$$
\begin{array}{c}
\hline
\text{Formulas for } PV \text{ and } FV \\
FV = PV(1+i)^t = PV\left(1 + \frac{i^{(m)}}{m}\right)^{mt} \\
PV = FV\left(1 + \frac{i^{(m)}}{m}\right)^{-mt} \qquad (2.24) \\
\hline
\end{array}
$$

Example 2.33 Find the accumulated amount if \$4,500 is invested at an effective monthly rate of .175% monthly for three years.

Solution: We have $m = 12$, $\frac{i^{(12)}}{12} = .00175$ and $t = 3$

$$FV = PV\left(1 + \frac{i^{(m)}}{m}\right)^{mt} = 4500(1.00175)^{36} = \$4,792.36$$

TI BA II Plus solution. Since we have an effective monthly rate, we can count in months without having to convert the interest. In this case I/Y is really interest per month (Table 2.32).

TABLE 2.32

N	I/Y	PV	PMT	FV
36	.175	4,500	0	CPT = 4,792.356898

Example 2.34 In the situation of Example 2.33, how much will the account be worth in ten years? Use the **TI BA II Plus**.

Solution. All we need to do is to change N to 120 (Table 2.33).

TABLE 2.33

N	I/Y	PV	PMT	FV
120				CPT = 5,550.532454

Example 2.35 X is placed in an account which carries a nominal annual interest of 2.5% compounded monthly. After five years the accumulated value is placed in an account which earns a nominal annual interest of 3.2% compounded quarterly. The value of this account eight years after inception is \$10,000. Find X.

Solution: We use equation 2.24 over the first period (five years) to obtain

$$FV = X \cdot \left(1 + \frac{.025}{12}\right)^{60}$$

We now accumulate this amount through the final three years or twelve quarters of the account's life at the second interest rate of 3.2% per quarter

$$FV = 10,000 = X \cdot \left(1 + \frac{.025}{12}\right)^{60} \cdot \left(1 + \frac{.032}{4}\right)^{12}$$

Solving for X yields $X = \frac{10,000}{\left(1+\frac{.025}{12}\right)^{60}\left(1+\frac{.032}{4}\right)^{12}} = \$8021.27$

TI BA II Plus solution. We start with the final (future value) of \$10,000 and work backwards to the present value in two steps. The account earned 2.5% per month for five years and then 3.2% per quarter for the final three years.

Step One: find the amount in the account at the start of the final three years. Since interest in the final three years is computed in quarters, we use $N = 3 \times 4 = 12$ (Table 2.34).

TABLE 2.34

N	I/Y	PV	PMT	FV
12	$\frac{3.2}{4} = .8$	$CPT = -9,088.110831$	0	10000

Step Two: Convert the amount in the account at the end of five years (\$9,088.11) to a FV by pressing the *FV* key and then compute the starting balance. Enter the new interest rate and then compute the *PV* (Table 2.35):

TABLE 2.35

N	I/Y	PV	PMT	FV
60	$\frac{2.5}{12} = .2083333$	$CPT = 8,021.27$	0	-9,088.110831

2.9   Discount Rates

In some cases, interest is paid at the **inception** of the loan, rather than at the end. PayDay loan establishments often operate in this fashion. This decreases the default risk (sometimes called the exposure) of the loan. Here are two scenarios:

Loan 1: \$500 is borrowed for a year at an effective rate of interest of 6%. At the end of the year, the borrower pays the lender \$530.

Loan 2: \$500 is borrowed for a year, but the borrower pays the interest at the loan's inception. The borrower receives \$470 and must repay \$500 at the end of one year.

In both cases the interest paid is \$30. However in the second case the interest is paid on a loan of only \$470. Loan 2 is an example of computing interest as a *discount* and is said to have an **effective rate of discount** of 6%.

The effective rate of discount is a measure of interest which is paid at the **beginning** of the period. The symbol for the rate of discount is d. Note that d computes an amount which is paid at the beginning but is **based on the amount at the end of the period**. In our example Loan 2 required a payment of \$30 which is 6% of the \$500 loan but 6.3829% of the \$470 which the borrower receives.

Suppose we have a loan of \$1 at an effective interest rate of i and compare this to the same loan with an effective discount of d. In the each case, we compute the interest rate as the ratio of the amount of interest paid (i or d) to the amount of the principal. In the first case, the principal is 1, while in the second case it is $1 - d$. We thus obtain the fundamental relationship between i and d

$$\frac{i}{1} = \frac{d}{1-d} \tag{2.25}$$

In our example, the effective rate of interest for the second loan is

$$i = \frac{d}{1-d} = \frac{.06}{.94} = .06383 = 6.38\%$$

We can compute i an alternative way by considering the loan as a loan of \$470 (PV) to be paid off in one year by a payment of \$500 (FV). In that case we have

$$FV = (1+i)PV$$
$$500 = (1+i)470$$
$$i = \frac{500}{470} - 1 = .06383$$

The **TI BA II Plus** will perform this calculation as well. We use $N = 1$ (Table 2.36)

TABLE 2.36

N	I/Y	PV	PMT	FV
1	$CPT = 6.382978723$	470	0	−500

We can obtain two other versions of Equation 2.25 by solving for i and $v = \frac{1}{1+i}$. Multiplying both sides of 2.27 by $1 - d$ and solving for d yields

$$i(1 - d) = i - id = d$$
$$i = d + id = d(1+i)$$
$$d = \frac{i}{1+i}$$

Since $v = \frac{1}{1+i}$ we have $v + iv = 1$ or $i = \frac{1-v}{v}$ and hence

$$d = i\frac{1}{1+i} = \frac{1-v}{v}v = 1-v \qquad (2.26)$$

This is perhaps most useful in the form $v = 1 - d$. We can now compute the relationships between PV and FV in terms of compound discount d or compound interest i.

Compound Interest and Compound Discount

$$PV = FVv^t = FV(1+i)^{-t} = FV(1-d)^t$$
$$FV = (1+i)^t = PVv^{-t} = PV(1-d)^{-t} \qquad (2.27)$$

Note: The FM exam often requires that you know these formulas.

Using the TVM Keys for Discount Problems.

NOTE: The form of each equation is the same except that the signs are reversed when we use d instead of i. This means we can use the TVM keys for computing with discount rates so long as we enter $-d$ for I/Y and $-N$ for N.

Example 2.36 Compute the accumulated amount (FV) of \$560 deposited at

a) An effective annual rate of interest of 5.6%.

b) An effective annual rate of discount of 5.6%.

c) What is the effective annual rate of interest for part b)?

d) What is the effective annual rate of discount for part a)?

Solution:

a) Table 2.37.

TABLE 2.37

N	I/Y	PV	PMT	FV
1	5.6	560	0	$CPT = -591.36$

b) Table 2.38. We enter both N and I/Y as negative numbers

TABLE 2.38

N	I/Y	PV	PMT	FV
-1	-5.6	560	0	$CPT = -593.22$

c) Table 2.39. We change to $N = 1$ and compute the I/Y

TABLE 2.39

N	I/Y		PV	PMT	FV
1	$CPT = 5.93220339$				

c) Table 2.40. We use the future value obtained in part a) and enter $N = -1$

TABLE 2.40

N	I/Y		PV	PMT	FV
-1	$CPT = -5.30303030$		560		-591.36

We ignore the minus sign: $i = 5.303\%$
We can also use Equations 2.25 and 2.26

c) $i = \frac{d}{1-d} = \frac{.056}{1-.056} = .05932203 = 5.93\%$

d) $d = \frac{i}{1+i} = \frac{.056}{1+.056} = .053$

2.9.1 Nominal Rate of Discount

If interest is computed at discount and is compounded p times per interest period we denote the nominal rate of discount per period as $d^{(p)}$ and have the relation

$$\left(1 - \frac{d^{(p)}}{p}\right)^p = 1 - d \qquad (2.28)$$

We can combine the information so far into a single equation which enables one to convert to and from nominal rates of discount to nominal rates of interest

$$\left(1 + \frac{i^{(m)}}{m}\right)^m = 1 + i = \frac{1}{v} = \frac{1}{1-d} = \left(1 - \frac{d^{(p)}}{p}\right)^{-p} \qquad (2.29)$$

Finally, we have the following relationships among discounts and interest rates compounded at any rate.

$$\boxed{\begin{array}{c}
\text{Present and Future Value Using } i \text{ or } d \\[2mm]
PV = FV\left(1 - \dfrac{d^{(p)}}{p}\right)^{pt} = FV\left(1 + \dfrac{i^{(m)}}{m}\right)^{-mt} \\[5mm]
FV = PV\left(1 - \dfrac{d^{(p)}}{p}\right)^{-pt} = PV\left(1 + \dfrac{i^{(m)}}{m}\right)^{mt} \qquad (2.30)
\end{array}}$$

The simplest equation which allows for easy conversion between discount and interest rates is:

$$\left(1 - \frac{d^{(p)}}{p}\right)^{-pt} = \left(1 + \frac{i^{(m)}}{m}\right)^{mt} = 1 + i \qquad (2.31)$$

Equation 2.31 is sometimes called the all in one formula.

Example 2.37 Compute the nominal rate of discount **compounded quarterly** which is equivalent to a nominal rate of interest of 5.6% annually **compounded monthly**.

Solution: We have $p = 12$, $m = 4$, $i^{(12)} = 0.56$. and are to compute $d^{(4)}$. Using Equation 2.31 we have

$$\left(1 - \frac{d^{(4)}}{4}\right)^{-4} = \left(1 + \frac{.056}{12}\right)^{12}$$

This gives us

$$\frac{d^{(4)}}{4} = 1 - \left(1 + \frac{.056}{12}\right)^{-3} = .01387034$$

As a result $d^{(4)} = .0554813746 = 5.548\%$. A nominal interest rate of 5.6% converted every quarter is the same as a nominal discount of 5.548% converted every month. As requested by the problem, we need to report the nominal annual rate $d^{(4)}$. With rare exceptions nominal rates are stated as annual rates.

TI BA II Plus solution. Recall that we can use the TVM keys to compute with discount by entering $I/Y = -d$ and entering the period as a negative number as well. We first compute the FV of $1 at the stated interest rate (Table 2.41):

TABLE 2.41

N	I/Y	PV	PMT	FV
12	$\frac{5.6}{12}$	1	0	$CPT = -1.057459928$

To compute the value of $\frac{d^{(4)}}{4}$ we merely change the value of N to -4 and compute I/Y (Table 2.42)

TABLE 2.42

N	I/Y	PV	PMT	FV
-4	$CPT = -1.387034356$			

Note that we obtained a negative number. We ignore the sign and obtain

$$\frac{d^{(4)}}{4} = 1.387034356$$
$$d^{(4)} = 5.548137025$$

Note: The ICONV worksheet will not work for discount. Entering a negative value for I/Y results in an error message.

Example 2.38 Find the present value of $1,000 to be repaid at the end of six years at a nominal 6% per year paid in advance and convertible semiannually. How much interest is paid?

Solution. We have $FV = 1,000$, $p = 2$ and $d^{(2)} = .06$. Because the interest is paid in advance it is a discount rate and we use d not i. Finally, $t = 6$. We use the first of equations 2.30 to obtain:

$$PV = 1,000 \left(1 - \frac{.06}{2}\right)^{2 \times 6} = 1,000(.97)^{12} = \$693.84$$

For this transaction, the borrower would receive $693.84 and be obligated to repay $1,000 in six years. The interest paid is thus $1,000 - 693.84 = \$306.16$.

TI BA II Plus solution. We can proceed in one of two ways.
  I: Convert d to i. $i = \frac{d}{1-d} = \frac{.03}{.97} = 0.0309278351$ (Table 2.43).

TABLE 2.43

N	I/Y	PV	PMT	FV
12	3.09278351	$CPT = 693.84$	0	$-1,000$

As we noted earlier the compound discount formula is the same as the formula for compound interest except for changes in sign on the exponent and the d. Hence, we can use the **TI BA II Plus TVM** worksheet to compute discount results by entering **negative values** for d (into the I/Y register) and N (Table 2.44).

TABLE 2.44

N	I/Y	PV	PMT	FV
-12	-3	$CPT = 693.84$	0	$-1{,}000$

2.10 Forces of Interest and Discount

Force of interest is defined as the instantaneous relative rate of change of either the accumulation or amount function. It is computed as (note that $a'(t)$ is the derivative of $a(t)$ with respect to t)

$$\delta_t = \frac{a'(t)}{a(t)} = \frac{A'(t)}{A(t)} \tag{2.32}$$

We can establish a formula for $a(t)$ in terms of δ_t by integration

$$\int_0^t \delta_r dr = \int_0^t \frac{A'(r)}{A(r)} dr = \ln(A(r))|_0^t$$
$$= \ln(A(t)) - \ln(A(0))$$
$$= \ln\left(\frac{a(t)}{a(0)}\right) = \ln(a(t)) \tag{2.33}$$

In the above calculation we assumed $a(0) = 1$. We then have

$$\ln(a(t)) = \int_0^t \delta_r dr$$
$$a(t) = e^{\int_0^t \delta_r dr} \tag{2.34}$$

Note: To compute the integral in 2.33, make the substitution $u = A(r)$, $du = A'(r)dr$ to obtain $\int \frac{1}{u} du = \ln(u)$.

Computing $a(t)$ from δ_r

$$a(t) = e^{\int_0^t \delta_r dr} \tag{2.35}$$

In the case of **compound interest**, we have

$$\delta_t = \frac{a'(t)}{a(t)} = \frac{\frac{(d(1+i)^t)}{dt}}{(1+i)^t} = \frac{(1+i)^t \ln(1+i)}{(1+i)^t} = \ln(1+i) = -\ln(v) \tag{2.36}$$

Here we used the fact that $\frac{du^t}{dt} = u^t \ln(u)$ and that $\ln(1+i) = \ln(\frac{1}{v}) = -\ln(v)$.

Force of Interest for Compound Interest

$$\delta_t = -\ln(v) = \ln(1+i) \qquad (2.37)$$

If the force of interest has a constant value of δ, we have $a(t) = e^{\int_0^t \delta dr} = e^{\delta t}$. This case is often referred to as continuous compounding.

Formulas for Continous Compounding

$$a(t) = e^{it}$$
$$A(t) = A(0)e^{it}$$
$$\boldsymbol{FV = PVe^{it}} \qquad (2.38)$$

There is a good reason for the use of the term continuous compounding. If we compound m times per year, the formula for $a(t)$ is given by

$$a(t) = \left(1 + \frac{i}{m}\right)^{mt} \qquad (2.39)$$

We now consider what happens to this expression as m approaches infinity

$$\lim_{m \to \infty} a(t) = \lim_{m \to \infty} \left(1 + \frac{i}{m}\right)^{mt} = \left(\lim_{m \to \infty} \left(1 + \frac{i}{m}\right)^m\right)^t \qquad (2.40)$$

The expression $\lim_{m \to \infty} \left(1 + \frac{i}{m}\right)^m$ in Equation 2.40 may look familiar; it's value is simply e^i. This results in the following

$$\lim_{m \to \infty} a(t) = e^{it} \qquad (2.41)$$

NOTE: Continuously compounded interest at $\ln(1+i)$ yields an effective annual rate of interest of i. For example, an effective annual rate of interest of 5.65 will be the same as a continuously compounded rate of $\ln(1.056) = .05448 = 5.45\%$. This means that we can use effective an effective annual rate of i or a continuous rate of $\ln(1+i)$, whichever is easiest.

Example 2.39 Bank A offers a savings account paying a nominal annual rate (APR) of 3.2% compounded monthly. Bank B offers the same rate, but compounds continuously. If you deposit \$4500 for a period of five years how much more will you earn with Bank B?

Solution: In each case, $PV = 4,500$ and $i = .032$. For Bank A, we have

$$FV = PV\left(1 + \frac{i}{12}\right)^{12.5} = 4,500(1.00266666667)^{60} = \$5,279.674$$

TI BA II Plus solution for Bank A (Table 2.45)

TABLE 2.45

N	I/Y	PV	PMT	FV
60	$\frac{3.2}{12} = .2666667$	4500	0	$CPT = -5,279.67$

For Bank B, we have

$$FV = 4500e^{.032 \cdot 5} = \$5,280.80$$

The difference is very small: $5,280.80 - 5,279.67 = \$1.13$!

Force of Discount

We can define a force of discount, δ_t' by using the discount function, $\frac{1}{a(t)}$:

$$\delta_t' = -\frac{\frac{d}{dt}\frac{1}{a(t)}}{\frac{1}{a(t)}} = -\frac{-\frac{1}{a^2(t)}\frac{da(t)}{dt}}{\frac{1}{a(t)}} = \frac{\frac{da(t)}{dt}}{a(t)} = \delta_t \qquad (2.42)$$

Thus, the force of discount is equal to the force of interest.

We can also compute the force of interest in the case of simple interest

Force of Interest for Simple Interest
$$\delta_t = \frac{\frac{d}{dt}a(t)}{a(t)} = \frac{\frac{d}{dt}(1+it)}{1+it} = \frac{i}{1+it} \qquad (2.43)$$

Note the similarity to the formula for the effective rate of interest in the case of simple interest which is $i_{eff} = \frac{i}{1+(n-1)i}$.

Example 2.40 An account is earning **simple** interest. At time 3, the force of interest is .045 and the acount has a balance of \$5,345. At this time the account is transferred to a bank which pays annual compound interest at the same rate as the rate of simple interest earned previously. How long will it take for the account to accumulate to \$7,000?

Solution: We need to compute the rate of interest. To do so, we use Equation 2.43. We are given $\delta_t = .045$ and $t = 3$ and must solve for i.

$$.45 = \frac{i}{1+3i}$$
$$i = .045 + .135i$$
$$.865i = .045$$
$$i = .052023121 = 5.2023121\%$$

Now we use the **TI BA II Plus** to solve for N. We are given $PV = 5,345$ and $FV = 7,000$ and $I/Y = 5.2023121$ (Table 2.46)

TABLE 2.46

N	I/Y	PV	PMT	FV
$CPT = 5.319$	5.2023121	5,345	0	$-7,000$

2.10.1 Varying Rates of Interest

If the force of interest is given as a function which we can integrate, the most direct means of calculating the accumulation function uses Equation 2.35

$$a(t) = e^{\int_0^t \delta_r\, dr} \tag{2.44}$$

NOTE: This assumes we are starting at $t = 0$. If we don't start at $t = 0$, we need to use a slightly different technique. See Example 2.44 for the details.

Example 2.41 Given that $\delta = \frac{1}{1+t}$, find an expression for $a(t)$.

Solution: We know that $a(t) = e^{\int_0^t \delta_r\, dr}$. In this case

$$\int_0^t \delta_r\, dr = \int_0^t \frac{1}{1+r}\, dr = \ln(1+t)$$

Hence $a(t) = e^{\ln(1+t)} = 1 + t$. Note that $a(0) = 1$ and that $a'(t) = 1 > 0$ so our accumulation function satisfies the basic criterion we outlined earlier in this chapter.

Example 2.42 Find an expression for $a(t)$ if the force of interest is given by $\delta_r = r^2$.

Solution: We have

$$\int_0^t \delta_r\, dr = \int_0^t r^2\, dr = \frac{r^3}{3}\Big|_0^t = \frac{t^3}{3}$$
$$a(t) = e^{\int_0^t \delta_r\, dr} = e^{\frac{t^3}{3}}$$

Example 2.43 An account earns a force of interest of $\delta_r = r^2$ starting at time $t = 0$. The account has a value of $500 at $t = 3$. What is the account balance at $t = 10$?

Solution. From Example 2.42 we know that $a(t) = e^{\frac{t^3}{3}}$. We have $A(1) = P_0 a(1) = P_0 e^{\frac{1}{3}} = 500$. Hence $P_0 = \frac{500}{e^{\frac{1}{3}}} = 358.2656553$. We want $A(10) = 358.2656553 \cdot e^{\frac{4^3}{3}} = \$659,407,865,000$. Clearly this is not a realistic force of interest.

Example 2.44 The force of interest is $\frac{2t}{t^2+1}$. If $A(2) = 300$, find a formula for $A(t)$.

Solution: We have

$$\int_0^t \delta_r dr = \int_0^t \frac{2t}{t^2+1} dt = \ln(t^2+1)$$

Hence $a(t) = e^{\int_0^t \delta_r dr} = e^{\ln(t^2+1)} = t^2 + 1$. $A(t) = P_0 \cdot a(t)$ and $A(2) = 300$, so

$$300 = P_0(2^2 + 1) = 5P_0$$
$$P_0 = 60$$
$$A(t) = 60 \cdot (t^2 + 1)$$

We can also use Equation 2.17:

$$A(t) = A(2)\frac{a(t)}{a(2)} = 300 * \frac{t^2+1}{5} = 60 \cdot (t^2 + 1)$$

Example 2.45 The force of interest is $\frac{2t}{t^2+1}$ and the accumulated amount at $t = 3$ is 10, what is the accumulated amount at time $t = 4$?

Solution: From the previous example, we know that $a(t) = t^2 + 1$. We can then use Equation 2.17

$$A(t_2) = A(t_1)\frac{a(t_2)}{a(t_1)}$$
$$= 10 \cdot \frac{4^2+1}{3^2+1}$$
$$= 17$$

Alternative Solution: We are interested in the amount accumulated between $t = t_1$ and $t = t_2$. We can compute the accumulation function (call it $b(t)$) directly as

$$b(t) = e^{\int_{t_1}^{t_2} \frac{2t}{t^2+1} dt} \tag{2.45}$$

We then obtain

$$A(t_2) = A(t_1)b(t) = A(t_1)\frac{t_2^2+1}{t_1^2+1}$$

Note that this is the same result we obtained with our first method.

Interest Which Varies, But Is Constant Over Each Period

In many cases, the interest varies but is constant over each of several periods. These rates are called **one-year forward rates**. We will discuss this in more detail in Chapter 7.

If we use the notation i_n for the interest in the nth period (recall that this is the period from $t = n - 1$ to $t = n$). In Chapter 7 we will consider more general forward rates and use the notation $i_{n-1,n}$. We can compute the value of $a(t)$ as follows:

$$a(t) = (1 + i_1)(1 + i_2) \cdots (1 + i_n) = \prod_{k=1}^{n}(1 + i_k) \qquad (2.46)$$

This leads to the following equations relating PV and FV in the case of interest which varies by period:

$$FV = PV \cdot \prod_{k=1}^{n}(1 + i_k)$$

$$PV = FV \cdot \prod_{k=1}^{n}(1 + i_k)^{-1} \qquad (2.47)$$

Example 2.46 A deposit of $5,500 earns interest $i_1 = .04$, $i_2 = .05$, $i_3 = .02$, $i_k = .03$ if $k \geq 4$. What is the accumulated value of this deposit after ten periods?

Solution: We are given $PV = 5,500$ and must find FV. We use the first of the two equations above

$$FV = 5,500(1.04)(1.05)(1.02)(1.03)^7$$
$$= \$7,534.35$$

TI-30XS solution: The expression above can be evaluated directly on the **TI-30XS**. This is the fatest way to proceed in this case.

TI BA II Plus solution: We start with $5,500 = PV$ and compute each term, setting the value to PV and computing FV (Table 2.47):

TABLE 2.47

N	I/Y	PV	PMT	FV
1	4	5,500	0	$CPT = -5,720$

Convert the computed FV to a PV by pushing the PV key and then enter in the period and interest rate for the second interest rate (Table 2.48):

TABLE 2.48

N	I/Y	PV	PMT	FV
1	5	-5,720	0	$CPT = 6,006$

Convert the computed FV to PV and proceed as above (Table 2.49):

TABLE 2.49

N	I/Y	PV	PMT	FV
1	2	6,066	0	$CPT = -6,126.12$

N	I/Y	PV	PMT	FV
7	3	$-6,126.12$	0	$CPT = 7,534.35$

Example 2.47. What rate of interest, paid over the entire period of the previous example, would yield the same accumulated value? This is called the **average rate of return** for the transaction.

Solution: If the interest is i over the entire period, we have

$$FV = 5,500(1+i)^{10} = 5,500(1.04)(1.05)(1.02)(1.03)^7$$

We then have

$$(1+i)^{10} = (1.04)(1.05)(1.02)(1.03)^7$$
$$1+i = \sqrt[10]{(1.04)(1.05)(1.02)(1.03)^7} = 1.031973$$

Hence, our average rate of return $i = .031973 = 3.1973\%$

TI-30XS Solution. Divide 7534.35 by 5500 and store it as x then compute $x(\wedge).1$ (ENTER)1.031973

NOTE: If we have a PV and a FV at two different times which differ by Δt, the ratio $\frac{FV}{PV}$ represents the interest rate over the period Δt. The interest rate per period is then given by

$$1+i = \left(\frac{FV}{PV}\right)^{\frac{1}{\Delta t}} \tag{2.48}$$

TI BA II Plus solution. Starting with the last line in Example 2.42, we enter $PV = -5500$ and $N = 10$ and solve for I/Y (Table 2.50)

TABLE 2.50

N	I/Y	PV	PMT	FV
10	$CPT = 3.197.00836$	$-5,500$	0	7,534.354884

Note that the "average" we obtain here is not the numerical average of the interest rates. The average we computed in this example is called the geometric average. The geometric average of the numbers $x_1, x_2, \ldots, x_n$ is defined as

$$\mathbf{Geometric Average} = \sqrt[n]{x_1 \cdot x_n \cdots x_n} \tag{2.49}$$

Exercises for Chapter 2

1) Suppose that the accumulation function for an account is $a(t) = (1+it)$. You invest $500 in this account today. Find i if the account's value twelve years from now is $1,250.

2) It is known that $a(t)$ is of the form $at^2 + b$. If $100 invested at time 0 accumulates to $172 at time 3, find the accumulated value at time 20 of $200 invested at time 5.

3) At a certain rate of simple interest, $1,000 will accumulate to $1,110 after a certain period of time. Find the accumulated value of $100 at a rate of simple interest half as great over a period five times as long.

4) Suppose that $a(t) = at^2 + 10b$. If $\$x$ invested at time 0 accumulates to $1,000 at time 10 and to $2,000 at time 20, find x.

5) At a certain rate of compound interest 100 will increase to 200 in x years, 200 will increase to 300 in y years, and 300 will increase to 1,500 in z years. If 600 will increase to 1,000 in n years, find an expression for n in terms of x, y, and z.

6) Find the effective rate of compound interest and compound discount in each case

   a) A deposit of $340 accumulates to $400 at the end of one year.

   b) A deposit of $550 earns $23 interest in one year.

7) $450 is deposited at 5.6% annual simple interest.

   a) Write an expression to compute the accumulated value of the account at any time t.

   b) Graph this expression. What sort of function is it?

   c) Find an expression for the effective rate of interest earned in period n.

   d) Evaluate your expression c) for $n = 1, 2, 3,$.

   e) Graph the expression in c) over the range $n = 1$ to $n = 20$.

8) $500 is deposited at 6.5% annual simple interest for ten years. What rate of annual compound interest will accumulate to the same amount as this account over this same time period?

9) $1 is deposited in each of two accounts for a period of one year.

Account 1 earns 5% annual simple interest.

Account 2 earns 5% compound interest.

a) Write an expression for the accumulated amount in Account 1 minus the accumulated amount in Account 2. Graph this expression for $0 \leq t \leq 1$. At what (approximate) point in time is the difference a maximum?

b) Use calculus to find the exact point at which the difference is a maximum.

c) Find an expression for the maximum difference in the case of an interest rate of $i = .05$. Use calculus.

10) How much must be deposited now in order to accumulate to $4,000 under each of the following scenarios?

     a) The account earns 5% simple interest and is left on deposit for five years.

     b) The account earns 5% compound interest and is left on deposit for five years.

     c) The account earns 5% compound interest for the first three years and then earns 5% simple interest for two more years.

11) Find the accumulated amount of a deposit of $600 in each case.

     a) The account earns 4.5% simple interest for seven years.

     b) The account earns 4.5% compound interest for seven years.

12) What monthly rate of interest is equivalent to the following?

     a) A quarterly rate of 4%.

     b) A weekly rate of 6% (assume four weeks exactly per month).

     c) An annual rate of 5%.

13) What is the effective rate of annual interest on a car loan advertised as 1.5% per month?

14) Compute the amount of interest paid in each case.

     a) $500 is borrowed for one year at 6% compound interest.

     b) $500 is borrowed for one year at 6% discount.

15) Convert as indicated

     a) Find i if $d = .07$.

     b) Find d if $i = .07$.

16) A borrower receives $560 and must repay $600 at the end of one year.

     a) What is the effective annual rate of discount?

     b) What is the effective annual rate of interest?

17) What nominal rate of interest, converted monthly is equivalent to a nominal rate of discount of 5.6% converted quarterly?

18) Find $d^{(4)}$ if $i^{(3)} = 2\%$.

19) a) Find an expression for $a(t)$ if $\delta_t = \frac{t}{1+t^2}$.
    b) If $A(4) = 30$, find $A(2)$.

20) Find an expression of $a(t)$ if $\delta_t = \frac{1}{(t+3)\ln(t+3)}$.

21) A deposit of $100 earns interest $i_1 = .07$, $i_2 = .05$, $i_3 = .056$, $i_4 = .08$, $i_5 = i_6 = \cdots i_{10} = .03$.

   a) What is the accumulated amount in the account at the end of period 10?

   b) What is the effective annual rate of compound interest for this account?

   c) What rate of annual simple interest will yield the same accumulated value?

22) Verify that $\delta_t = \frac{i}{1+it}$ results in the correct formula for $a(t)$ in the case of simple interest.

23) A CD increases in value from $4,000 to $5,432.66 in five years.

   a) What is the effective annual rate of compound interest?

   b) What is the effective rate of simple interest?

   c) What is the effective annual rate of compound discount?

24) How much must you deposit now in order to have $10,000 in seven years?

   a) If interest is a nominal 5% compounded monthly.

   b) If interest is 7% simple interest.

   c) If interest is a nominal 4% compounded quarterly.

25) An insurance company earns 7% on their investments. How much must they have on reserve on January 1, 2008, to cover the claims for the next three years if they expect claims of $500,000 at the end of 2008, $300,000 at the end of 2009, and $250,000 at the end of 2010?

26) What is the effective annual interest over eight years for an account which grows at 5% simple interest for the first four years and then 5% compounded monthly for the next four years? What if the situtation is reversed? (compound the first years, simple the second four years).

27) The amount of interest earned on $X for one year is $562.50. The equivalent amount of discount on $X is $500. What is X?

28) At a constant force of interest, f, \$200 increases to \$240 over two years. Find f.

29) A car loan is paid off in forty-eight equal monthly payments of \$365. The total interest paid over the life of the loan was \$2520. Find the original amount borrowed.

30) Consider the function $a(t) = t^2 + 3t + 1$.

    Verify that $a(t)$ satisfies the three properties of an accumulation function.

31) (a) Prove that $A(n) - A(0) = I_1 + I_2 + \cdots + I_n$.

    (b) Verbally interpret the result from part (a).

    (c) Is it true that $a(n) - a(0) = i_1 + i_2 + \cdots + i_n$? Explain.

32) Joe has \$653 at Bank A and \$429 at Bank B. He just received a bank statement in forming him that for a past year he earned \$30 in interest at Bank A and \$24 in interest at Bank B.

    (a) Find the effective interest rate for each bank.

    (b) Which bank would you recommend that Joe should use for future saving?

33) Assume that $A(t) = 100 + 6t$.

    (a) Find i_4.

    (b) Find i_8.

34) Assume that $A(t) = 100(1.085)^t$.

    (a) Find i_4.

    (b) Find i_8.

35) At an unknown simple rate of interest, \$150 will accumulate to \$170 over a period of time. Find the accumulated value of \$700 at a simple rate of interest three fourths as great over double the period of time.

> **For Problems 36–39, solve assuming first simple interest and then compound interest.**
>
> 36) A local bank is advertising that you can double your money in eight years if you invest with them. Suppose you have \$1,000 to invest. What interest rate is the bank offering?

37) Steve plans to put his graduation money into an account and leave it there for four years while he goes to college. He receives $1,250 in graduation money that he puts it into an account at Cardinal Credit Union that earns 5.65% interest. How much will be in Steve's account at the end of four years?

38) (a) At what rate will $2,500 accumulate to $3,200 in three and a half years?

(b) In how many years will $2,500 accumulate to $3,500 at 9.6%?

39) (a) In how many years will $1,000 accumulate to $1,500 at 11.8% interest?

(b) At what rate of interest will $1,000 accumulate to 1500 in five years?

40) It is known that $500 invested for two years in Dan Green's Generous Index Fund will earn $220 in interest. Assuming continued performance of this fund, find the accumulated value of $2000 invested at the same rate of compound interest for three years.

41) Find the accumulated value of $5,000 invested for ten years, if the compound interest rate is 8% for the first three years, 2% for the fourth year, -3% the fifth year, and then 12% the last five years. Hint: use $i = -.03$ for year 5.

42) A deposit of $654 accumulates to $781 in four years what is the rate of

(a) Simple interest.

(b) Compound interest.

43) At what rate of interest per month will $1,500 invested at an interest rate of i per month **compounded monthly** accumulate to $4,000 in ten years?

44) What deposit is required to accumulate $6,000 in five years at 5.41% simple interest?

45) Nathan had $25,000 in a Money Market Account on January first of this year.

(a) Assuming a compound interest rate of 5.7%, what was the present value of the account when it was opened eight years ago?

(b) Nathan wants to buy a $150,000 condo and needs a 20% down payment to avoid private mortgage insurance charges. Assuming he makes no contribution to his current savings, how long does he have to wait to achieve his savings goal?

Excel Projects For Chapter 2

1) Construct an Excel spreadsheet which performs the following conversions. See Table 2.51.

TABLE 2.51

Use enters	Output	Output	Output
i, m	$i^{(m)}$	$d^{(m)}$	
$i^{(m)}, m, p$	i	$d, d^{(p)}$	
$d^{(p)}, m, p$	$i^{(m)}$	d	i
$i^{(m)}, m, n$	d	i	

Example: (Table 2.52) Spreadsheet is on the left, formulas are on the right. To convert your worksheet to the format on the right click on the Formulas tab and then chose Show Formulas. Click again to return normal view.

TABLE 2.52

Enter i	1.45%	Enter i	0.0145
Enter m	4	Enter m	4
$i^{(m)} =$	1.442%	$i^{(m)} =$	$= B2 * ((1 + B1)^{(1/B2)} - 1)$
$d^{(m)} =$	1.437%	$d^{(m)} =$	$= B2 * (1 - (1 + B1)^{-(1/B2)})$

2) Construct an Excel spreadsheet which performs the following calculations assuming a nominal rate of interest of $i^{(m)}$: See Table 2.53.

TABLE 2.53

User Enters	Output
$PV, n, m, i^{(m)}$	FV
$FV, n, , m, i^{(m)}$	PV

Table 2.54 is a sample of how it might look:

TABLE 2.54

ENTER THESE VALUES				Computed Value
PV	n	m	i	FV
1000	4	3	5.60%	$1,248.49

Chapter 3

Solving Problems in Interest

3.1 Introduction

In this chapter, we extend the ideas in Chapter 2 to more complicated situations. Each problem will involve the same quantities:

1. The principal originally invested (PV).

2. The length of the investment period (t) (N for the **TI BA II Plus**).

3. The rate of interest (i). (I/Y for the TI BA II Plus).

4. The accumulated value of the investment at the end of the investment period (FV).

5. The payment (PMT) which will be assumed to be 0 in this chapter. We will consider cases where the payment is not 0 in Chapters 4 and 5.

Assuming that the $PMT = 0$ knowledge of any three of the first four quantities is sufficient to determine the fourth quantity. The **TI BA II Plus** TVM keys allow you to enter any three of these and compute the value of the fourth.

3.2 Measuring Time Periods: Simple Interest and Fractional Time Periods

In all of our examples so far we have simply stated a value of t. In most practical problems, time is expressed in terms of a date of inception and a date of termination. Thus, we might have an investment which is initiated on January 12, 2012, and terminates on September 11, 2014. In order to use our formulas we need to convert these dates into a time period. There are two methods in common use for calculating the value of t corresponding to the time between two specified dates.

Method One: Exact

This method uses the exact number of days and calculates t based on the assumption that there are 365 days in a year.

Method Two: Ordinary Simple Interest

This method assumes that each month has 30 days, resulting in a year of 360 days. If our two dates are M_1, D_1, Y_1 and M_2, D_2, Y_2 the formula for computing the number of days between two dates are M_1, D_1, Y_1 and M_2, D_2, Y_2 days for ordinary simple interest is

$$360(Y_2 - Y_1) + 30(M_2 - M_1) + (D_2 - D_1) \tag{3.1}$$

In both of these two cases, the easiest way to calculate the number of days between two dates is to use the **DATE** Worksheet on the **TI BA II Plus**.

We invoke this worksheet by keying in 2ND DATE (it's above the 1). As for other worksheets, we move from entry to entry using the ↑, ↓. Dates are entered as mm.ddyy. Thus, March 5, 2007, would be entered as 03.0507. It will display as 3.0507 until you hit the ENTER key, at which point it will display as 03-05-2007.

There are four variables which are accessed by using the ↑ or ↓ keys. DT1 is the first date and is always assumed to be prior to DT2, the second date. DBD is the number of days between the two dates. The final variable is either ACT (for exact number of days) or 360 (using the ordinary simple interest method). This last variable is toggled by using 2ND SET. Here's how to compute the number of days between March 3, 2008, and October 5, 2009, using both the ordinary simple interest rule and the exact number of days (Table 3.1).

2ND Date

TABLE 3.1

	Key in	Display
$DT1$	3.0314 ENTER	3-03-2014 ↓
$DT2$	10.0515 ENTER	10-05-2015 ↓↓
$ACT/360$	Use 2*ND Set* to toggle this	ACT ↑
DBD	CPT	581 ↓
ACT	$2ndSet$	360 ↑
DBD	CPT	572

Since there are 581 days between March 3, 2014, and October 5, 2015, using the exact method we obtain $t = \frac{581}{365} = 1.592$. For the ordinary simple interest method we have $t = \frac{572}{360} = 1.588$.

Example 3.1 Compute the interest earned on a loan of \$4,500 at simple interest of $i = 5.76\%$ over the period from March 3, 2014, and October 5, 2015, using both methods

Solution:
Method One: $t = 1.592$ and so we obtain $.0576 \cdot 1.592 \cdot 4,500 = \$412.65$
Method Two: $t = 1.588$ yielding $.0576 \cdot 1.588 \cdot 4,500 = \$411.60$

3.3   Fractional Time Periods

The equation $FV = PV(1+i)^t$ can be evaluated directly for any value of t. For example, if we want the value of an investment of \$5,456 at six and a third years earning compound annual interest of 5.6%, we have (Table 3.2)

$$FV = 5456(1.056)^{6\frac{1}{3}} = 5456 \cdot 1.412119603 = \$7,704.52$$

TABLE 3.2

N	I/Y	PV	PMT	FV
6.3333	5.6	5,456	0	$CPT = -7,704.52$

This method gives the exact value of the accumulated amount over any time period. In some cases the interest earned over the fractional portion of a time period is computed using simple interest. Historically this practice developed due to the difficulties involved in computing the value of $(1+i)^t$ if t is not an integer.

As an example, simple interest is sometimes used to compute the interest due at inception for a home mortgage. Home mortgages are timed to start on the first day of the month subsequent to the date of inception of the loan. If you close on a home loan on a day other than the first of the month, you are assessed simple interest on the loan amount for the number of days remaining in that month at the inception of the loan. After that, interest is computed using compound interest. A loan which closes on May 12 will be assessed interest on the remaining days in May as part of the closing costs. The loan will begin June 1, and the first regular payment will be due at the end of June.

We can calculate the FV of an account using simple interest over the fractional period as follows. Suppose that $t = n + q$ where n is a whole number and q represents the fractional portion of time $(0 < q < 1)$ we would compute the FV using simple interest for the fractional portion of time as

$$FV = PV(1+i)^n(1+iq) \tag{3.2}$$

Using this method in our previous example gives us

$$FV = 5456(1.056)^6 \left(1 + .056 \left(\frac{1}{3}\right)\right)$$
$$= 5456 \cdot 1.41258 = \$7,707.08$$

Notice that this method always yields a larger value than the exact calculation. Unless otherwise stated, we will always use the exact calculation.

Example 3.2 The Piersons purchase a home for $350,000 and finance $275,000. The interest rate is a nominal 8% convertible quarterly. The Piersons close on July 12, 2007, and must pay simple interest on the amount financed for the remaining days in July. How much do they owe at closing?

Solution: There are 19 days remaining in July. If we use the exact number of days formula the interest due is

$$275000 \cdot (.08) \cdot \left(\frac{19}{365}\right) = \$1,145.21$$

Computing FV Over Non-Integer Time Periods $t = n + q$
     Exact Answer

$$FV = PV(1+i)^t \tag{3.3}$$

Using Simple Interest for the Fractional Time Period q

$$FV = PV(1+i)^n(1+iq) \tag{3.4}$$

3.4   Equations of Value at any Time

We will often need to compare two or more amounts of money at different points in time. If the rate of interest is not zero, this is only makes sense if we compute the values of each amount at a common point in time. This common date is called the **comparison date**. We can choose any date we like. An amount which occurs prior to the comparison date must be accumulated to that date while an amount which occurs after the comparison date must be discounted. It is often helpful to use a number line to place each amount at the appropriate time. This is called a time diagram. The equation which equates two different methods of computing the value of a transaction is called the **equation of value**. An amount which will be available n periods after the comparison date is discounted[1] by the factor v^n Recall that $v = \frac{1}{1+i}$. An

---

[1]Note that this is not the rate of discount (d) which was discussed earlier.

amount which will be available m periods prior to the comparison date is accumulated by a factor of $(1+i)^m$.

Example 3.3 In return for the promise of a payment of $600 at the end of eight years, a person agrees to pay $100 now, $200 at the end of five years, and to make one further payment at the end of ten years. What is the required final payment if the nominal rate of annual interest is 6.2% compounded semi-annually?

Solution. We will measure time in six-month intervals. This gives us $i = \frac{.062}{2} = .031$ and $v = \frac{1}{1.031}$. We will use **the inception date as our comparison date**. We then have the following situation for each amount of money using X for the (unknown) value of the final payment. See Figure 3.1.

$600: is not available for eight years or sixteen periods. Its value at inception is thus $600v^{16}$

$100: is available now. Its value is thus 100

$200: is not available for five years or ten periods. Its value at inception is thus $200v^{10}$

$X: is not available for ten years or twenty periods. Its value at inception is thus xv^{20}

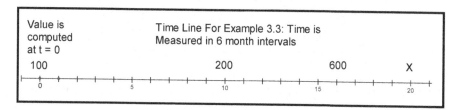

FIGURE 3.1

By assumption the value at inception of the $600 payment is equal to the sum of the value at inception of the other three amounts. This gives us the following **equation of value:**

$$600v^{16} = 100 + 200v^{10} + xv^{20}$$

Solving this equation for x gives us

$$X = \frac{600v^{16} - 100 - 200v^{10}}{v^{20}} = \$222.38$$

TI-30XS Solution.

We store $v = \frac{1}{1.031}$ as x and then compute directly

$$(600 \times x(\wedge)16 - 100 - 200 \times x(\wedge)10)/x(\wedge)20(ENTER)222.3767$$

TI BA II Plus solution.

We write the equation of value as

$$600v^{16} - 100 - 200v^{10} = xv^{20}$$

The three numbers on the left are present values, while x is a future value. We compute the sum of the three present values and then compute the future value that they correspond to. We will need to store the two computed values on the left (Table 3.3).

TABLE 3.3

N	I/Y	PV	PMT	FV
16	3.1	$CPT = 368.136591$	0	-600

We then press STO 1. This stores the \$368.1396591 into storage register 1 (Table 3.4).

TABLE 3.4

N	I/Y	PV	PMT	FV
10	3.1	$CPT = -147.3816287$	0	200

Now subtract 100 to obtain -247.3816287 and then add RCL 1 $=$ 368.1396591 to obtain 120.75880334, the present value of the left hand side. We want to convert this to a future value at period 20 (Table 3.5):

TABLE 3.5

N	I/Y	PV	PMT	FV
20	3.1	120.7580334	0	$CPT = -222.3767288$

Note: be careful with signs. The **TI-30XS** is clearly the easier method here. You can enter the formula directly into the **TI BA II Plus** as well, but it is very easy to make a mistake since you can't see what's being entered as you go along.

As a general rule, you should use the **TI-30XS** for all strictly arithmetic calculations and the **TI BA II Plus** when the TVM keys or other worksheets are required.

Alternative Solution: We compute the value of each monetary amount at the time of the final payment. See Figure 3.2.

$600: has been available for two years or four periods. Its value is thus $600(1 + i)^4$

$100 has been available for ten years or twenty periods. Its value is thus $100(1 + i)^{20}$

$200 has been available for five years of ten periods. Its value is this $200(1 + i)^{10}$

Since X is being valued at the time it becomes available, its value is thus just X.

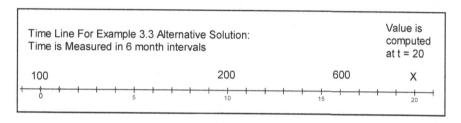

Time Line For Example 3.3 Alternative Solution:
Time is Measured in 6 month intervals

Value is computed at t = 20

100   200   600   X

FIGURE 3.2

Using this comparison date our equation of value becomes

$$600(1 + i)^4 = 100(1 + i)^{20} + 200(1 + i)^{10} + X$$

Solving for x gives us

$$X = 600(1 + i)^4 - 100(1 + i)^{20} - 200(1 + i)^{10}$$

This is easily calculated using the **TI-30XS**. Store 1.031 as variable x and then enter the formula just as written

$600 \times x(\wedge)4$ $- 100 \times x(\wedge)20$ $- 200 \times x(\wedge)10(ENTER)$

222.3767

While we can create the equation of value at any comparison date, the inception or termination dates are most commonly used. In this example, the termination date was easier to use. In some cases several payments occur on a particular date. In that case it's often easier to compute the equation of value on that date. Here's an example.

Example 3.4 Find the interest rate at which payments of $200 at the end of five years and $300 at the end of ten years have the same value as a single payment of $400 at the end of five years. See Figure 3.3.

Solution. Since two of the three amounts occur at the end of five years, this is the most convenient comparison date since we need only adjust the $300 payment. The values at the comparison date are (Table 3.6):

TABLE 3.6

Amount	Time Available	Value at the end of year 5
200	End of year 5	200
300	End of year 10	$300v^5$
400	End of year 5	400

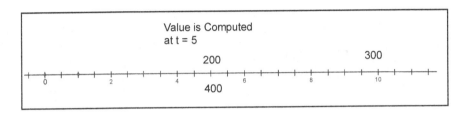

FIGURE 3.3

Our equation of value computed at the end of year 5 is

$$200 + 300v^5 = 400$$

We first solve for v and then for i. We have $v^5 = \frac{400-200}{300} = \frac{2}{3}$ so that $v = \left(\frac{2}{3}\right)^{\frac{1}{5}} = .92211$ and hence $i = \frac{1}{v} - 1 = .08447$ or 8.447%.

    TI BA II Plus solution. We re-write the equation of value as $200 = 300v^5$. $200 is a present value, while $300 is the future value. We have $N = 5$ and need the value of I/Y (Table 3.7):

TABLE 3.7

N	I/Y	PV	PMT	FV
5	$CPT = 8.44717712$	200	0	-300

TI-30XS solution.

$$1 + i = \left(\frac{300}{200}\right)^{\frac{1}{5}} = 1.084471, \; i = .084471$$

3.5   Unknown Time

Problems involving an unknown value of t often lead to equations which can only be solved approximately using numerical methods. The **TI BA II**

Plus, Excel, and MAPLE all have approximation algorithms. Of the three, Excel seems to be the least accurate. While the differences are slight, for large amounts of money, Excel might lead to significant errors.

A common problem concerns replacing a sequence of payments (called an annuity – to be discussed in greater detail later) with a single payment equal to the sum of the numerical values of the other payments. The problem – determine the time at which the single payment is to be made in order that the sequence of payments is equal in value to the single payment.

As our first example, suppose that payments in the amounts $p_1, p_2, p_3, \ldots, p_n$ are made at times $t_1, t_2, t_3, \ldots, t_n$. Suppose further that interest is compounded at an effective rate of i per period. We are to find a time, t, at which a single payment of $p_1 + p_2 + p_3 + \cdots + p_n = \sum_{k=1}^{n} p_k$ has the same value as this annuity. We compute our equation of value at the inception of the transaction resulting in the following equation of value. The left side of Equation 3.4 is the value of the single payment at time $t = 0$, while the right side is the value of the sequence of payments at this same time

$$\left(\sum_{k=1}^{n} p_k \right) v^t = \sum_{k=1}^{n} p_k v^{t_k} \tag{3.5}$$

Solving this equation for t yields the following equation:

$$t = \frac{\ln \left(\frac{\sum_{k=1}^{n} p_k v^{t_k}}{\left(\sum_{k=1}^{n} p_k \right)} \right)}{\ln(v)} \tag{3.6}$$

Example 3.5 Suppose that the rate of interest is a nominal 5.04% annually and that interest is compounded monthly. Find the time at which a single payment of $1000 is equivalent in value to the following sequence of payments displayed in Table 3.8.

TABLE 3.8

Payment	Date of Payment	Period in Months
400	End of year 2	24
300	End of year 5	60
200	End of year 7	84
100	End of year 8	96

Solution:

Method One: Since interest is compounded monthly we measure time in months. Our interest rate is then $i = \frac{.0504}{12}$ and $v = \frac{5000}{5021} = .9958$.

Using Equation 3.5 we have

$$t = \frac{\ln \left(\frac{400v^{24} + 300v^{60} + 200v^{84} + 100v^{96}}{1000} \right)}{\ln(v)} = 52.48454$$

The single payment is to be made in month 52.48454 or year 4.37371167. This would be four years later in month $(.37.71167)(12) = 4.48454$. The payment will thus be made in April (month 4). Since April has thirty days in it, the payment will be made on $30(.48457) = 14.536$. Thus, the single payment is due on April 14 four years from the present. If your bank is really greedy, we could compute the exact time of day the payment would be due!!

The calculation is easiest on the **TI-30XS** store v as x and then compute the value of the expression inside ln in the numerator. Store this as y and then compute $\frac{\ln(y)}{\ln(x)}$. Note that this is not the same as $\ln(\frac{y}{x}) = \ln(y) - \ln(x)$.

Method Two: Using years as the unit of measure. We must first compute the effective annual interest rate corresponding to a nominal annual rate of 5.04% compounded monthly. We have

$$1 + i = \left(1 + \frac{.0504}{12}\right)^{12} = 1.051580694$$

So $i = .051580694$ and $v = .950949371$. Our expression for t in years, is

$$t = \frac{\ln\left(\frac{400v^2 + 300v^5 + 200v^7 + 100v^8}{1000}\right)}{\ln(v)} = 4.3737120470$$

Note the slight difference in the two values. Using monthly interest, we obtained year 4.35477308.

We can also use the **TI BA II Plus** Cash Flow Worksheet. We will use this worksheet again in Chapter 6. The **TI BA II Plus** Cash Flow Worksheet is accessed by the CF key (it's not a 2 ND function). As with all worksheets start by entering $2nd\ CLRWORK$ to delete any previous data. You then enter in each cash flow (using plus signs for contributions and negative signs for withdrawals). When all the deposits/withdrawals have been entered, the keystrokes IRR, CPT will compute the internal rate of return. Note that you must include every period - even those in which no contribution or withdrawal occurs.

To illustrate how CF works, we create a table representing the payments in Example 3.5. We will measure time in years and use the effective annual interest rate. See Table 3.9.

To begin enter CF, $2ndCLRWORK$ to clear out any previous entries in the CF worksheet. This is critical as all previous work is retained otherwise.

The display will show CFo = 0. CFo is the initial cash flow (at time $t = 0$, hence CFo). If there is no cash flow the first period, enter 0. We have 0 as CFo so press 0 ENTER. **Each entry in this worksheet must be confirmed by use of the ENTER key.** The calculator will display CFo = 0. If you don't see the = sign, you forgot to press the ENTER key.

Each subsequent contribution has both an amount ($CO1, CO2$, etc) and a frequency ($FO1, FO2$, etc.). This saves time when the same amount is deposited/withdrawn for two or more consecutive periods. The frequency represents the number of periods its associated deposit/withdrawal is made. We

TABLE 3.9

Year(s)	Payment
0	0
1	0
2	400
3,4	0
5	300
6	0
7	200
8	100

must account for every period even if there is no activity. This means we often need to enter 0 for multiple years.

These later entries are accessed using the ↓ key. You can retrace your steps using the ↑ key. C01 represents the deposit/withdrawal at time $t = 1$. Pressing the ↓ once after entering CFo = 0 will result in a display of CO1 = 0.00000. In our example, this amount is 0, so we press 0 and then enter. The display now shows CO1 = 0. We use the ↓ to get to F01 = 1.00. This is the default setting and represents that fact that the deposit of 0 is only made at the end of period one. That is correct for this example, so we don't need to change it.

We now use the ↓ key to enter each subsequent cash flow and its associated frequency. We must account for each year, so if there is no deposit or withdrawal in a period it is entered as 0. There is a deposit of $400 in period 2. We thus enter CO2 = 400 and FO2 = 1.

Table 3.10 is a chart showing the values which are entered into the **TI BA II Plus** CF worksheet for this example. Note that each entry must be followed by pushing the ENTER key.

TABLE 3.10

Time	Symbol	Amount	Frequency
0	CFo	0	
1	C01	0	$F01 = 1$
2	C02	400	$F02 = 1$
3,4	C03	0	$F03 = 2$
5	C04	300	$F04 = 1$
6	C05	0	$F05 = 1$
7	C06	200	$F06 = 1$
8	C07	100	$F07 = 1$

We want to find the time t at which a deposit of \$1,000 has the same present value as the sequence of deposits listed above. The CF worksheet will tell us the present value of the sequence if we provide the interest rate. Since we are counting in years, we use the effective annual interest rate of .05158694 = 5.1580694%. The Cash Flow Worksheet uses NPV for the (net) Present Value. We compute that as follows:

1) Press the NPV key

2) You will see I - key in the interest rate and press ENTER

3) Press ↓. You will see NPV.

4) Press CPT and you will see the NPV = 802.54

This tells us that a single payment of \$802.54 at time $t = 0$ has the same PV as the sequence of payments in this problem. We want to know the time N at which the value is \$1,000. We use The TVM keys to compute that N as follows

1) Transfer the NPV value to the TVM keys by pressing PV.

2) Exit the CF Worksheet by pressing 2ND, QUIT.

3) Enter the Interest rate and desired future value and compute the required value of N (Table 3.11).

TABLE 3.11

N	I/Y	PV	PMT	FV
$CPT = 4.37$	5.1580694	802.54	0	$-1,000$

We can use the CF Worksheet to solve most of the other problems in this section as well. As an example we re-work Example 3.4.

Example 3.4: (using the CF keys) Find the interest rate at which payments of \$200 at the end of five years and \$300 at the end of ten years have the same value as a single payment of \$400 at the end of five years.

Solution: In this case we need to compute an interest rate. The CF Worksheet uses the notation IRR(internal rate of return) for the interest rate. It computes IRR as the interest rate at which the NPV(net present value) is \$0. That means we must have withdrawals as well as payments. We can make this work for us by entering the payments of \$200 and \$300 as negative and the payment of \$400 as positive[2]. That means we have we have a net positive

---

[2]We always attach signs to the entries so that the initial entry is positive. As we will see in Chapter 6, this is required if we want to ensure that the value for IRR is unique. Note that the **TI BA II Plus** will not warn you about this.

payment of $200 in year 5 and a negative payment of $300 in year 10. Here is how it's entered into the CF Worksheet. Make sure you press 2ND CLR WORK before you begin and 2ND QUIT when you are finished (Table 3.12).

TABLE 3.12

Time	Symbol	Amount	Frequency
0	CFo	0	
1, 2, 3, 4	C01	0	4
5	C02	200	$F01 = 1$
6, 7, 8, 9	C03	0	$F03 = 4$
10	C04	-300	$F03 = 1$

We now press IRR and then CPT resulting in $IRR = 8.44717712$. This is the same value we obtained earlier.

Method of Equated Time

There is a reasonably good approximation to the solution to the single payment problem known as the **method of equated time**. The method of equated time uses a time-weighted average of the payments to compute an estimate for the value of t. In the equation below p_k represents a payment (or withdrawal) at time t_k.

$$\bar{t} = \frac{\frac{\sum_{k=1}^{n} p_k t_k}{n}}{\sum_{k=1}^{n} p_k} \qquad (3.7)$$

The method of equated time dates to the period before calculators had the capability to compute logarithms and is little used today.

3.6 Doubling Time

We are often interested in determining how long it will take a fixed deposit to increase in value by a given amount. In the case of compound interest the time required for a deposit to increase by any given factor is independent of the amount which is currently on deposit. The amount of time a given amount takes to double in value is known as the doubling time.

We begin by deriving a formula for the doubling time as a function of the rate of interest in the case that interest is compounded.

We have $A(t) = P_0 a(t) = P_0(1+i)^t$. Starting with any initial time, t_0 we want to find t so that $A(t) = 2A(t_0)$. This gives us

$$P_0 a(t) = 2P_0 a(t_0)$$
$$P_0(1+i)^t = 2P_0(1+i)^{t_0}$$
$$(1+i)^t = 2(1+i)^{t_0} \tag{3.8}$$

Taking logarithms of both sides we obtain

$$t\ln(1+i) = \ln(2) + t_0 \ln(1+i)$$
$$(t - t_0) = \frac{\ln(2)}{\ln(1+i)} \tag{3.9}$$

An easy generalization of this gives the following formula for the time required for an investment to increase by a factor of k at a given rate of interest i is given by

$$\Delta t = \frac{\ln(k)}{\ln(1+i)} \tag{3.10}$$

We can use Equation 2.17 to obtain this same solution. We have

$$A(t_1) = kA(t_0) = \frac{a(t_1)}{a(t_0)} A(t_0)$$
$$\frac{a(t_1)}{a(t_0)} = k = \frac{(1+i)^{t_1}}{(1+i)^{t_0}} = (1+i)^{t_1-t_0}$$
$$\ln(k) = (t_1 - t_0)\ln(1+i) \tag{3.11}$$

We then solve for $(t_1 - t_0)$ as before.

Time for an Investment to Increase by a Factor of k

$$t_k = \Delta t = \frac{\ln(k)}{\ln(1+i)} \tag{3.12}$$

To use the **TI BA II Plus**, we use the TVM keys. Use $PV = 1$, $FV = -k$ and solve for n. We enter PV as negative because of the **TI BA II Plus** convention that PV and FV must have opposite signs.

Example 3.6 Find the time required to increase an investment by a factor of 7 if the interest rate is $i = .076$.

Solution: Using Equation 3.12, $t_7 = \frac{\ln(7)}{\ln(1.076)} = 26.565$.

TABLE 3.13

N	I/Y	PV	PMT	FV
$CPT = 26.57$	7.6	1	0	-7

TI BA II Plus solution

We can solve Equation 3.11 for i to find the interest rate required to obtain an increase by a factor of k in time t_k (Table 3.13).

Interest Rate Required to Increase an Investment by a Factor of k in time t_k

$$\ln(1+i) = \frac{\ln(k)}{t_k}$$
$$1+i = e^{\frac{\ln(k)}{t_k}}$$
$$i = e^{\left(\frac{\ln(k)}{t_k}\right)} - 1 \qquad (3.13)$$

Example 3.7 What interest rate is required in order that an investment of $500 will grow to $988 in three and a half years?

Solution

$$k = \frac{988}{500} = 1.976, \ t = 3.5 \Rightarrow i = e^{\frac{\ln(1.976)}{3.5}} - 1 = 1.21481614 - 1 = 21.48\%$$

TI BA II Plus solution (Table 3.14):

TABLE 3.14

N	I/Y	PV	PMT	FV
3.5	$CPT = 21.48161429$	500	0	-988

Alternatively, we can compute the interest as follows using the growth factor rather than the absolute amounts (Table 3.15).

TABLE 3.15

N	I/Y	PV	PMT	FV
3.5	$CPT = 21.48161429$	1	0	$\frac{-988}{500} = -1.976$

Example 3.8 How long will it take for an investment earning a nominal 7.88% compounded semiannually to increase from $1,456.78 to $2,345.89?

Solution; We will measure time in half-years so that $i = \frac{.0788}{2}$ and $k = \frac{2345.89}{1456.78}$ This gives us:

$$\Delta t = \frac{\ln\left(\frac{2345.89}{1456.78}\right)}{\ln\left(1 + \frac{.0788}{2}\right)} = 12.32897683$$

This number represents the number of half-years, so the time in years is $6.164488414 \approx 6.16$ years.

TI BA II Plus solution (Table 3.16):

TABLE 3.16

N	I/Y	PV	PMT	FV
$CPT = 6.164488$	$\frac{.0788}{2}$	1	0	$-\frac{2345.89}{1456.78}$

3.6.1   The Rule of 72

This is an archaic estimate for the doubling time of an investment. If an investment earns $x\%$ effective annual interest it will double in approximately $\frac{72}{x}$ years. For example an investment earning 7.2% interest would double in approximately $\frac{72}{7.2}$ = ten years. Using the **TI BA II Plus** TVM keys we obtain 9.969 years, so the estimate is pretty good in this instance. See Chapter 10 for more on the Rule of 72.

3.7   Finding the Rate of Interest

Problems involving an unknown rate of interest often require an approximation technique. We work several examples to give the flavor of the sorts of issues which arise. As usual the TVM keys will be our best friend!

Example 3.9 At what nominal rate of annual interest, convertible monthly, will \$456 accumulate to \$500 in three years?

Solution: We will use months as our unit of measurement to avoid an interest-conversion problem.

We use Equation 2.25 and the **TI-30XS**

$$FV = PV\left(1 + \frac{i^{(m)}}{m}\right)^{mt}$$

We have $mt = 36$ (months), $PV = 456$, and $FV = 500$

$$500 = 456 \left(1 + \frac{i^{(12)}}{12}\right)^{36}$$

$$\left(1 + \frac{i^{(12)}}{12}\right) = \left(\frac{500}{456}\right)^{\frac{1}{36}}$$

$$\frac{i^{(12)}}{12} = \left(\frac{500}{456}\right)^{\frac{1}{36}} - 1 = .002562034$$

Thus $i^{(12)} = 12(.002562034) = 0.03074440799 = 3.0744\%$.
TI BA II Plus solution (Table 3.17):

TABLE 3.17

N	I/Y	PV	PMT	FV
36	$CPT = .256203444$	456	0	-500

This gives the monthly rate of .002562034, so $i = 12 \cdot .00256204 = .03074440799 = 3.0744\%$

NOTE: We didn't use the $i^{(m)}$ conversion formula since the problem requested nominal annual rate of interest and not the effective annual rate of interest.

Example 3.10 At what effective rate of interest will the sum of the present value of \$2,000 at the end of two years and \$3,000 at the end of four years be equal to \$4,000?

Solution: Figure 3.4 displays the time line for this problem.

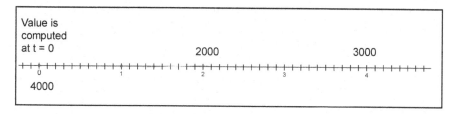

FIGURE 3.4

The equation of value at the inception of the transaction is

$$4000 = 2000v^2 + 3000v^4$$

We can re-write this as $3000v^4 + 2000v^2 - 4000 = 0$ or $3v^4 + 2v^2 - 4 = 0$. At this point we can use several methods to find v and thus i.

1) Cute, but doesn't work very often: Let $u = v^2$. This results in the quadratic equation $3u^2 + 2u - 4 = 0$ with solutions

$$v^2 = u = \frac{-2 \pm \sqrt{52}}{6}$$

Since $v^2 > 0$, only the positive solution is viable. We thus have $v^2 = \frac{-2+\sqrt{52}}{6}$. Giving us $v^2 = .86852$, $v = .93194$ and $i = \frac{1}{v} - 1 = .07303 = 7.303\%$.

NOTE: The FM exam often has a problem like this. Being aware of this little trick can make it easy to work. If you see one exponent which is twice another exponent, consider the possibility that you can use this technique.

2) Using the CF Worksheet. We find the interest rate at which an initial deposit of \$4,000 has the same NPV as deposits of \$2,000 in year 2 and \$3,000 in year 4. The interest is computed by pressing the IRR key. The IRR function calculation assumes a NPV of 0. As a result we treat the deposits in years 2 and 4 as withdrawals. Table 3.18 displays the entries to be made in the CF Worksheet.

TABLE 3.18

Time	Symbol	Amount	Frequency
0	CF0	4000	
1	C01	0	$F01 = 1$
2	C02	−2,000	$F02 = 1$
3	C03	0	$F03 = 1$
4	C04	−3,000	$F04 = 1$

Now we press IRR and then CPT resulting in $IRR = 7.30274083$

Note: We could have entered $CF0 = -4,000$, $C02 = 2,000$ and $C04 = 3,000$ and obtained the same answer. That being said, it is **recommended that you enter the initial transaction as a positive number.** As we shall see in a later chapter, the yield rate (IRR) is not always unique unless the account balance is always positive.

Example 3.11 If interest is compounded quarterly deposit will double in ten years. What is the effective annual interest rate?

Solution: The investment must double in forty quarters (Table 3.19).

TABLE 3.19

N	I/Y	PV	PMT	FV
40	$CPT = 1.74796921$	1	0	−2

We can now compute the effective annual interest by changing N to 4 and computing the FV. Since PV $= 1$, we have $i = FV - 1$ (Table 3.20).

TABLE 3.20

N	I/Y	PV	PMT	FV
4				$CPT = -1.0717736463$

As usual, we ignore the minus sign and obtain $i = .071773463$.

Example 3.12 At what nominal rate of interest compounded semiannually would investments of \$1,000 now and \$2,000 in three years accumulate to \$5,000 ten years from now?

Solution: Let $j = \frac{i^{(2)}}{2}$. We compute the equation of value at the end of ten years or twenty, six-month intervals

$$5000 = 1000(1 + j)^{20} + 2000(1 + j)^{14}$$

If we let $u = 1 + j$, this simplifies to $2u^{14} + u^{20} - 5 = 0$ which is not easy to solve. We employ the CF worksheet. We enter the first two deposits as positive and the last as negative (Table 3.21).

TABLE 3.21

Time	Symbol	Amount	Frequency
0	CF0	1,000	
1,2,3,4,5	C01	0	$F01 = 5$
6	C02	2,000	$F02 = 1$
7-19	C03	0	$F03 = 13$
20	C04	-5,000	$F04 = 1$

We press IRR, CPT and obtain $IRR = 3.217767$.

Here is the logic behind this calculation: the CF worksheet computes IRR under the assumption that the Net Present Value (NPV) of sequence of payments/withdrawals is \$0. If the payments of 1,000 and 2,000 accumulate to \$5,000 then a withdrawal of \$5,000 will create a NPV of 0.

HINT: When you use the CF worksheet you must have an entry for every period. This means that **the sum of all the frequencies must equal the total period of time involved in the problem.** Note that the intitial deposit (CFo) occurs at period 0 and is not included in the calculation. In the last example the period of the investment was ten years or twenty half-years. We have

$$5 + 1 + 13 + 1 = 20$$

The tricky ones are the periods with no deposit. If we look at the above example, we need to total six at the time C02 is made. Since we have $F02 = 1$, we must have $F01 = 5$. Likewise, we must total twenty at $C04$, so $F03 = 19 - 5 - 1 = 13$. We can add a column to the CF worksheet template to keep track of the running total of periods (Table 3.22).

TABLE 3.22

Time	Symbol	Amount	Frequency	Total Periods
0	CFo	1,000		0
1,2,3,4,5	C01	0	$F01 = 5$	5
6	C02	2,000	$F02 = 1$	6
7-19	$C03$	0	$F03 = 13$	19
20	$C04$	$-5,000$	$F04 = 1$	20

3.8   Deposits and Withdrawals: Cash Flow Problems

We now look at a class of problems where the CF Worksheet is virtually required. Cash flow problems involve situations where money is deposited and/or withdrawn at various points during the time period of interest. Many of the problems we have just finished looking at involve a sequence of deposits and withdrawals over a specified period of time. Such a sequence of deposits/withdrawals is called an **investment scheme**. We make no assumptions as to the interest earned by these deposits and withdrawals. They may or may not earn interest and the interest rate earned may or may not be constant.

We are often interested in a number called the internal rate of return (IRR). As you may recall from our earlier work:

The internal rate of return (IRR) is a single interest rate which represents the effective or average rate of interest earned by the investment scheme. If the IRR is used as the (constant) rate of interest for a given investment scheme the final balance will match the actual final balance.

If we assume that the account is closed by withdrawing all funds available on the date of closure the sequence would have a net present value at inception of $0. The IRR is the (constant) interest rate which would also result in a NPV of $0. The IRR is used by managers to compare the returns on various schemes. Schemes which yield higher IRR values are deemed to be more profitable than those with smaller IRR values.

We begin with situations where we know the interest rate and then proceed to problems which involve calculating the IRR. Our first examples are just slightly more complicated versions of the problems we have been solving.

Example 3.13 On January 1, 2007, an account paying 7.5% annual interest compounded annually is opened with a deposit of $1,000. One year later a withdrawal of $500 is made. Two years after that, a deposit of $1,500 is made. What will be the balance in the account on January 1, 2011?

Solution 1: We work through the balance year by year. Just after the $500 withdrawal the balance will be $1000(1.075) - 500 = \$575$. That amount earns interest for two years until the $1,500 deposit is made. The balance just after that deposit is $575(1.075)^2 + 1500 = 2,164.48$. That amount then earns interest for the remaining year resulting in a final balance of $2164.48(1.075) = \$2,326.82$.

Solution 2: We keep track of each deposit/withdrawal separately and add the results remembering to enter withdrawals as negative. The equation of value at the end of the sequence of transactions is then:

Ending Balance $= 1000(1.075)^4 - 500(1.075)^3 + 1500(1.075)^2 = \$2,326.82$.

Solution 3: TI BA II Plus solution. We are interested in a future value. The CF Worksheet uses the notation NFV for this number. It is accessed by pressing NPV and using the ↓ key. Table 3.23 displays the entries we need to make.

TABLE 3.23

Time	Symbol	Amount	Frequency
0	CFo	1,000	
1	C01	−500	$F01 = 1$
2	C02	0	$F02 = 1$
3	C03	1,500	$F03 = 1$
4	C04	0	$F04 = 1$

We now press $NPV, I = 7.5, \downarrow \downarrow NFV \; CPT$ obtaining $NFV = 2,326.82$.

NOTE: The **TI BA II Plus Professional Edition** provides NFV but the **TI BA II Plus** does not. To use the **TI BA II Plus**, compute NPV and then use the TVM keys to compute the NFV. Using the **TI BA II Plus** we first compute NPV and then press the PV key. Next exit the CF worksheet $(2nd \; QUIT)$ and proceed as follows (Table 3.24):

TABLE 3.24

N	I/Y	PV	PMT	FV
4	7.5	1,742.46	0	$CPT = -2,326.82$

A generalization of the technique used in **Solution 2** allows us to treat this situation in a very general way. Let B_j represent the balance in an account at time j. If the life of the account is t periods we then have $B_0 =$ Opening Balance and $B_t =$ Ending Balance. Suppose we have a sequence of deposits/withdrawals C_i made at times t_i. We assume $C_0 = B_0$. The deposit (withdrawal) C_i will be on account for a time period of $t - t_i$ and thus accumulate to $C_i(1+i)^{t-t_i}$ by the end of the investment. We then just add up all the pieces:

$$B_t = C_0(1+i)^t + C_1(1+i)^{t-t_1} + \cdots + C_n(1+i)^{t-t_n} \qquad (3.14)$$

If $C_t < 0$ we are subtracting the value of C_t plus its accumulated value. We could have added/subtracted at each time t_i, but it is easier to treat each deposit/withdrawal as a separate investment and then just add up the results. We will revisit this formula in Chapter 6. In that context, we will be trying to find the effective annual interest i given the account balances at various points in time. We will use Equation 3.14 in Chapter 6 when we consider the question of the uniqueness of the yield rate. It turns out that some investment schemes don't have a unique yield rate – a disturbing possibility if you are a manager! For now, we continue our survey of the sorts of problems which arise under the Cash Flow label.

Example 3.14 An account is opened with a deposit of \$1,000. At the end of two years, \$500 is withdrawn. At the end of four years, \$300 is deposited. If the rate of interest is 4% what is the amount on deposit at the end of seven years? Note that we have three years at the end with no deposits or withdrawals.

Solution: TI BA II Plus Solution using CF worksheet (Table 3.25).

TABLE 3.25

Time	Symbol	Amount	Symbol	Frequency
0	CFo	1000		
1	C01	0	F01	1
2	C02	−500	F02	1
3	C03	0	F03	1
4	C04	300	F04	1
5, 6, 7	C05	0	F05	3
NPV	I	4	↓ NFV	$CPT = 1,045.064$

Using the **TI BA II Plus** we compute NPV, press PV, *2nd QUIT* and then compute FV (Table 3.26)

TABLE 3.26

N	I/Y	PV	PMT	FV
7	4	794.163	0	$CPT = 1,045.06$

Example 3.15 At what effective annual rate of interest will the following account show a balance of $454.19 at the end of year 10? See Table 3.27.

TABLE 3.27

Year	Deposit	Withdrawal
0	1000	
3		300
4	250	
7		1,000
10		454.19 (closes account)

Solution: We use the CF worksheet which computes the effective annual interest (IRR) rate given that the net present value of the cash flow is 0. That means that we need to enter the final value of $454.19 as a withdrawal. Table 3.28 is a table showing the entries required to solve this problem.

TABLE 3.28

Time	Symbol	Amount	Frequency
0	CFo	1000	
1, 2	C01	0	F01 = 2
3	C02	−300	F02 = 1
4	C03	250	F03 = 1
5, 6	C04	0	F04 = 2
7	C05	−1,000	F05 = 1
8, 9	C06	0	F06 = 2
10	C07	−454.19	F07 = 1

After this is entered, enter $IRR\ CPT = 5.6\%$.

Example 3.16 In return for payments of $500 at the end of year 3 and $800 at the end of year 6 an investor agrees to make an initial $(t = 0)$ deposit of $1,000 and deposit an additional $40 at the end of year x. If the rate of interest if 5% find x.

Solution: See Table 3.29. We can use the cash flow worksheet to compute the NPV (net present value) of the sequence of cash flows without the $40 payment. That is the PV of $40 to be paid in year x. We then use the TVM keys to find x so that the FV is $40.

Note that it is negative – this indicates that the investor has withdrawn more than is warranted by the deposit of $1,000 at an interest rate of 5%. To make up for that, an additional amount needs to be invested. This amount is indicated as $40 in the problem. Since that exceeds the actual deficit as

TABLE 3.29

Time	Symbol	Amount	Frequency
0	CFo	1000	
1, 2	C01	0	F02 = 2
3	C02	−500	F02 = 1
4,5	C03	0	F03 = 2
6	C04	−800	F04 = 1
	NPV	$I = 5$	
	↓	NPV	$CPT = -28.89$

computed at time $t = 0$, the \$40 payment can be made at a later date which we will now find. We transfer the 28.89 to the TVM keys by pressing PV and then exit the CF worksheet (2ND QUIT). We enter the I/Y, PMT, and FV and then compute N (Table 3.30).

TABLE 3.30

N	I/Y	PV	PMT	FV
$CPT = 6.67$	5	28.8912	0	−40

The final payment of \$40 should be made at $t = 6.67$ to balance the account.

Example 3.17 (a modification of Example 3.16) In return for an investment of \$1,000 at time 0 and \$250 at the end of year 2 an investor will receive a payment of \$800 at the end of year six and an additional payment of \$$x$ at the end of year 3. If the interest rate is 5%, find x.

Solution: See Table 3.31. We compute the NPV of the cash flow without the payment in year three. That is the present value of the required payment. We then use the TVM keys to compute the FV for $t = 3$

TABLE 3.31

Time	Symbol	Amount	Frequency
0	CFo	1,000	
1	C01	0	F01 = 1
2	C02	250	F02 = 1
3-5	C03	0	F03 = 3
6	C04	−800	F04 = 1
	NPV	$I = 5$	
	↓	NPV	$CPT = 629.78$

Press *PV* to transfer the NPV to the TVM keys and then *2nd* Quit (Table 3.32)

TABLE 3.32

N	I/Y	PV	PMT	FV
3	5		0	$CPT = -729.05$

The additional payment received at the end of year 3 is $729.05.

Example 3.18 A deposit of $1,000 is made into an account earning 8% effective annual interest rate. At the end of year 3, $200 is withdrawn. At the end of five years an additional deposit of $100 is made. At the end of eight years the amount on deposit is transferred to an account earning 4% effective annual interest rate. At the end of the first three years in the new account an additional $1,000 is deposited. At the end of the first seven years in that account an additional $500 is deposited.

a) How much is on deposit at the end of eight years in the second account?

b) What was the effective annual interest rate (IRR in **TI BA II Plus** terms) for the entire investment scheme?

Solution: See Tables 3.33 and 3.34. a) We compute the NFV of the cash flow in the first account. This amount is the initial investment in the second account. We then compute the NFV of that account.

Account One

TABLE 3.33

Time	Symbol	Amount	Frequency
0	CFo	1,000	
1,2	C01	0	F01 = 2
3	C02	-200	F02 = 1
4	C03	0	F03 = 1
5	C04	100	F04 = 1
6, 7, 8	C05	0	F05 = 3
	NPV	I = 8	
	↓↓	NFV	$CPT = 1,683.04$

Account Two

TABLE 3.34

Time	Symbol	Amount	Frequency
0	CFo	1,683.035495	
1, 2	C01	0	F01 = 2
3	C02	1,000	F02 = 1
4, 5, 6	C03	0	F03 = 3
7	C04	500	F04 = 1
8	C05	0	F05 = 1
	NPV	I = 4	
	↓↓	NFV	$CPT = 4,040.00$

b) See Table 3.35. To compute the IRR for the entire period we treat the cash flows as being made at a single interest rate with a final value (NFV) of \$4,040.00362 and compute the effective annual interest rate. To do this we enter the final amount as a withdrawal. This is because the CF Worksheet computes the effective annual interest rate under the assumption that the ending balance is \$0.

TABLE 3.35

Time	Symbol	Amount	Frequency
0	CFo	1,000	
1, 2	C01	0	F01 = 2
3	C02	−200	F02 = 1
4	C03	0	F03 = 1
5	C04	100	F04 = 1
6-11	C05	0	F05 = 5
11	C06	1,000	F06 = 1
12, 13, 14	C07	0	F07 = 3
15	C08	500	F08 = 1
16	C09	−4,040.0036	F09 = 1
	IRR	CPT = 5.68	

Study Tip: If you are planning on taking Exam 2/FM you will need to be very proficient in using the **TI BA II Plus** and the **TI-30XS**. Each calculator has its own set of advantages and disadvantages. By exam time you shouldn't have to think about which one to use for a given type of problem.

Exercises for Chapter 3

1) Compute the interest earned on a loan of \$5,650 at simple interest of $i = 6.54\%$ over the period from January 18, 2007, to September 15, 2009, using the day counting methods discussed in this chapter

2) Compute the interest earned on \$5,700 at 7.8% annual compounded interest over a period of eight and a quarter years

   a) Using the exact calculation.

   b) Using simple interest over the fractional period.

3) \$500 is invested at 6.4% interest per year. Find the accumulated amount using each of the following three methods for the time periods indicated.

   i) Simple interest.

   ii) Compound interest.

   iii) Compound interest over integral time periods and simple interest for the fractional portion.

      a) After one year.

      b) After five years.

      c) After a half year.

      d) After five years and three months.

      e) After twelve years and eight months.

4) Compute the value of FV in the case of a fractional time period as follows: Define $h(x) = PV(1+i)^x$

   a) Find the equation of the line through $(n, h(n))$ and $(n+1, h(n+1))$ as $y = mx + b$.

   b) Compute y for $x = n + q$, $0 < q < 1$.

Show that your result in b) is the same as that obtained using simple interest for the fractional period.

5) Use the result of 4) to show that using simple interest always overstates the accumulated value. (Hint, what is the concavity of the compound interest curve?) Use calculus to find the value of q (fractional portion of time) which maximizes the difference between the result using simple interest and the exact value. That is, find q which maximizes

$$f(q) = (1+i)^n(1+iq) - (1+i)^{n+q}$$

Note: n is a constant and the domain of f is $0 < q < 1$.

6) In return for the promise of a payment of \$900 at the end of eight years, a person agrees to pay \$50 now, \$200 at the end of three years and make one final payment at the end of eleven years. If the nominal rate of interest is 7.2% compounded monthly what is the value of the final payment?

7) Find the interest rate at which payments of \$500 at the end of four years and \$300 at the end of twelve years have the same value as a single payment of \$750 at the end of four years.

8) Suppose that the rate of interest is a nominal 5.34% annually and is compounded quarterly. Find the time at which a single payment of \$5,000 is equivalent in value to the following sequence of payments displayed in Table 3.36.

TABLE 3.36

Payment	Date of Payment
450	end of month 16
3000	end of year 2
550	end of month 48
1000	end of year 7

9) Compute the doubling time for an investment at a nominal 6% annual interest under each scenario.

   a) Interest is compounded annually.

   b) Interest is paid at simple interest.

   c) Interest is compounded for integer time periods and simple interest is paid over any fractional time period.

   d) Interest is compounded continuously

10) At what effective rate of interest will the sum of the present value of $3,000 in two years and $2,000 in one year be equal to $4,000 now?

11) In return for payments of $2,000 at the end of four years and $5,000 at the end of ten years, an investor is willing to pay $3,000 immediately and make an additional payment at the end of three years. If $i^{(4)} = .08$, find the value of this additional payment.

12) Georgia invests $550 at 3% annual interest. At the end of the first year, she withdraws $100, at the end of the second year, she withdraws $300. At the end of the third year she deposits $50. She closes the account at the end of year 4. How much is in the account at the end of year 4?

13) Bob needs to have $10,000 to pay off his student loan at the end of five years. He plans to make a payment of $x in year 2 and $2x in year 3. What is the value of x if interest is $i^{(12)} = .04$?

14) Alice is saving for a new car she will purchase in four years. She anticipates that the price of the car will be $20,000. She is putting her money into an account which pays 5.5% effective. At the **beginning** of this year she deposited $4,000. She plans to deposit $5,000 at the beginning of the second year and again at the beginning of year 3. How much more will she need at the end of four years to make the purchase?

15) An account earns 5% for three years and then 7% for an additional four years. What is the effective rate of interest over the life of the account?

16) Find t if, after t years the accumulated value of $1,000 at 6% effective annual interest rate compounded quarterly is exactly twice the accumulated value of $1000 at 4% nominal compounded annually. Hint: Convert to quarters or to years, take your pick.

17) At a varying force of interest $\delta_r = kr$ the doubling time is ten years. Find k.

18) The present value of two payments: $100 at the end of n years, and $100 at the end of $2n$ years is $100.

   a) If $i = .08$, find n,

   b) If $n = 10$, find i.

19) An investor will pay $3,000 now and an unknown additional amount in three years. In return the investor will receive $2,000 at the end of two years and $5,000 at the end of ten years. If $i^{(4)} = .06$, find the amount of the payment which must be made at the end of three years.

20) Find a function $f(i, k)$ which computes the length of time it takes money to increase by a factor of k at an annual effective rate of i. Graph $f(i, k)$ for $i = .01, .02, .03, .06$.

21) You are purchasing a TV and can pay for it in one of two ways.

   a) Pay now and receive a 30% discount off full retail.

   b) Pay in six months and receive a 25% discount off full retail.

   At what effective rate of annual interest are these two payment methods equivalent?

22) I win a prize at the local state gambling emporium and can claim my winnings in one of two ways.

   a) $50,000 now and an additional $50,000 in three years.

   b) $80,000 now.

   At what interest rate are these two methods equivalent?

23) $1,000 accumulates at $i^{(6)} = .06$ for six years and then accumulates an additional five years at $\delta_t = \frac{t}{1+t^2}$. What is the accumulated value of this investment?

24) $10,000 accumulates a 5% for three years, an additional two years at 6% and an additional five years at an unknown rate of interest, i. Find i if the final amount is $18,243.09

25) At what rate of nominal annual interest, compounded monthly will $500 increase to $850 in three years?

26) A deposit of $100 earns 5% interest for ten years and then an unknown interest for an additional twenty years. At the end of the thirty years it has accumulated to $630.33. What is the effective annual interest rate during the last twenty years?

27) Mable needs to double her money in five years. What effective annual interest rate is required to achieve her goal? What if Mable can earn interest compounded continuously?

28) An account is opened with $6,000. At the end of two years, $4,000 is withdrawn. At the end of four years the account is closed with a withdrawal of $3,000. What is the effective annual rate of compound interest?

29) On January 1, 2005, Susan put $5,000 into a bank account at Stingy Bank which pays a nominal annual interest rate of 3.5% compounded twice a year. On July 1, 2006, she withdrew $500. On January 1, 2007, she deposited an additional $700. How much will be in the account on January 1, 2010?

30) The present value of $2,500 one year from now and $1,000 two years from now is $3,000, what is the effective annual interest?

31) A savings account pays a nominal 7.3% compounded daily (assume 365 days in a year). The account is opened with a deposit of $1,000. A deposit of $1,500 is made at the end of two years. If no further deposits or withdrawals are made what is the balance in the account at the end of eight years?

32) Account A earns interest at an effective annual rate of 5%, while Account B earns interest compounded continuously at 5%. Compute the ratio of the doubling time for Account A to that for Account B.

33) An account is opened with a deposit of $4,000. At the end of the first year, $250 is withdrawn. A deposit of $500 is made at the end of year 4 and a deposit of $200 is made at the end of year 6. The account is closed at the end of year 11 with a withdrawal of $6,571.68. What was the effective rate of annual interest for this account?

34) An account earning 5% annual effective interest is opened with a $500 deposit. At the start of year 2, $200 is withdrawn, at the start of year 3, $300 is deposited. What is the amount of the account at the start of year 4?

35) The account in problem in problem 34) is transferred to a new account at the start of year 4. The accumulated amount plus $500 is deposited at this time. $200 is deposited at the start of year 5 and the account shows a balance of $1,717.34 at the end of year 7. What effective rate of interest did the account earn in years 4 through 7?

Chapter 4

Annuities

4.1 Introduction

Definition: An **annuity** is a sequence of payments made at agreed upon intervals of time.

We will begin our study of annuities with the case in which the payments are equal in value and are made at equal periods of time for an agreed upon period of time. For example, an annuity which pays $100 every month for three years. This is called a **fixed term level payment annuity**. The word term refers to the total length of time payments are to be made. The term of an annuity can be stated in years or as the number of payments. Thus an annuity which provides monthly payments of three years has a term of three years or thirty-six months. The appropriate term to use depends on the conversion period for the interest rate.

Rent, mortgage payments, and retirement plans are all examples of annuities. Additionally, bonds often provide periodic interest payments. Thus, a bond can be thought of as a form of annuity. While annuities, loans, and bonds are very similar, the sorts of questions we need to solve are very different. As a result, each instrument will get its own chapter. We will consider general annuities in this chapter, mortgage payments in Chapter 5, and bonds in Chapter 7.

If an annuity consists of payments which are certain to be made for a proscribed period, it is called an annuity-certain. Mortgage payments or rent are examples of annuities-certain. In other cases an annuity is only paid under certain conditions. For example, a retirement plan typically only provides payments during the life of its beneficiary. Such an annuity is known as a contingent annuity. The value of a contingent annuity depends on the expected life of the beneficiary of the annuity and is thus more complicated to compute. We will restrict our discussion to annuities-certain.

An annuity usually has a fixed number of payments to be made at specified intervals of time known as the payment period. The duration of the annuity in payment periods is known as the **term** of the annuity. An annuity which offers a (potentially) infinite sequence of payments is known as a **perpetuity**. Thus, an annuity which provides payments each month for fifteen months would have a monthly period and a term of 15 (months).

Annuities typically offer payments at either the start or the end of each period.

Types of Annuities

Annuities-Immediate

An annuity in which payments are made at the end of each period is called (contrary to intuition) an **annuity-immediate**.

Annuities-Due

An annuity in which payments are made at the beginning of each period is known as an **annuity-due**.

If the annuity starts at time $t = 0$, an annuity-immediate provides the first payment at time $t = 1$, and an annuity-due provides the first payment at $t = 0$. An annuity-immediate starting at $t = 0$ is the same thing as an annuity-due starting at $t = 1$. As a result an annuity-immediate is just an annuity-due delayed one period. Mortgage payments are typically made at the end of each period and are therefore best thought of as annuities-immediate. By contrast, state lotteries are paid out as annuities-due[1].

Our goal for this chapter will be to analyze the techniques used to compute the value of an annuity of either type at any point in time. We begin with annuities-immediate with fixed terms and constant (often called level) payments. These annuities pay a fixed amount, R at the end of each period for a pre-specified number of payments known as the term of the annuity.

4.2 Fixed Term Annuities-Immediate with Constant Payments

We start our analysis with an annuity-immediate which provides n payments of $1 at the end of each period. We assume further that the effective interest rate[2] is i per period. For example, if the payments are made every month, we will assume that interest is compounded monthly as well. Later we will deal with cases where the payment period and interest conversion periods are not the same. For example, an annuity with monthly payments is valued using an interest rate convertible quarterly.

The Present Value of an Annuity

We compute the value of a fixed term annuity-immediate which pays $1 each period at the time of its inception using techniques developed in Chapter 1:

Since the first payment occurs one period after inception it has a value at inception ($t = 0$) of : $1 \cdot \frac{1}{1+i} = \frac{1}{1+i} = v$. The second payment will be delayed two periods and so has a value of v^2, and so forth. The total value at inception

---

[1] Although Illinois failed to pay anything to big winners for a period of time during the state "budget crisis."

[2] Compound interest is assumed unless stated otherwise.

of this annuity is given by the sum $v + v^2 + \cdots + v^n$. This expression occurs so often that it has a special symbol: $a_{\overline{n}|,i}$ If we are clear about the interest rate we will write it as just $a_{\overline{n}|}$. Using this notation we have:

$$a_{\overline{n}|,i} = v + v^2 + v^3 + \cdots v^n = \sum_{i=1}^{n} v^i \qquad (4.1)$$

We can simplify this expression by using the fact that it is a geometric series. We can then use Equation 1.1 to compute its sum. Here is Equation 1.1

$$a + ar + ar^2 + ar^3 + \cdots + ar^{n-1} = a\frac{1 - r^n}{1 - r} \qquad (1.1)$$

Comparing Equation 4.1 with Equation 1.1 we see that $a = v$, $r = v$ and have:

$$\begin{aligned} a_{\overline{n}|,i} &= v + v^2 + v^3 + \cdots v^n \\ &= v(1 + v + v^2 + \cdots + v^{n-1}) \\ &= v\frac{1 - v^n}{1 - v} \\ &= v\frac{1 - v^n}{iv} \\ &= \frac{1 - v^n}{i} \qquad (4.2) \end{aligned}$$

Present Value (PV) of an Annuity Immediate of \$1 per period for n periods

$$a_{\overline{n}|,i} = \frac{1 - v^n}{i}$$

To compute the value of $a_{\overline{n}|,i}$ using the TVM keys we use the PMT key. Since payments cease at the end of the term of the annuity the future value is \$0. If we enter PMT $= -1$ the PV will be reported as positive (Table 4.1).

TABLE 4.1

N	I/Y	PV	PMT	FV
n	i	$CPT =$	-1	0

Table 4.2 shows the computation for $a_{\overline{10}|,.03}$:

TABLE 4.2

N	I/Y	PV	PMT	FV
10	3	$CPT = 8.5302$	-1	0

If the level payment is \$R, we simply multiply by R to obtain the present value of our annuity:

The Value at Inception(PV) of an Annuity-Immediate of n Payments (made at the end of each period) of \$R at Interest Rate i

$$PV = Ra_{\overline{n}|,i} = R\frac{1 - v^n}{i} \tag{4.3}$$

We can use this to easily compute the accumulated value of all the payments of an annuity just after the final payment is made. We use the formula we developed earlier:

$$FV = PV(1 + i)^n$$

In this case, $a_{\overline{n}|,i}$ represents the Present Value and we want to compute the future value.

The symbol $s_{\overline{n}|,i}$ is used to denote the accumulated value of a \$1 annuity. The symbol $s_{\overline{n}}$ is used if the interest rate is not in doubt. With this notation we have:

The Accumulated Value (FV) of an Annuity-Immediate of n Payments of \$R at Interest Rate i just after the last payment is made.

$$FV = Rs_{\overline{n}|,i} = Ra_{\overline{n}|,i}(1 + i)^n \tag{4.4}$$

With a little algebra we can derive another relationship between $a_{\overline{n}|,i}$ and $s_{\overline{n}|,i}$. First, note that since $v^n = \frac{1}{(1+i)^n}$

$$s_{\overline{n}|,i} = a_{\overline{n}|,i}(1 + i)^n = \frac{1 - v^n}{i}(1 + i)^n = \frac{(1 + i)^n - 1}{i} \tag{4.5}$$

We then have

$$\frac{1}{s_{\overline{n}|,i}} + i = \frac{i}{(1 + i)^n - 1} + i$$
$$= \frac{i + i(1 + i)^n - i}{(1 + i)^n - 1}$$
$$= \frac{\frac{i}{v^n}}{\frac{1}{v^n} - 1}$$
$$= \frac{i}{1 - v^n} = \frac{1}{a_{\overline{n}|,i}} \tag{4.6}$$

To summarize:

$$s_{\overline{n}|,i} = (a_{\overline{n}|,i})(1 + i)^n$$
$$\frac{1}{s_{\overline{n}|,i}} + i = \frac{1}{a_{\overline{n}|,i}} \tag{4.7}$$

This last expression is useful in cases in which we do not know the term of the annuity in question. It enables conversion between $a_{\overline{n}|,i}$ and $s_{\overline{n}|,i}$ using only the value of i. See Example 4.3 below.

TI BA II Plus (Table 4.3)

The TVM keys will compute values of both $a_{\overline{n}|,i}$ and $s_{\overline{n}|,i}$. The PMT key allows us to enter the value of a periodic payment. Note that the payments must be equal and occur at equal intervals[3].

TABLE 4.3

To compute $a_{\overline{n}|,i}$

N	I/Y	PV	PMT	FV
n	i	$CPT = a_{\overline{n}|,i}$	-1	0

To compute $s_{\overline{n}|,i}$

N	I/Y	PV	PMT	FV
n	i	0	-1	$CPT = s_{\overline{n}|,i}$

Note: We enter $FV = 0$ when we compute $a_{\overline{n}|,i}$ and $PV = 0$ when we compute $s_{\overline{n}|,i}$.

If you need to use these values later, use STO to store them. Enter STO and then a digit from 1 to 9 to place the computed value in one of the nine storage registers. You can retrieve your result by pressing RCL and then the number of the register you stored your value in.

While the **TI BA II Plus** will solve many problems involving annuities, there are others for which you need to know the formulas for $a_{\overline{n}|,i}$ and $s_{\overline{n}|,i}$. Note also that the **TI-30XS** can easily evaluate the expression in Equation 4.3.

Example 4.1: Compute the present and accumulated value (just after the last payment) of a sequence of twenty-five payments of $1 each if the interest rate is 3.5% per payment period.

Solution: Present Value (Table 4.4)

TABLE 4.4

N	I/Y	PV	PMT	FV
25	3.5	$CPT = 16.48$	-1	0

Accumulated (FV) Value: If you have just computed $a_{\overline{n}|,i}$ you can set $PV = 0$ and compute FV (Table 4.5).

TABLE 4.5

N	I/Y	PV	PMT	FV
25	3.5	0	-1	$CPT = 38.95$

---

[3]For payments which occur at unequal intervals, we use the CF (cash flow) worksheet.

Example 4.2: Find the present and accumulated value of an annuity which pays \$450 at the end of each quarter for ten years. The rate of interest is 4.5% per quarter converted quarterly.

Solution: Since interest is converted quarterly we count in quarters. We then have $v = \frac{1}{1.045}$ and $n = 10 * 4 = 40$. Hence:

$$PV = 450 \cdot a_{\overline{n}|,i} = 450 \cdot \frac{1 - v^n}{i} = 450 \cdot \frac{1 - \left(\frac{1}{1.045}\right)^{40}}{.045} = \$8,280.712989$$

TI BA II Plus: Present Value (Table 4.6)

TABLE 4.6

N	I/Y	PV		PMT	FV
40	4.5	$CPT = 8,280.71$		-450	0

$$FV = 450 \cdot s_{\overline{n}|,i} = 450 \cdot \left(1 + \frac{1}{1.045}\right)^{40} \cdot (1.045)^{40} = \$48,163.64$$

TI BA II Plus: Future Value (Table 4.7)

Enter 0 for PV and then compute FV. You don't need to enter any other numbers as they are retained from the previous calculation.

TABLE 4.7

N	I/Y	PV	PMT	FV
		0		$CPT = -48,163.65$

There is one real-life example of the concept of future and present value that most of us encounter most every day. State lotteries announce a total prize, but pay the money out in annual installments. In Example 4.2 above, the total prize would have been announced as $450 \cdot 40 = \$18,000$. The real value of the prize (at 4.5% interest per quarter) is the present value of the annuity the state provides: \$8,280.16. The moral is simple: lottery tickets are a bad buy! If you insist on gambling go to Vegas.

Example 4.3: An annuity-immediate has a present value of \$12. If the interest rate is $i = .05$ what is the accumulated value of this annuity just after the final payment is made?

Solution: We are given $a_{\overline{n}|,i} = 12$, $i = .05$ and want to determine $s_{\overline{n}|,i}$. We don't know n. But we don't need it! We use the second form of Equation 4.7

$$\frac{1}{s_{\overline{n}|,i}} + i = \frac{1}{s_{\overline{n}|,i}}$$

$$\frac{1}{s_{\overline{n}|,i}} + .05 = \frac{1}{12}$$

$$s_{\overline{n}|,i} = \frac{1}{\frac{1}{12} - .05} = 30$$

Note that the **TI BA II Plus** is not of any help at all in this case! That's because, we only know the value of three of the FV variables in the TVM worksheet. The TVM keys will solve all problems (and only those problems) where we know four of the five numbers: N, I/Y, PV, PMT, FV.

Example 4.4 You are given that $v^n = .6139133$ and that the present value of an annuity-immediate of \$1 for n periods is 7.72. What is the interest rate?

Solution: We have

$$7.72 = \frac{1 - v^n}{i} = \frac{.3860867}{i} \text{ hence } i = \frac{.380867}{7.72} = .05001, \ i = 5\%$$

Note that we did not need to find n.

The Price of an Annuity

Many financial institutions sell annuities. These are often purchased in advance as part of a retirement plan. Suppose we wish to purchase an annuity which pays \$R at the end of each period for n periods at an effective interest rate of i per period. If the annuity is purchased at the start of the first period its purchase price (not including any profit for the financial institution) is the present value of the annuity:

$$\text{Price} = PV = Ra_{\overline{n}|,i} \tag{4.8}$$

We can compute the payment of an annuity in terms of its price (PV) as well:

$$R = \frac{PV}{a_{\overline{n}|,i}} \tag{4.9}$$

In most cases an additional fee will be charged by the financial institution. The fee may be built into the PV calculation or billed up front. In the case of home loans these up front fees are called points. Each point is valued at 1% of the loan amount. A loan of \$200,000 with an up front charge of 1 point would incur a fee of $.01 \cdot 200,000 = \$2,000$ to be paid at closing. Since this fee is considered interest, it is usually a deductible expense for income tax purposes. We will discuss this further when we consider Truth in Lending Laws in Chapter 5.

Example 4.5 A person has $10,000 to invest and wishes to purchase an annuity which will provide constant monthly payments for a period of ten years. If the interest rate used to price the annuity is a nominal rate of annual interest of 12% compounded monthly, what will the monthly payment be?

Solution: We have $i = \frac{.12}{12} = .01$, $n = 120$, and $PV = 10,000$. Using Equation 4.9, we obtain $R = \frac{10000}{a_{\overline{120}|,.01}} = \$143.47$.

TI BA II Plus (Table 4.8)

TABLE 4.8

N	I/Y	PV	PMT	FV
120	1	$-10,000$	$CPT = 143.47$	0

The annuity will provide 120 monthly payments of $143.47.

Example 4.6: Maria will retire in ten years and is saving $300 every month to purchase a thirty-year monthly annuity to provide income in retirement. Her savings account pays a nominal rate of 8% annual interest converted monthly while the annuity is priced based on a nominal annual interest rate of 15% compounded monthly. What will Maria's monthly payment be?

Solution: Maria's saving account is an annuity as well. The value of her savings account at the time she wishes to purchase the annuity can be found using Equation 4.4:

$$FV = Rs_{\overline{n}|,i} = Ra_{\overline{n}|,i}(1+i)^n = 300a_{\overline{n}|,i}(1+i)^n$$

In this case $n = 12 \cdot 10 = 120$, $i = \frac{.08}{12} = .006667$ and we obtain $FV = \$54,883.81055$. This number now becomes the present value of the annuity Maria will purchase. We use Equation 4.9 to determine the payment, R. She will receive $R = \frac{\$54,883.81}{a_{\overline{n}|,i}}$.

We now have $n = 12 \times 30 = 360$ and $i = .15/12$. This results in a monthly payment of $693.98.

TI BA II Plus solution: (Table 4.9) This can all be done in one sequence of keystrokes! After we compute the future value of Maria's savings plan, we convert it to the present value of the annuity she will purchase and then have the **TI BA II Plus** compute the payment.

TABLE 4.9

N	I/Y	PV	PMT	FV
120	$\frac{8}{12}$	0	300	$CPT = 54,883.81$

Now we push the PV key to convert the FV to a present value for our annuity and then key in $N = 360$ and $I/Y = 15/12$, $FV = 0$ and compute PMT (Table 4.10)

TABLE 4.10

N	I/Y	PV	PMT	FV
360	1.25	54,883.51	$CPT = -693.97$	0

Example 4.7: David donates to North Central College the remaining payments of his thirty-year annuity with monthly payments to North Central to begin upon his death. Based on current statistics, David is likely to live an additional ten years. Thus, North Central can expect to start receiving payments starting with payment 121. David will report this donation on his tax return as a charitable donation. If the annuity pays $500 per month, what is the value of this donation?

Solution: We need an interest rate to compute the answer. Since the payments begin in the future, we can't use prevailing rates. In this case we would have to consult the IRS as to what rate is acceptable. Let's assume the IRS allows us to use a nominal 8% annual rate of interest compounded monthly.

The value of the annuity at start of the payments the college will receive is $Ra_{\overline{n}|,i}$ with $R = 500$, $i = \frac{.08}{12}$ and $n = 240$ (David will get the first 120 of the 360 total payments, leaving 240 for North Central). This gives us a value of $59,777.15. Since North Central doesn't start getting any money for ten years (120 periods), this number is a future value. We use $PV = FVv^t$ to compute the present value of this amount. In this case $FV = 59,777.15$, $t = 120$, and $v = \frac{1}{1+i} = \frac{1}{1+\frac{.08}{12}}$. This gives us a present value of $26,931.00.

This was calculated as $26,931.00662 = 59,777.15 \cdot \left(\frac{1}{1+\frac{.08}{12}}\right)^{120}$.

TI BA II Plus (Table 4.11) Again, we can do the calculations all at once. This time we compute the present value (at the end of year 10) of the annuity and then rename that the FV and compute its present value at the start of year 1.

TABLE 4.11

N	I/Y	PV	PMT	FV
240	.666666	$CPT = 59,777.15$	500	0

Now push the FV key to convert this to a future value and enter in $N = 120$, $PMT = 0$ and compute the present value (Table 4.12).

TABLE 4.12

N	I/Y	PV	PMT	FV
120		$CPT = 26,931.00$	0	59,777

This is a very nice deal for David – he is assured of his monthly payments for as long as he lives and is able to take a one time deduction of $26,931.00. This sort of donation is so attractive to both sides that many non-profit organizations sell such annuities. You might find it interesting to visit the Development Office at your college to see how they handle these kinds of donations.

Example 4.8: A group of workers wins $40,000,000 in the Illinois lottery. The state pays this off in twenty annual payments of $2,000,000. If the prevailing interest rate is 6% compounded annually, how much will the state spend to pay off this group of lottery winners?

Solution: We are asked to find the present value (PV) of a twenty-payment annuity with payments $R = \$2,000,000$ at the end of each year. We have

$$PV = Ra_{\overline{n}|,i} = \$22,939,842.44$$

TI BA II Plus (Table 4.13)

TABLE 4.13

N	I/Y	PV	PMT	FV
20	6	$CPT = 22,939,842.44$	$-2,000,000$	0

In most cases lottery winners have the option of taking this amount as a single lump sum payment. The choice of lump sum or annuity depends on the prevailing interest rates.

Example 4.9 A loan of $1,200 is to be repaid over a period of twelve years. The effective annual interest rate of interest is 5.6% per year. Find the total amount of interest paid under each of the following repayment methods and explain why they differ.

1) The entire loan, including accrued interest, is paid off in one lump-sum at the end of twelve years.

2) The interest is paid each year and the loan amount ($1,200) is paid off at the end of twelve years.

3) The loan is paid off using a fixed payment annuity-immediate with payments each year. The payments are sufficient to cover the interest due and also reduce the principal. Note that this is the standard repayment method for home and auto loans.

Solution

1) At the end of twelve years the value of the loan will be $1200(1.056)^{12} =$ \$2,307.534827. The total interest paid is then $2,307.53 - 1200 =$ \$1,107.53.

   Using the TVM keys (Table 4.14):

TABLE 4.14

N	I/Y	PV	PMT	FV
12	5.6	$-1,200$	0	$CPT = 2,307.53$

2) The interest due each year is $1200(.056) = \$67.20$ and so the total interest paid is $12 \times 67.20 = \$806.40$.

3) The present value of the annuity which pays off the loan must equal the present value of the loan, which is \$1,200. We use the TVM keys to compute the required payment (Table 4.15).

TABLE 4.15

N	I/Y	PV	PMT	FV
12	5.6	1,200	$CPT = -140.01$	0

The payments total $12 \times 140.01035515 = \$1,680.12$ so the total interest paid using this method of repayment is $1,680.12 - 1,200 = \$480.12$.

The differences in the amount of interest paid in these three scenarios is a nice example of what is called the time value of money (hence the abbreviation TVM on the **TI BA II Plus**) – the longer you have to wait to be paid, the more money you are entitled to. In the first method no payments are made until the end of the loan period. In the second case only the interest is paid, so the lender must wait until the end of the period to recover the principal. In the third case, each payment is a mixture of interest and principal. These methods are equivalent in that the present value of the payments paid is the same for each scenario.

As a way of seeing that methods 1) and 2) are equivalent we will compute the value of each sequence of payments as measured at the inception of the loan. In the first case, the lender receives $1,107.53 at the end of twelve years. The value of this amount at the inception of the loan is $1107.53 \cdot v^{12} = \$575.95$. In the second case, the interest is paid as an annuity of $67.20 every year for twelve years. The value at inception of this annuity given by $PV = 67.20 a_{\overline{n}|,i} = \$575.95$ (Table 4.16).

TABLE 4.16

N	I/Y	PV	PMT	FV
12	5.6	$CPT = 575.96$	67.2	0

Methods 1) and 2) are thus equivalent.

Suppose we computed the value of the interest paid at the end of the loan period? This makes method one easy to compute – the value of the interest is just $1,107.53. In the case of the second method we are looking for the accumulated value of an annuity of $67.20 which is given by $FV = 67.20 s_{\overline{n}|,i} = 67.20 \times 16.48117301 = \$1,107.53$. Again, the methods are seen to be equivalent.

It's more difficult to compare scenario 3) to the others since each payment is a mixture of principal and interest. The interest in this case is a varying payment annuity. The process of determining the interest and principal portions of a payment is known as amortization and will be discussed in Chapter 5. It is clear that the interest paid using the third method is less than that in either of the other two methods, since the total interest paid is $480.12 and hence its present value can't exceed this number.

The third method is the least risky for the lender since the principal is reduced with each payment. The second method is more risky for the lender than the third, since only interest is paid , but less risky than the first method since in the first method the lender receives no money until the end of the loan period. Note that the interest paid varies increases with increased risk to the lender. The amount of money outstanding is sometimes called the exposure of a loan.

Another measure of the risk of the risk associated with loans or other financial transactions is a number called the duration. Using the first payment method, the entie loan is paid off after twelve years and (not surprisingly) the duration is twelve years. The other two methods pay off some of the obligation each period and hence have shorter computed durations. More about that in Chapter 9.

4.3   Fixed Term Annuities-Due

An annuity-due consists of a sequence of payments which commence immediately and are then made at the start of each subsequent period. Recall that annuities-immediate provide payments at the end of the each period. We will use $\ddot{a}_{\overline{n}|,i}$ for the present value of an annuity-due and $\ddot{s}_{\overline{n}|,i}$ for the accumulated value of an annuity-due. Since an annuity-due starts one term earlier than an annuity-immediate there is a simple relationship between the present values of these two types of annuities:

Present Value of an Annuity-Due

$$\ddot{a}_{\overline{n}|,i} = a_{\overline{n}|,i}(1+i) \qquad (4.10)$$

Accumulated Value of an Annunity-Due

$$\ddot{s}_{\overline{n}|,i} = s_{\overline{n}|,i}(1+i) \qquad (4.11)$$

We can also compute these values directly from their definition. For the present value (PV) of an annuity-due we have:

$$\begin{aligned}
\ddot{a}_{\overline{n}|,i} &= 1 + v + v^2 + v^3 + \cdots v^{n-1} \\
&= \frac{1-v^n}{1-v} \\
&= \frac{1-v^n}{iv} \\
&= \frac{1-v^n}{d} \qquad (4.12)
\end{aligned}$$

Likewise, for the accumulated value we have:

$$\begin{aligned}
s_{\overline{n}|,i} &= (1+i) + (1+i)^2 + \cdots + (1+i)^n \\
&= (1+i)\frac{(1+i)^n - 1}{(1+i) - 1} \\
&= \frac{(1+i)^n - 1}{iv} \\
&= \frac{(1+i)^n - 1}{d} \qquad (4.13)
\end{aligned}$$

Recall that d is the discount rate and that $d = \frac{i}{1+i}$.

NOTE The FM exam often contains questions which require that you know the various versions of these equations.

Example 4.10 Which of the following is equal to d?

I: $i - i^2 + i^3 - i^4 \ldots$

II: $\delta - \frac{\delta^2}{2!} + \frac{\delta^3}{3!} - \frac{\delta^4}{4!} \ldots$

III: $1 - \left(1 - \frac{d^{(m)}}{m}\right)^{-m}$

A) I only.

B) II only.

C) III only.

D) I and II only.

E) The correct answer is not given by A), B), C), or D).

Solution. We need to compute each of the three expressions and see if it simplifies to d.

I: Is a geometric series with $a = i$, $r = -i$.

$$\frac{i}{1-(-i)} = \frac{1}{1+i} = d$$

II: The Taylor Series for $e^{\delta} = 1+\delta+\frac{\delta^2}{2!}+\frac{\delta^3}{3!}\cdots$, so $e^{-\delta} = 1-\delta+\frac{\delta^2}{2!}-\frac{\delta^3}{3!}\cdots$
Hence

$$II = 1 - e^{-\delta} = 1 - v = d$$

III: Recall the conversion rule to convert d and $d^{(m)}$
Hence I and II are correct and the correct choice is D).

NOTE: The type of annuity we are dealing with depends entirely on the time at which we begin the analysis. An annuity-due with annual payments commencing in 2007 is exactly the same thing as an annuity-immediate commencing in 2006.

Example 4.11: Georgia is saving up for college and needs to have $20,000 in twelve years. She plans to make an equal deposit at the end of each year with the final payment to be made one year prior to entering college (the end of the investment period). What is the size of the payment she needs to make each year if the interest earned on her account is 5% per year?

Solution (assuming a payment of R each year) We will analyze this problem as an annuity-due and as an annuity-immediate.

a) As an annuity-immediate: she will make eleven payments ending in year 11. At that point the value of the annuity will be $R \cdot s_{\overline{n}|,i}$. This amount will accumulate for one more year, yielding a total value of $R \cdot s_{\overline{n}|,i} \cdot (1+i)$

b) As an annuity-due: she will make eleven payments at the start of each of years 2 through 12 for an accumulated value of $R\ddot{s}_{\overline{n}|,i} = R \cdot s_{\overline{n}|,i} \cdot (1+i)$.

As a result, whichever way we look at the problem, we have:

$$R = \frac{20,000}{s_{\overline{n}|,i}(1+i)}$$

Since $n = 11$ and $i = .05$, we obtain R = \$1,340.74.

TI BA II Plus Solution: We compute R using the annuity-immediate function (TVM worksheet) and divide by $1 + i$ (Table 4.17).

TABLE 4.17

N	I/Y	PV	PMT	FV
11	5	0	$CPT = -1,407.78$	20000

Now we divide by 1.05 to obtain $1,340.74.

To set the **TI BA II Plus** up to convert annuities-due directly enter 2ND BGN. You should see END (the setting for annuities-immediate). Enter 2ND SET and you should see BGN, the setting for annuities-due. Now enter 2ND QUIT. You will see a small BGN in the upper right corner of the viewing window to remind you that you are in annuities-due mode (Table 4.18).

TABLE 4.18

				BGN
N	I/Y	PV	PMT	FV
11	5	0	$CPT = 1,340.74079$	20000

To re-set your **TI BA II Plus** to annuities-immediate mode enter 2ND BGN 2ND SET 2ND QUIT.

Note: The risk in converting to annuity-due is that you will forget to reset your calculator. The result will be answers that are reasonable, but wrong!

Example 4.12: An investor wishes to accumulate $50,000 in ten years. Payments will be made at the end of each month with a nominal annual interest rate of 12% converted monthly. Payments in the last four years will be twice those in the first six years. What is the monthly payment in the first six years?

Solution: There are several ways to solve this problem. Here are two. In each case the monthly interest rate is 1%.

Method One:

Let R be the payment in the first six years, so that $2R$ is the payment in the final four years. We compute the value of each annuity (six years at R followed by four years at $2R$) at the end of the investment period. The sum of these two future values (FV) must equal $50,000.

a) The first six years. We have seventy-two payments of R each. The value at the end of the four years is given by $Rs_{\overline{72}|,.01}$. This value will accumulate at 1% per month for the final four years or forty-eight months. Hence, its value is $Rs_{\overline{72}|,.01}(1.01)^{48}$.

b) The accumulated value of the final forty-eight payments of $2R$ is given by $2Rs_{\overline{48}|,.01}$

We then have $50,000 = Rs_{\overline{72}|,.01}(1.01)^{48} + 2Rs_{\overline{48}|,.01}$ Factoring out R and solving give us

$$R = \frac{50,000}{s_{\overline{72}|,.01}(1.01)^{48} + 2s_{\overline{48}|},.01} = \$171.67$$

TI BA II Plus solution: We compute and store each term in the denominator and then compute the needed payment (Table 4.19).

TABLE 4.19

$s_{\overline{72}|,i}:$	N	I/Y	PV	PMT	FV
	72	1	0	-1	$CPT = 104.7099312$

Now convert the result to a present value by pushing the PV key, set PMT $= 0$, $N = 48$, and compute the new FV (Table 4.20)

TABLE 4.20

N	I/Y	PV	PMT	FV
48	1	104.7099312	0	$CPT = -168.8160817$

Convert to a positive number and store the result

$$\pm STO\ 1$$

Now we compute the second term in the denominator (Table 4.21):

TABLE 4.21

$s_{\overline{48}|,i}:$	N	I/Y	PV	PMT	FV
	48	1	0	-2	$CPT = 122.4452$

Now we add these two results, compute $\frac{1}{x}$ and multiply by 50,000:

$$+RCL\ 1\frac{1}{x} = \times 50000 = 171.67$$

Method Two

We can think of the investment scheme as consisting of ten years of payments of R per year plus a second sequence of four years of payments of R. The accumulated value is then given by $Rs_{\overline{120}|,.01} + Rs_{\overline{48}|,.01} = 50,000$. This gives us

$$R = \frac{50000}{s_{\overline{120}|,.01} + s_{\overline{48}|,.01}} = \$171.67$$

TI BA II Plus (Table 4.22)

TABLE 4.22

$s_{\overline{120}|,i}$:

N	I/Y	PV	PMT	FV
120	1	0	-1	$CPT = 230.0386895$

STO 1

$s_{\overline{48}|,i}$:

N	I/Y	PV	PMT	FV
48	1	0	-1	$CPT = 61.22260777$

$$+\text{RCL } 1 = \frac{1}{x} \times 50000 = \$171.67$$

Example 4.13 Lakeesha wins the Illinois lottery with an announced prize of $40,000,000 to be paid in twenty equal annual installments of $2,000,000 with the first installment payable immediately. She wants her money up front. The fine folk at the lottery offer her a lump sum payment of $21,000,000. What interest rate did they use to compute this number?

Solution: Table 4.23. The state is equating a twenty-payment annuity-due with payments of $2,000,000 per year with a lump sum of $21,000,000. This means that $21,000,000 is the present value of the annuity. **TI BA II Plus** solution: Since we are solving for i we can't just divide by $1 + i$ to convert to an annuity due. Hence, we will convert to BGN mode: 2ND BGN, 2ND, SET (**Result:** the letters BGN appear), 2ND QUIT.

Now we compute the interest rate using the TVM keys. Note that one of PV or PMT must negative or you will get an error message.

TABLE 4.23

N	I/Y	PV	PMT	FV
20	$CPT = 8.151236605$	21,000,000	-2,000,000	0

Note: Make sure you return the calculator to annuity-immediate mode after you are finished. If you see BGN in the upper right hand corner of the display, you are still in annuity-due mode.

Which payment method should you choose? That depends on the interest rate you can earn. If you can deposit the $21,000,000 at an interest rate of exactly 8.15%, it will purchase an annuity of exactly $2,000,000 so the two methods are exactly the same. If, on the other hand, you can earn better than 8.15%, you are better off taking the lump sum.

For example, if Lakeesha can purchase an annuity with a price based on 8.5%, her $21,000,000 will purchase twenty payments of 2,045,244.67. On the other hand, if the prevailing interest rate is less than the rate the state used, she is better off taking the annuity payments. Note that we are assuming that all payments are certain to be made and that the interest rates will not change. Interest rates are seldom constant and Illinois became the first state to default on lottery payments in 2015. Once, again, only death and taxes are certain.

4.4   The Value of an Annuity at any Date

Our formulas thus far give us the value of an annuity at its inception $(a_{\overline{n}|,i})$ and at its conclusion $(s_{\overline{n}|,i})$. We will now see how to compute:

1) The value of an annuity one or more periods before its inception.

2) The value of an annuity one or more periods after the final payment is made.

3) The value of an annuity at any time between the first and last payment dates.

4.4.1   The Value of an Annuity prior to its Inception

In many cases, persons purchase an annuity as a source of retirement income. In the case of a divorce it is often necessary to determine a value of an annuity prior to its inception since the annuity will likely be part of the settlement. As we discussed earlier, persons sometimes donate the remaining payments of an annuity to nonprofit organizations with payments to begin for the nonprofit upon the death of the person making the donation. In order to determine the tax implications of such a donation, it is required to determine the value of such a deferred annuity at the time of its donation.

To compute the value of an annuity prior to its inception, we merely discount the value of the deferred annuity at the time payments actually begin by the number of years remaining until payments begin. The value at inception (the year payments begin) is $a_{\overline{n}|,i}$. Hence the value m years prior to inception is given by $v^m a_{\overline{n}|,i}$. We can also compute this as an annuity of $m+n$ payments minus the first m payments giving us

$$v^m a_{\overline{n}|,i} = a_{n+\overline{m}|,i} - a_{\overline{m}|,i} \qquad (4.14)$$

Example 4.14 An annuity immediate will consist of ten years of monthly payments of $500 each. What is the value of this annuity seven months prior to its first payment if the nominal annual interest rate is 6% compounded quarterly?

Solution: We first observe that the period for compounding interest does not match the period for payments. The rate of interest per quarter is $\frac{.06}{4} = .015$. We need to convert this to a monthly rate so that the compounding period is equal to the payment period. Let i be the monthly interest rate. We then have:

$$(1+i)^3 = 1.015$$

Solving for i, we obtain $i = .004975206$. We can now keep track of time in months.

Since we are interested in the value seven months prior to inception $m = 7$. The annuity pays monthly for ten years so $n = 120$. The required value is then

$$500 \cdot v^7 \cdot a_{\overline{120}|,.00497} = \$43,557.61$$

TI BA II Plus Solution: Tables 4.24, 4.25, and 4.26. We first compute the interest rate per month by entering the accumulated value of $1 after a quarter (1.015) as the future value and $N = 3$. We then compute the value of I/Y

TABLE 4.24

N	I/Y	PV	PMT	FV
3	$CPT = .4975$	-1	0	1.015

Now we just enter $N = 120$, $PMT = 500$, $FV = 0$, and compute the present value. The advantage of this method is that we do not have to enter the value of I/Y again.

TABLE 4.25

N	I/Y	PV	PMT	FV
120		$CPT = -45,097.39$	500	0

Finally, we convert the computed PV to a future value and compute the final answer. Press the FV key to convert our computed answer to a FV and change N to 7. Then compute the PV.

TABLE 4.26

N	I/Y	PV	PMT	FV
7		$CPT = 43,557.61$	0	$-45,097.39$

Helpful Hint: If you complete a TVM calculation with the calculator in BGN, you can reset it to END and then return to the TVM calculation. You don't need to enter the numbers again.

4.4.2   The Value of an Annuity after the Final Payment Is Made

Suppose that an annuity consists of n payments with an interest rate per payment period of i and we want the value of the annuity m periods after the final payment is made. We can do this in at least three different ways.

a) Accumulating the value of the annuity at inception $(R \cdot a_{\overline{n}|,i})$ for $m + n$ periods

$$FV = Ra_{\overline{n}|,i}(1 + i)^{n+m} \tag{4.15}$$

b) Accumulating the accumulated value of the annuity just after the last payment $(R \cdot s_{\overline{n}|,i})$ for m periods

$$FV = Rs_{\overline{n}|,i}(1 + i)^{m} \tag{4.16}$$

c) Treating the annuity as if the payments had continued for the entire period $(m + n)$ and then subtracting the value of the missing payments

$$FV = R(s_{\overline{n}|n+m,i} - s_{\overline{n}|,i}) \tag{4.17}$$

Example 4.15 An annuity consists of level payments of \$5,000 at the end of each year for twenty years. If the prevailing interest rate is a nominal rate of annual interest of 8% per year compounded monthly, how much must be deposited in five years as a single payment in order that the accumulated value of the annuity and that of the single deposit are equal at the end of thirty years? That is: accumulated value of annuity = accumulated value of the single payment when measured at thirty years.

Solution. The annual interest rate is found on the **TI BA II Plus** (Table 4.27) using

The effective annual interest rate is $8.2999507 = 8.3\%$.

The accumulated value of a single deposit of \$$x$ made at the end of year 5 after thirty years is $x(1 + i)^{25}$. The accumulated value of the annuity (using Equation 4.15) is $5000a_{\overline{20}|,.08299}(1.0829995)^{30}$.

TABLE 4.27

N	I/Y	PV	PMT	FV
12	8/12	1	0	$CPT = -1.0829995$

N	I/Y		PV	PMT	FV
1	$CPT = 8.299$				0

Equating these expressions we obtain

$$x(1.083)^{25} = 5000a_{\overline{20}|,.08299}(1.083)^{30} \tag{4.18}$$

Solving yields $x = \dfrac{5000a_{\overline{20}|,.082999507}(1.0829995)^{30}}{(1.0829995)^{25}} = 5000a_{\overline{20}|,.0829995}(1.0829995)^{5}$
$= \$71,533.54$ **TI BA II Plus** solution.

We compute $5000a_{\overline{20}|,.082999507}$ and then compute its future value at $t = 5$ (Table 4.28).

TABLE 4.28

N	I/Y	PV	PMT	FV
20	8.29995	$CPT = -48,014.06$	5000	0

We now set the $PMT = 0$ and $N = 5$ to compute the final value of x as a future value at $t = 5$ (Table 4.29).

TABLE 4.29

N	I/Y	PV	PMT	FV
5			0	$CPT = 71,533.54$

Notice that this expression for X (Equation 4.18) can be interpreted as the value of the annuity five years into its life cycle. This equation thus ends up comparing the single payment and the annuity at a point in time five years after the inception of the annuity.

4.4.3 The Value of an Annuity at any Time between the First and Last Payments

We suppose that the annuity consists of n payments with an interest rate per payment period of i and want to compute the value of the annuity m

periods after the inception of the annuity. In this case, we assume further that $m < n$. We also assume that we are computing the value of **all payments, not just those remaining at time of our calculation.** We have several methods:

a) Accumulate the value of the annuity at inception for m periods to obtain

$$Ra_{\overline{n}|,i}(1+i)^m \tag{4.19}$$

b) Add the accumulated value of the payments already made (m of them) to the present value of the payments yet to be made ($n - m$ of these).

$$R(s_{\overline{m}|,i} + a_{\overline{n-m}|,i}) \tag{4.20}$$

c) Deflate the accumulated value of the annuity just after the final payment by $n - m$ years

$$Rs_{\overline{n}|,i}v^{n-m} \tag{4.21}$$

We can also compute the value of the remaining payments (as opposed to the value of all the payments). If we want the value of the remaining $n - m$ payments just after the mth payment is made, we use the formula for the value at inception of an annuity of $m - n$ payments: $a_{\overline{n-m}|,i}$.

Example 4.16 Bob and Carol are divorcing. Among other assets which must be split up is a forty-year annuity-immediate which they purchased ten years ago with level payments of \$1,000 per month. Bob would like a lump sum payment while Carol would prefer to continue to receive monthly payments for the thirty years remaining on the annuity. If the prevailing interest rate is a nominal 11% annually compounded monthly, what lump sum payment to Bob would be fair? What will Carol's new payments be?

Solution: The present value of the remaining payments is $1000a_{\overline{360}|,\frac{.11}{12}} = $ \$105,006.35

Since they are to split this as a lump sum to Bob plus an annuity for Carol, the present value for each of them must be one-half of the total or \$52,503.17. Since Bob has elected a lump sum, this is the amount he receives.

Carol has elected to continue to receive monthly payments - how much should she get each month? That's easy! Had they elected to share the annuity payments equally, they would each receive \$500 per month. Since the scheme proposed also shares the value equally, Carol's annuity should be the same in either case and hence her share must be \$500 per month - no calculations required.

4.5 Perpetuities

Definition: A **perpetuity** is an annuity in which the payments are assumed to continue forever. A perpetuity-immediate begins payments at the end of the first period, while a perpetuity-due begins payments immediately.

While no one person could benefit from such a scheme, colleges and other nonprofit institutions often receive gifts in the form of perpetuities. The simplest example is a savings account which pays a fixed interest rate each period. If you withdraw only the interest each period, the principal remains the same and the payments can continue forever.

The symbol $a_{\overline{\infty}|}$ is used to denote the value of a perpetuity-immediate consisting of payments of \$1 at each payment period. We compute the present value at inception of such a scheme by going back to first principals using Equation 1.2:

$$
\begin{aligned}
a_{\overline{\infty}|} &= v + v^2 + v^3 + \cdots \\
&= \frac{v}{1-v} = \frac{v}{1-\frac{1}{1+i}} = \frac{v}{\frac{i}{1+i}} \\
&= \frac{v}{iv} \\
&= \frac{1}{i}
\end{aligned}
\tag{4.22}
$$

In the case of a perpetuity-due we have

$$
\ddot{a}_{\overline{\infty}|} = a_{\overline{\infty}|}(1+i) = \frac{1+i}{i} = \frac{1}{d}
\tag{4.23}
$$

Perpetuities

Present Value of a perpetuity-immediate with Payments of \$R

$$
R \cdot \frac{1}{i}
\tag{4.24}
$$

Present Value of a perpetuity-due

$$
R \cdot \frac{1}{d} = R\frac{1+i}{i}
\tag{4.25}
$$

Example 4.17 Albert leaves his estate of \$1,000,000 to his son Harry, his daughter Grace, and his dear alma mater North Central College in the form of a perpetuity paid by the interest on the value of the estate. The payments are to be paid monthly based on a nominal annual interest rate of 8% compounded monthly. The payments are to be divided as follows:

Harvey will receive the first ten years of payments.

Grace will receive the next twenty years of payments.

North Central College will receive all further payments.

What is the value of Albert's gift to each party at the time he makes it?

Solution: Harvey's share is an annuity consisting of 120 monthly payments. The value of each payment is the interest earned on the estate: $\left(\frac{.08}{12}\right)(1,000,000) = \$6,666.67$ per month. Harvey's share consists of ten years (120 payments and is thus valued at

$$6666.66667 a_{\overline{120}|,\frac{.08}{12}} = \$549,476.54$$

Harry's share using the TI BA II Plus (Table 4.30)

TABLE 4.30

N	I/Y	PV	PMT	FV
120	$\frac{8}{12}$	$CPT = -549,476.54$	6666.67	0

Grace's share is a 240-payment annuity with payments of $6,666.67 per month. But she must wait 120 periods to get it, so the value to Grace must be reduced by a factor of v^{120}:

$$6666.67 \cdot a_{\overline{240}|} \frac{.08}{12} v^{120} = \$359,080.09$$

Grace's share using the TI BA II Plus (Table 4.31)

We first compute the value of the 240-payment annuity at its inception. Since we are using the same payments and the same interest rate, we just need to input $N = 240$.

TABLE 4.31

N	I/Y	PV	PMT	FV
240		$CPT = 797,028.53$		

Push the FV key to convert this to a future value, enter $PMT = 0$ and $N = 120$ and re-compute the present value (Table 4.32):

TABLE 4.32

N	I/Y	PV	PMT	FV
120		$CPT = 359,080.05$	0	$-797,028.53$

Finally, North Central College (NCC) will receive all remaining payments. NCC's share is thus a perpetuity of $6666.67 per month which is deferred for 360 months. Its value at the time of the gift is thus:

$$6666.66666 \cdot \left(\frac{1}{i}\right) \cdot v^{360} = \$91,443.37$$

NCC's share using the TI BA II Plus (Table 4.33)

Note that the term $6666.66666 \cdot \left(\frac{1}{i}\right) = 1,000,000$ so we are computing the present value of 1,000,000 to be paid in 360 months. We enter $FV = 1,000,000$ and $N = 360$ and compute the PV

TABLE 4.33

N	I/Y	PV	PMT	FV
360		$CPT = -91,443.37$		1,000,000

Note: Since the interest rate was the same for all three parts of this problem, it did not have to be re-entered. So long as its memory is not cleared, the **TI BA II Plus** retains all entries in the TVM keys. Not sure if your entries are correct? Use the RCL key at any time to check what is currently in a given TVM register.

NCC's share, easy way.

The total value at inception is $1,000,000. We subtract the value of Harry's and Grace's shares to obtain the value of NCC's share. In fact, we can compute any two of the three shares and subtract from $1,000,000 to get the value of the third share.

The next two examples illustrate the need to know the formula for $a_{\overline{n}|,i} = \frac{1-v^n}{i}$. Problems of this sort can't be solved using just a calculator.

Example 4.18: Annuity A pays X at the end of each year for n years and has a present value of 493. Annuity B pays $3X$ at the end of each year for $2n$ years and has a present value of 2,748 at the same effective rate of annual interest. Determine the value of v^n.

Solution: We have:

$$493 = Xa_{\overline{n}|,i}$$
$$2748 = 3Xa_{\overline{2n}|,i}$$

Dividing the second equation by the first gives us

$$\frac{3a_{\overline{2n}|,i}}{a_{\overline{n}|,i}} = \frac{2748}{493}$$

Recalling that $a_{\overline{n}|,i} = \frac{1-v^n}{i}$ allows us to simplify this to

$$\frac{1-v^{2n}}{1-v^n} = \frac{2748}{3\cdot 493}$$

Now we use an old factoring fact: $1 - v^{2n} = (1-v^n)(1+v^n)$. This results in

$$1 + v^n = \frac{2748}{3\cdot 493}$$

Solving gives $v^n = .85801$.

Example 4.19: Find i given that $\frac{a_{\overline{18}|,i}}{a_{\overline{9}|,i}} = 1.25$

Solution Using the same factoring technique as in Example 4.18 we obtain $1 + v^9 = 1.25$. Hence $v^9 = .25$, $v = .85724$, $i = 16.65\%$.

4.6 Non Integer Time Periods

Thus far we have considered only the case where the time period, t, is a whole number. In a few rare cases it is necessary to compute the value of annuity which has a lifetime of $n + k$ where $0 < k < 1$. For example, wrongful death lawsuits are often settled by computing the lifetime earnings expectancy of the decedent. The process involves calculating the present value of an annuity whose payments are based on the expected annual earnings of the decedent and whose duration represents the decedent's life expectancy at the time of death. This last number is often not an integer (and greed is a common human trait) so we need to compute a reasonable value for the symbol $a_{\overline{n+k}|,i}$ – the present value of an annuity of n years and fractional duration k. The difficulty is that payments are made at integer times, while we want a present value at a fractional time. We can get around this by noting that even though our formula in the case for $a_{\overline{n}|,i}$ was derived under the assumption that n is an integer, its value can be computed for any value of n

$$a_{\overline{n+k}|,i} = \frac{1 - v^{n+k}}{i}$$
$$= \frac{1 - v^n + v^n - v^{n+k}}{i} = \frac{1-v^n}{i} + \frac{v^n - v^{n+k}}{i}$$
$$= a_{\overline{n}|,i} + v_{n+k}\frac{(1+i)^k - 1}{i} \tag{4.26}$$

The first term is just the present value of the n "regular" payments. The second term can be thought of as the value of a payment of $\left(\frac{(1+i)^k - 1}{i}\right)$ delayed by

$n+k$. This may seem a bit strange, but if we expand $(1+i)^k$ using the binomial series the first two terms are $1+ki$. Using this as a first approximation gives us

$$\frac{(1+i)^k - 1}{i} \approx \frac{1+ki-1}{i} = k \qquad (4.27)$$

This makes more sense! Since we have a period of k and payments of \$1 per whole period, it makes sense to think of a final payment of $\frac{k}{1} = k$. If payments of \$500 are made on a monthly basis and we want the value at 15.5 months, it makes sense to assume a payment of \$250 at the .5 month period. At any rate, a common solution to the fractional time conundrum is to use the value obtained in Equation 4.26 for the value of the payment for the fractional portion of the life of the annuity, namely

$$R \cdot \frac{(1+i)^k - 1}{i} \qquad (4.28)$$

4.7   Unknown Time

We can compute the time needed to accumulate a required amount if we know the payments which will be made and the interest rate in effect. This calculation is almost always approximate, but the **TI BA II Plus** can provide a very accurate estimate. Since the solution is rarely an integer, the last payment is usually not the same as the others.

Example 4.20 An investor wishes to accumulate \$4,000 with payments of \$200 at the end of each month for as long as needed. If the interest rate is a nominal 8% per year compounded monthly, how many regular payments will be made? Find the amount of the final payment using each of the three methods described above.

Solution: The interest rate is $\frac{.08}{12} = .00666667 = i$. We need to find n so that $200s_{\overline{n}|,i} = 4000$.

TI BA II Plus (Table 4.34)

TABLE 4.34

N	I/Y	PV	PMT	FV
$CPT = 18.84$	$\frac{8}{12}$	0	-200	4,000

We now compute the amount of the single final payment. At the end of year 18, a total of $200s_{\overline{18}|,.0066666} = \$3,811.43$. will be accumulated. The investor then has the choice of paying the remaining $4,000 - $3,811.43 = $188.56 at the end of year 18 for a total payment of $388.56 or waiting until the end of year 19 to make the needed fractional payment. The fractional payment was "due" at 18.84 years and so will accumulate to slightly larger amount at time 19 (Table 4.35).

TABLE 4.35

N	I/Y	PV	PMT	FV
.8369837		188.56	0	$CPT = 189.61$

The investor can thus make an additional payment of $188.56 at time 18 or $189.61 at time 19. The payment of $188.56 is known as a drop payment while the $189.61 payment is a balloon payment.

4.8   Unknown Rate of Interest

We now consider the situation where the period of the annuity, the amount of each payment, and either the present or future value is known and we wish to calculate the rate of interest. As an example consider the case of the Illinois State Lottery. In most cases, the prize is announced as a single number. However, the announced number is never paid out at one time. Instead, the winner is given a choice of a smaller single payment or an annuity with total payments (not present value!) equal to the total prize. As the lucky winner, you need to decide between the annuity and the single fixed payment. Which of these is the better deal depends on the interest rate the state uses to compute the value of the single payment. Here's an example.

Example 4.21   You win $10,000,000 and are offered the choice of twenty annual payments of $500,000 (paid at the start of each year) or a single payment of $7,000,000. At what interest rate is the present value of these two alternative prizes equal?

Solution: We have the equation of value

$$\$7,000,000 = \$500,000 a_{\overline{20}|,i}(1+i) \tag{4.29}$$

which we must solve for i. Note that the term $1+i$ is due to the fact that the annuity is paid at the start of each year and is thus most easily thought of as an annuity-due.

TI BA II Plus solution: (Table 4.36) Since we are solving for i, we must convert to an annuity-due.

Note: Recall that this is a toggle variable (BGN or END indicating payments which start at the beginning(BGN) or end (END) of a period. To do the conversion the following keystrokes are needed: 2ND, BGN, 2ND, SET, 2ND QUIT. A tiny BGN will show up in the upper right-hand corner of the screen.

We now compute I/Y after entering all the other values.

TABLE 4.36

N	I/Y	PV	PMT	FV
20	$CPT = 4.120974059$	7,000,000	$-500,000$	0

The interest rate is 4.121%.

Remember to re-set the **TI BA II Plus** to annuity-immediate mode: 2ND BGN 2ND SET 2ND QUIT as soon as you are finished.

Example 4.22 At what interest rate will an annuity-immediate paying $500 per month at the end of each month for ten years have a present value of $30,000?

Solution: TI BA II Plus (Table 4.37)

TABLE 4.37

N	I/Y	PV	PMT	FV
120	$CPT = 1.322001322$	30,000	-500	0

The interest rate is 1.32%.

4.9 Varying Rates of Interest

We now consider the case where the rate of interest varies from period to period. We will use the symbol i_j for the interest rate in period j. Be alert to the fact that period j is the period from $t = j - 1$ to $t = j$. There are two possibilities.

a) The interest rate $i_k = f_k$ applies in period k for all payments. These are known as **forward rates** and will be discussed in more detail in Chapter 7.

b) The interest rate $i_k = s_k$ applies to the payment made in period k and for all prior periods. Each payment has its own private interest rate! These are known as **spot rates** and will also be discussed in Chapter 7.

4.9.1 Present Value of Annuities with Varying Interest Rates

We consider a payment of $1 made at the end of each period for a total of n payments. There are two common ways the problem of varying interest is approached. One uses forward rates and the second uses spot rates.

a) The interest rate in period k is f_k.

$$PV = (1 + f_1)^{-1} + (1 + f_1)^{-1}(1 + f_2)^{-2}$$
$$+ (1 + f_1)^{-1}(1 + f_2)^{-2}(1 + f_3)^{-1} \cdots$$
$$= \sum_{t=1}^{n} \left(\prod_{s=1}^{t} (1 + f_s)^{-1} \right) \tag{4.30}$$

b) The interest paid on the payment made in period k is s_k during the period from 0 to k.

$$PV = (1 + s_1)^{-1} + (1 + s_2)^{-2} + \cdots = \sum_{t=1}^{n} (1 + s_t)^{-t} \tag{4.31}$$

Example 4.23 Compute the present value of an annuity-immediate paying $500 per year over four years with interest rates of 5%, 4%, 6%, and 7% respectively. Use both methods.

Solution

a) Treating the given interest rates as forward rates and using Equation 4.30:

$$PV = 500[(1.05)^{-1} + (1.05)^{-1}(1.04)^{-1} + (1.05)^{-1}(1.04)^{-1}(1.06)^{-1}$$
$$(1.05)^{-1}(1.04)^{-1}(1.06)^{-1}(1.07)^{-1}] = \$1,769.72$$

b) Treating the given interest rates as spot rates and equation 4.31

$$PV = 500[(1.05)^{-1} + (1.04)^{-2} + (1.06)^{-3} + (1.07)^{-4}] = \$1,739.73$$

Both of these calculations are fairly easy using the **TI-30XS**.

4.9.2 Accumulated Value of Annuities under Varying Interest Rates

We have the same two possibilities. We will use an annuity-due so that all terms enter the formula.

a) If the forward interest rate in period k is f_k we compute the accumulated value of an annuity-due as

$$(1 + f_n) + (1 + f_n)(1 + f_{n-1}) + \cdots = \sum_{t=1}^{n} \left(\prod_{s=1}^{t} (1 + f_{n-s+1}) \right) \quad (4.32)$$

b) If the spot interest rate for the payment made at time k is s_k we compute the accumulated value of an annuity-due as

$$\sum_{t=1}^{n} (1 + s_{n-t+1})^t \quad (4.33)$$

Example 4.24 Find the accumulated value of a ten-year annuity-immediate paying \$150 per year if the effective annual interest rate is 5% in the first three years, 6% in the next three years and 7% for the final four years.

Solution. Since the interest doesn't change every year, it's easier to use our standard formulas and adjust the interest rate. We compute the accumulated value of the payments grouped by the interest rate in effect when the payments were made.

The first three payments accumulate to $s_{\overline{3}|.05}$. This amount then accumulates for an additional three years at 6% and then for four years at 7%. The total accumulated value of this portion of the payments is thus

$$150 s_{\overline{3}|.05}(1.06)^3 (1.07)^4 = \$738.24$$

TI BA II Plus solution for these payments (Table 4.38)

TABLE 4.38

N	I/Y	PV	PMT	FV
3	5	0	-150	$CPT = 472.87$

Now convert this to a present value, set the PMT $= 0$, N $= 3$, $I/Y = 6$ and compute FV (Table 4.39)

TABLE 4.39

N	I/Y	PV	PMT	FV
3	6	472.875	0	$CPT = -563.20$

Convert this to a present value again and then compute its future value with $N = 4$ and $I/Y = 7$ (Table 4.40)

TABLE 4.40

N	I/Y	PV	PMT	FV
4	7	−563.20	0	$CPT = 738.24$

Now store this result: STO 1
Likewise, the next three years of payments accumulate to

$$150s_{\overline{3}|,.06}(1.07)^4 = \$625.96$$

TI BA II Plus solution for these three payments (Table 4.41)

TABLE 4.41

N	I/Y	PV	PMT	FV
3	6	0	-150	$CPT = 477.54$

Convert the FV to a PV, set PMT = 0 compute the future value with $N = 4$, $I/Y = 7$ (Table 4.42)

TABLE 4.42

N	I/Y	PV	PMT	FV
4	7	477.54	0	$CPT = -625.96$

Store this result : ± STO 2
Finally, the last four years of payments accumulate to $150s_{\overline{4}|,.07} = \$665.99$
TI BA II Plus solution for the final four years: (Table 4.43)

TABLE 4.43

N	I/Y	PV	PMT	FV
4	7	0	−150	$CPT = 665.99$

Now add up the three results: + RCL 1 + RCL 2 = $2,030.19.

4.10 Annuities Payable at Different Frequencies than Interest Is Convertible

Method One: Convert the conversion period of the interest to the period of the payments.

The simplest way to deal with the case in which the annuity payments are made at a different frequency than the interest conversion frequency is to convert the interest to the equivalent rate convertible at the **period of the annuity payments**. After that is done, we do all of our calculations using the period of the annuity payments.

Example 4.25. Find the accumulated value of a ten-year annuity paying $500 at the end of each quarter if interest is paid at a nominal 12% compounded monthly.

Solution. The interest rate is 1% per month. We need to convert this to an equivalent rate of interest per quarter. If we call the quarterly rate j, we have

$$(1 + j) = (1.01)^3$$

$$j = (1.01)^3 - 1 = .030301$$

We now calculate the accumulated value of a forty-payment (ten years = forty quarters) annuity at the quarterly interest rate

$$FV = 500 s_{\overline{40}|,.030301} = \$37,958.93$$

TI BA II Plus solution: (Table 4.44) We compute the interest rate and the future value using the TVM keys.

TABLE 4.44

N	I/Y	PV	PMT	FV
3	1	-1	0	$CPT = 1.030301$

N	I/Y		PV	PMT	FV
1	$CPT = 3.0301$				

The advantage of this second method is that we can now compute the accumulated value requested by merely entering $N = 40$, $PV = 0$, and $PMT = -500$ (Table 4.45):

TABLE 4.45

N	I/Y	PV	PMT	FV
40		0	-500	$CPT = 37,958.93$

Example 4.26 Find the present value of an annuity consisting of fifty equal payments of $600 made at the end of each month if the effective annual interest rate is 11%.

Solution.   We need to convert the annual interest rate to an equivalent monthly rate. Note that the problem does not say a nominal interest rate, so we can't just divide by 12 to obtain the monthly interest rate. If j represents the monthly rate which is equivalent to an effective annual interest rate of 11% we have

$$(1+j)^{12} = 1.11 \tag{4.34}$$

$$j = (1.11)^{\frac{1}{12}} - 1 = .008734594 \tag{4.35}$$

We now calculate the present value in the usual way using this interest rate.

$$PV = 600a_{\overline{50}|,.008734594} = \$24,222.82$$

TI BA II Plus solution: (Table 4.46)

TABLE 4.46

N	I/Y	PV	PMT	FV
12	$CPT = .873459382$	−1	0	1.11

N	I/Y	PV	PMT	FV
50		$CPT = 24,222.82$	−600	0

Example 4.27 Find the effective annual rate of interest if payments of $100 at the end of every quarter accumulate to $2,500 at the end of five years.

Solution.   See Table 4.47. We find the quarterly interest rate and then convert this to an effective annual rate. There are a total of twenty payments of $100 so we are solving

$$100s_{\overline{20}|,i} = 2500$$

TABLE 4.47

N	I/Y	PV	PMT	FV
20	$CPT = 2.285398264$	0	−100	2,500

Now we convert this to an effective annual rate (Table 4.48)

TABLE 4.48

N	I/Y	PV	PMT	FV
4		1	0	$CPT = -1.09459777$

N	I/Y	PV	PMT	FV
1	$CPT = 9.45977746$			

We could also have used the ICONV worksheet. Multiply the quarterly rate multiply by 4 and proceed as below (Table 4.49):

TABLE 4.49

Key	Entry/Result
C/Y	4
EFF	$CPT = 9.45977$
NOM	9.1415931

4.11 Alternative Method: Annuities Payable Less Frequently than Interest Is Convertible

We can derive an expression for the present and accumulated values of an annuity without converting the interest rate. Suppose that there are k interest conversion periods in one payment period. Let n be the period of the annuity in **interest conversion periods**. Finally, suppose that i is the interest rate per interest conversion period. The number of payments made is $\frac{n}{k}$ which we will assume to be a whole number.

For example, suppose we have an annuity which pays $\$R$ every quarter but interest is converted monthly. If the life of the annuity is five years, we use $n = 12 \cdot 5 = 60$. The value of k is 3 and then number of payments is $\frac{60}{3} = 20$. We can also compute the number of payments directly. Since the annuity lasts five years with four payments per year, there are a total of twenty payments. If we make payments every three months for twelve months and interest is converted monthly, $k = 3$, $n = 12$ and the number of payments is $\frac{12}{3} = 4$.

We begin with an annuity-immediate paying $\$R$ at the end of each five periods for a total of n interest conversion periods. To compute the present value, we sum the present value of each payment. Since it is delayed by k interest conversion periods, its present value is v^k. Each subsequent payment is delayed by an additional k conversion periods. The present value of the annuity at inception is thus:

$$
\begin{aligned}
PV &= R(v^k + v^{2k} + \cdots v^{\frac{n}{k}k}) \\
&= Rv^k(1 + v^k + \cdots + v^{n-k}) \\
&= Rv^k \frac{1 - v^n}{1 - v^k} \\
&= R\frac{a_{\overline{n}|,i}}{s_{\overline{k}|,i}}
\end{aligned}
\tag{4.36}
$$

The accumulated value immediately after the last payment is made is computed by multiplying the expression in 4.36 by $(1+i)^n$

$$
FV = R\frac{a_{\overline{n}|,i}}{s_{\overline{k}|,i}}(1+i)^n = R\frac{s_{\overline{n}|,i}}{s_{\overline{k}|,i}}
\tag{4.37}
$$

We can use this same technique to find the present and future (accumulated) values of annuities for which payments are made at the beginning of each k interest conversion periods.

$$
PV = R(1 + v + v^5 + \cdots v^{n-k}) = R\frac{1 - v^n}{1 - v^k} = R\frac{a_{\overline{n}|,i}}{a_{\overline{k}|,i}}
\tag{4.38}
$$

$$
FV = R(1+i)^n \frac{a_{\overline{n}|,i}}{a_{\overline{k}|,i}} = R\frac{s_{\overline{n}|,i}}{a_{\overline{k}|,i}}
\tag{4.39}
$$

These formulas make it possible to use the **TI BA II Plus** to make these calculations by storing the numerator and denominator in memory and then performing the division. While converting the interest conversion period to the payment period is usually the most direct route there are questions on the FM exam for which the formulas just discussed are very helpful.

Example 4.28 A ten-year annuity consists of equal payments of $450 made at the end of every six months. Interest is a nominal annual rate of 15% compounded every two months. What is the accumulated value of this annuity five years after the last payment is made?

Solution. We have $n = 10 \cdot 6 = 60$ two-month intervals. Interest is paid every two months while payments are made every six months, hence $k = 3$. The two-month interest rate is $i = \frac{.15}{6} = .025$. We can compute the accumulated value of this annuity immediately after the last payment using Equation 4.39. We then accumulate this amount for an additional five years (thirty periods). We thus have:

$$
FV = 450 \cdot \frac{s_{\overline{60}|,.025}}{s_{\overline{3}|,.025}}(1.025)^{30} = \$41,735.646400
$$

Summary Table (Table 4.50)
Payments every k Interest Conversion Periods
In each case, the value is multiplied by R the payment)

TABLE 4.50

Type of Annuity	PV	FV
Annuity Immediate	$\frac{a_{\overline{n},i}}{s_{\overline{k},i}}$	$\frac{s_{\overline{n},i}}{s_{\overline{k},i}}$
Annuity Due	$\frac{a_{\overline{n},i}}{a_{\overline{k},i}}$	$\frac{s_{\overline{n},i}}{a_{\overline{k},i}}$

TI BA II Plus solution: we compute

a) $450(1.025)^{30}$

b) $s_{\overline{60},.025}$

c) $s_{\overline{3},.025}$

store these values and then compute the answer.

a) (Table 4.51)

TABLE 4.51

N	I/Y	PV	PMT	FV
30	2.5	−450	0	$CPT = 943.9054106$

STO 1

b) (Table 4.52)

TABLE 4.52

N	I/Y	PV	PMT	FV
60	2.5	0	−1	$CPT = 135.99159$

STO 2

c) (Table 4.53)

TABLE 4.53

N	I/Y	PV	PMT	FV
3				$CPT = 3.075625$

STO 3
Now we complete the calculation: RCL 1 × RCL 2 ÷ RCL 3 = \$41,735.646.

Example 4.29 An investment of \$1,000 is being used to fund payments of \$125 at the end of each year for as long as possible with a smaller final payment to be made at the time of the last regular payment. The nominal rate of annual interest is 6% compounded quarterly. How many regular payments will be made and what is the amount of the final payment (regular payment plus smaller amount)?

Solution: In this case, n is unknown, $k = 4$, $i = .015$. The present value of the annuity is $1,000, and the payments are $125. We can use Equation 4.43 to set up an equation of value:

$$\$1,000 = 125\frac{a_{\overline{n}|,.015}}{s_{\overline{4}|,.015}}$$

We solve for $a_{\overline{n}|,.015} = \frac{1000}{125}s_{\overline{4}|,.015} = 8 \cdot s_{\overline{4}|,.015}$

TI BA II Plus solution: (Tables 4.54 and 4.55) We compute $s_{\overline{4}|,.015}$, multiply by 8, and then use the TVM keys to solve for n.

TABLE 4.54

N	I/Y	PV	PMT	FV
4	1.5	0	−1	$CPT = 4.090903375$

$\times 8 = 32.727227$

TABLE 4.55

N	I/Y	PV	PMT	FV
$CPT = 45.34521454$	1.5	32.727227	−1	0

We must now determine the value of the additional payment required to clear out the account. At the end of forty-four interest conversion periods, the accumulated value of $1,000 would be $1000(1.015)^{44}$. This must be equal to the accumulated value of the eleven regular payments plus the additional payment. If the value of the additional payment is R we have the following equation of value:

$$R + 125\frac{s_{\overline{44}|,.015}}{s_{\overline{4}|,.015}} = 1000(1.015)^{44}$$

This implies that $R = \$40.3927539$. The total final payment (sometimes called a balloon payment) is $40.39 + 125 = \$165.39$.

To use the **TI BA II Plus**: (Tables 4.56, 4.57, and 4.58) we solve for R

$$R = 1000(1.015)^{44} - 125\frac{s_{\overline{44}|,.015}}{s_{\overline{4}|,.015}}$$

TABLE 4.56

N	I/Y	PV	PMT	FV
4	1.5	−1000	0	$CPT = 1,925.333019$

STO 1

TABLE 4.57

N	I/Y	PV	PMT	FV
44	1.5	0	−125	$CPT = 7,711.108493$

STO 2

TABLE 4.58

N	I/Y	PV	PMT	FV
4			−1	$CPT = 4.090903375$

STO 3

Now we complete the calculation: RCL 2 ÷ RCL 3 ± = + RCL 1 = 40.3927539. The total payment required is \$165.39.

4.12 Annuities Paid More Frequently than Interest Is Converted

We already know how to do these problems by converting the interest rate to the period of the annuity payments. In this section we develop an alternative method which avoids the conversion problem. We assume that m payments are made during each interest conversion period and that there are a total of n interest conversion periods. There are thus mn payments in all. We assume that the amount of each payment is $\frac{R}{m}$ (for a total of \$R each period!).

The symbol $a_{\bar{n}}^{(m)}$ we will used to denote the present value of an annuity-immediate in this situation. Note the similarity to the symbol $i^{(m)}$ which represents the nominal annual rate of interest compounded m times per year.

Remember: n **represents the period of the annuity in interest conversion periods and** m **is the number of payments per interest conversion period.**

We compute the present value by summing the appropriate geometric series:

$$a_{\overline{n}|}^{(m)} = \frac{1}{m}(v^{\frac{1}{m}} + v^{\frac{2}{m}} + \cdots + v^n)$$

$$= \frac{1}{m} \frac{v^{\frac{1}{m}} - v^{n+\frac{1}{m}}}{1 - v^{\frac{1}{m}}}$$

$$= \frac{1}{m} \frac{v^{\frac{1}{m}}}{v^{\frac{1}{m}}} \left(\frac{1 - v^n}{\left(\frac{1}{v^{\frac{1}{m}}}\right) - 1} \right)$$

$$= \frac{1 - v^n}{m((1+i)^{\frac{1}{m}} - 1)}$$

$$= \frac{1 - v^n}{i^{(m)}} = i \cdot \left(\frac{a_{\overline{n}|,i}}{i^{(m)}} \right) \tag{4.40}$$

Here we have used the symbol $i^{(m)} = m((1+i)^{\frac{1}{m}} - 1)$ from an earlier chapter. The accumulated value of this annuity immediately after the last payment is made is denoted by $s_{\overline{n}|}^{(m)}$ and is simply the value of $a_{\overline{n}|}^{(m)}$ accumulated over the n interest conversion periods.

$$s_{\overline{n}|}^{(m)} = a_{\overline{n}|}^{(m)}(1+i)^n$$

$$= \frac{(1+i)^n - 1}{i^{(m)}}$$

$$= i \cdot \left(\frac{s_{\overline{n}|,i}}{i^{(m)}} \right) \tag{4.41}$$

Example 4.30 Compute the present value and the accumulated value just after the last payment of an annuity which pays $300 each month for three years. Interest is a nominal 8% per year compounded quarterly.

Solution: (Tables 4.59, 4.60, and 4.61) The interest rate per quarter is $i = .02$. The annuity makes $m = 3$ payments per quarter. We have also that $\frac{R}{m} = 300$. The period of the annuity in interest conversion periods is $n = 3 \cdot 4 = 12$. We compute $i^{(m)}$ and $s_{\overline{n}|,i}$, and $a_{\overline{n}|,i}$ and then the present and accumulated values.

$\frac{i^{(m)}}{m}$

TABLE 4.59

N	I/Y	PV	PMT	FV
3	$CPT = .662270956$	-1	0	1.02

$i^{(m)}$

$\times 3 = \div 100 = .019868129$ STO 1

$a_{\overline{n}|,i}$

TABLE 4.60

N	I/Y	PV		PMT	FV
12	2	$CPT = 10.5734122$		-1	0

STO 2

$s_{\overline{n}|,i}$

TABLE 4.61

N	I/Y	PV	PMT	FV
		0		$CPT = 13.41208973$

STO 3

Now we compute

$$300 \cdot a_{\overline{n}}^{(m)} = 300 \cdot .02 \times RCL2 \div RCL1 = 3,193.659974$$

$$300 \cdot s_{\overline{n}}^{(m)} = 300 \cdot .02 \times RCL3 \div RCL1 = 4,050.332779$$

While these methods work, converting the interest rate to the period of the payments is usually easier than using Equations 4.40 or 4.41.

Annuities-Due

The present and accumulated values of annuities-due are merely the present and accumulated values of an annuity-immediate accumulated for $\frac{1}{m}$th of a period.

We use the symbols $\ddot{a}_{\overline{n}}^{(m)}$ and $\ddot{s}_{\overline{n}}^{(m)}$ for the present and accumulated values of and annuity-due respectively and have the following formulas

$$\ddot{a}_{\overline{n}}^{(m)} = a_{\overline{n}}^{(m)}(1+i)^{\frac{1}{m}} \tag{4.42}$$

$$= \frac{1 - v^n}{m[(1+i)^{\frac{1}{m}} - 1]}(1+i)^{\frac{1}{m}}$$

$$= \frac{1 - v^n}{m(1 - v^{\frac{1}{m}})}$$

$$= \frac{1 - v^n}{d^{(m)}} = \frac{ia_{\overline{n}|,i}}{d^{(m)}}$$

$$\ddot{s}_{\overline{n}|}^{(m)} = \ddot{a}_{\overline{n}|}^{(m)}(1+i)^n \tag{4.43}$$

$$= \frac{1-v^n}{d^{(m)}}(1+i)^n$$

$$= \frac{(1+i)^n - 1}{d^{(m)}}$$

$$= \frac{i \cdot s_{\overline{n}|,i}}{d^{(m)}} \tag{4.44}$$

Summary Table (Table 4.62)
Annuities-Payable m Times Per Interest Conversion Period
Payment is $\frac{1}{m}$ Per Interest Conversion Period
Multiply by R to obtain the required result.

TABLE 4.62

Annuity	PV	FV
Immediate	$a_{\overline{n}|}^{(m)} = \frac{1-v^n}{m((1+i)^{\frac{1}{m}}-1)} = i \cdot \left(\frac{a_{\overline{n}|,i}}{i^{(m)}}\right)$	$s_{\overline{n}|}^{(m)} = a_{\overline{n}|}^{(m)}(1+i)^n = i \cdot \left(\frac{s_{\overline{n}|,i}}{i^{(m)}}\right)$
Due	$\ddot{a}_{\overline{n}|}^{(m)} = a_{\overline{n}|}^{(m)}(1+i)^{\frac{1}{m}} = \frac{i a_{\overline{n}|,i}}{d^{(m)}}$	$\ddot{s}_{\overline{n}|}^{(m)} = \ddot{a}_{\overline{n}|}^{(m)}(1+i)^n = \frac{i s_{\overline{n}|,i}}{d^{(m)}}$

4.13 Perpetuities Paid More Frequently than Interest Is Convertible

Again, we go back to the basic equation of value. Assuming that the value of each payment is $\frac{1}{m}$ and using $a_{\infty}^{(m)}$ and $\ddot{a}_{\infty}^{(m)}$ for the present value of a perpetuity-immediate and a perpetuity-due we have

$$a_{\infty}^{(m)} = \frac{1}{m}(v^{\frac{1}{m}} + v^{\frac{2}{m}} + \cdots)$$

$$= \frac{1}{m}\frac{v^{\frac{1}{m}}}{1-v^{\frac{1}{m}}}$$

$$= \frac{1}{m(\frac{1}{v^{\frac{1}{m}}}-1)} = \frac{1}{d^{(m)}} \tag{4.45}$$

These equations are probably not worth memorizing. You can use the basic formulae for a geometric series to solve them. Here it is one more time:

$$a + ar + ar^2 + ar^3 + \cdots = \frac{a}{1-r}(\text{assuming } |r| < 1) \tag{4.46}$$

Example 4.31: A perpetuity pays \$600 per year with payments made at the end of the each month. If the effective annual rate of interest is $i = 5.6\%$ what is the present value of this annuity?

Solution: The monthly payment is \$50 and the formula for the present value is

$$50v^{\frac{1}{12}} + 50v^{\frac{2}{12}} + \cdots + 50v^{\frac{i}{12}} + \cdots$$

Using Equation 4.46 this is

$$\frac{50v^{\frac{1}{2}}}{1 - v^{1/12}} = 10,986.58$$

Example 4.32: A perpetuity pays \$10 at the end of each four-month period (three times per year) and has a present value of 609.89. What is the effective annual rate of interest used to value this annuity?

Solution: We have $\frac{10v^{\frac{1}{3}}}{1 - v^{\frac{1}{3}}} = 608.89$. Solving yields $v^{\frac{1}{3}} = .983868$, $v = .9523809$, $i = 5\%$.

4.14 Annuities with Varying Payments

It is often the case that the payments associated with an annuity vary. We consider several such cases below.

Payments Varying According to an Arithmetic Progression

We begin with an annuity-immediate of n payments at an interest rate per period of i. We assume that the first payment is P and that payments change by Q per period. The payments are thus P, $P+Q$, $P+2Q$, ..., $P+(n-1)Q$. Q can be either positive or negative, but we assume $P > 0$ and $P+(n-1)Q > 0$ so that all payments are positive. We will refer to this type of annuity as a $P - Q$ annuity. To compute the present value of a $P - Q$ annuity we return to our equation of value:

$$PV = Pv + (P + Q)v^2 + \cdots (P + (n-1)Q)v^n \tag{4.47}$$

This is the special case we considered in Chapter 1. The simplified version of this expression (Equation 1.8) is repeated below

$$PV = \frac{1}{i}(P(1 - v^n) + Q(a_{\overline{n}|,i} - nv^n))$$

$$= Pa_{\overline{n}|,i} + Q\frac{a_{\overline{n}|,i} - nv^n}{i} \tag{4.48}$$

The future value is computed using $FV = PV(1 + i)^n$

$$FV = Pa_{\overline{n}|,i}(1 + i)^n + Q\frac{a_{\overline{n}|,i}(1 + i)^n - nv^n(1 + i)^n}{i}$$

$$= Ps_{\overline{n}|,i} + Q\frac{s_{\overline{n}|,i} - n}{i} \qquad (4.49)$$

Computing the Present and Future Values of P–Q Annuities

$$PV = Pa_{\overline{n}|,i} + Q\frac{a_{\overline{n}|,i} - nv^n}{i}$$

$$FV = Ps_{\overline{n}|,i} + Q\frac{s_{\overline{n}|,i} - n}{i} \qquad (4.50)$$

Example 4.33: Compute the present and accumulated value of an annuity-immediate which has an initial payment of \$450. Payments increase by \$75 each subsequent period. The annuity provides thirty monthly payments and the interest rate is a nominal 8% compounded monthly.

Solution: (Tables 4.63, 4.64, and 4.65) We have $P = 450$, $Q = 75$, $n = 30$, and $i = .08/12$. We will compute $a_{\overline{n}|,i}$, $s_{\overline{n}|,i}$ and then do the calculation.

$a_{\overline{n}|,i}$

TABLE 4.63

N	I/Y	PV	PMT	FV
30	$\frac{8}{12}$	$CPT = 27.11$	-1	0

STO 1

$s_{\overline{n}|,i}$

TABLE 4.64

N	I/Y	PV	PMT	FV
		0		$CPT = 33.08885394$

STO 2

v^n

TABLE 4.65

N	I/Y	PV	PMT	FV
		$CPT = -.81927434$	0	1

$\pm$ STO 3

PV

$450 * \text{RCL1} = 12,198.98204 \text{ STO } 4 \; 30 \times \text{ RCL } 3 = \pm + \text{ RCL } 1 = \times 75 = \times 12 \div .08 = 28,469.4612 + RCL4 = 40,668.46$

FV

$450 * \text{ RCL } 2\pm = 14,889.98427 \text{ STO } 4 \text{ RCL } 2 - 30 = \times 12 \div .08 = \times 75 = 34,749.60683 + \text{ RCL } 4 = 49,639.61$

TI-30XS Solution: Since the **TI-30XS** will not compute $a_{\overline{n}|,i}$ or $s_{\overline{n}|,i}$ directly, the **TI BA II Plus** is probably the better choice here. If you chose to transfer values from one to the other, make sure you do not truncate them.

Payments Varying in a Geometric Progression

We assume an initial payment of P at the end of period 1 and that subsequent payments increase/decrease by rate g. The second payment is thus $P(1+g)$, the third is $P(1+g)^2$ and so on. For example, if payments increase by 5% each period, we have $g = .05$ and $1+g = 1.05$. If payments decrease by 5% each period we have $g = -.05$ and $1+g = .95$.

The present value of such an annuity is the sum of the geometric series

$$PV = \frac{P}{1+i} + \frac{P(1+g)}{(1+i)^2} + P\frac{(1+g)^2}{(1+i)^3} + \cdots P\frac{(1+g)^{n-1}}{(1+i)^n} \qquad (4.51)$$

We have $a = \frac{P}{1+i}$ and a common ratio of $r = \frac{1+g}{1+i}$. Using Equation 1.1 we have

$$PV = \frac{P}{1+i} \cdot \frac{1 - \left(\frac{1+g}{1+i}\right)^n}{1 - \left(\frac{1+g}{1+i}\right)} = P\frac{1 - r^n}{i - g} \qquad (4.52)$$

Present value of an annuity with payments varying geometrically

$$PV = P\frac{1 - r^n}{i - g} \qquad (4.53)$$

Note that the equation makes no sense if $g = i$. In that case the present value of each payment is $\frac{1}{1+i}$. If $g > i$, the present value of the payments actually increase!

Example 4.34 Burt has purchased a retirement annuity consisting of twenty annual payments of \$50,000 which was priced based on an interest rate of 7.5%. He went the extra distance and purchased an inflation protection plan which promises to increase the value of his annual payments by the amount of inflation. If inflation is expected to be 6% per year over the next twenty years what is a fair price for Bert's retirement annuity?

Solution: We have $i = .075$, $n = 20$, and $g = .06$. Hence, $r = \frac{1.06}{1.075} = .98605$

$$PV = 50,000 \cdot \frac{1 - (.98605)^{20}}{.015} = \$816,660.48$$

Example 4.35: A annuity consists of thirty payments made at the end of each year. The effective annual rate of interest is 7%. The initial payment is $400 and payments increase by 7% each year. What is the present value of this annuity?

Solution: We have $i = g$ so we can not use the formula. As mentioned above, the value of each payment is just $\frac{30}{1.07}$ and there are thirty payments so the total value is $\frac{900}{1.07} = \$841.12$.

Block Payments

In some cases annuities pay in "blocks" – for example, there might be ten payments of $500 followed by twenty payments of $600, and finally twelve payments of $1,000. There are several ways to approach this type of problem. The next example illustrates the method which is most efficient in terms of time spent.

Example 4.36 An annuity due provides annual payments and is valued at an effective annual rate of interest of 5.4%. The annuity provides five payments of $100, then ten payments of $135, then twenty-five payments of $500. What is the present value of this annuity?

Solution: This annuity consists of a five-payment annuity, a ten-payment annuity delayed five-periods, and a twenty-five-payment annuity delayed fifteen periods:

$$PV = 100a_{\overline{n}|,i} + 135a_{\overline{10}|,i} \cdot v^5 + 500a_{\overline{25}|,i} \cdot v^{15}$$

This method involves calculating values of $a_{\overline{n}|,i}$, and v^n. It's easier if we start with a forty-payment annuity of $500 and then subtract off terms. We obtain

$$500 \cdot a_{\overline{40}|,i} - (500 - 135) \cdot a_{\overline{15}|,i} - (135 - 100)a_{\overline{5}|,i}$$

This simplifies to

$$500a_{\overline{40}|,i} - 365a_{\overline{15}|,i} - 35a_{\overline{5}|,i} = 4,291.53$$

4.15   Varying Perpetuities

Perpetuities with an Arithmetic Increase In Payments

We will assume an initial payment of P and an increase of Q for each subsequent payment. A perpetuity-immediate has an infinite number of terms, so we compute its present value as the limit as $n \to \infty$ of a varying annuity with n terms.

$$\lim_{n\to\infty}\left(Pa_{\overline{n}|,i} + Q\frac{a_{\overline{n}|,i} - nv^n}{i}\right) \tag{4.54}$$

To compute this limit, we need to compute $\lim_{n\to\infty} a_{\overline{n}|,i}$ and $\lim_{n\to\infty} nv^n$. We first note that $v = \frac{1}{1+i} < 1$ so that $\lim_{n\to\infty} v^n = 0$. Using L'Hôpital's rule we can show that $\lim_{n\to\infty} nv^n = 0$ as well. Finally, since $a_{\overline{n}|,i} = \frac{1-v^n}{i}$ and $0 < v < 1$ we see that $\lim_{n\to\infty} a_{\overline{n}|,i} = \frac{1}{i}$. This last result makes sense since $\lim_{n\to\infty} a_{\overline{n}|,i}$ represents the present value of a perpetuity, which we already knew to be $\frac{1}{i}$.

Using these results we obtain

$$\lim_{n\to\infty} Pa_{\overline{n}|,i} + Q\frac{a_{\overline{n}|,i} - nv^n}{i}$$
$$= \frac{P}{i} + Q\frac{\frac{1}{i} - 0}{i} = \frac{P}{i} + \frac{Q}{i^2} \tag{4.55}$$

Example 4.37 A perpetuity-immediate has a first year payment of $100 and its annual payments increase by $50 in each subsequent year. If the present value of this perpetuity is $20,000, what is the interest rate?

Solution: From Equation 4.51 we obtain

$$22,000 = \frac{100}{i} + \frac{50}{i^2}$$
$$\Rightarrow 22000 \cdot i^2 - 100 \cdot i - 50 = 0$$
$$\Rightarrow i = \frac{1}{20} \text{ or } i = -\frac{1}{22}$$

Clearly, only $i = \frac{1}{20} = .05$ or 5% makes sense in this situation.

Perpetuities with a Geometric Increase in Payments

This situation only makes sense when $g < i$ so that $r = \frac{1+g}{1+i} < 0$. Hence $\lim_{n\to\infty} r^n = 0$. In this case Equation 4.52 reduces to the following formula for the present value of a perpetuity immediate in which payments increase at rate g:

$$PV = P\frac{1}{i-g} \tag{4.56}$$

Example 4.38 Suppose that a perpetuity provides an initial payment of $5,000 and that payments increase by 6% each year. Using an interest rate of 7.5%, what is the present value of this perpetuity?

Solution:
$$PV = 5000\frac{1}{.015} = \$333,333.33$$

There are two special cases of varying annuities which appear often enough that they have their own special notation.

Increasing Annuity of term n with $P = Q = 1$

The present value of this special type of varying annuity is denoted $(Ia)_{\overline{n}|,i}$. Its accumulated (future) value is denoted by $(Is)_{\overline{n}|,i}$. These can be computed directly from Equations 4.55 and 4.56, but we can get simpler versions of each with a little algebra.

$$
\begin{aligned}
(Ia)_{\overline{n}|,i} &= a_{\overline{n}|,i} + \frac{a_{\overline{n}|,i} - nv^n}{i} \\
&= \frac{1 - v^n + a_{\overline{n}|,i} - nv^n}{i} \\
&= \frac{\ddot{a}_{\overline{n+1}|,i} - (n+1)v^n}{i} \\
&= \frac{\ddot{a}_{\overline{n}|,i} - nv^n}{i}
\end{aligned}
\tag{4.57}
$$

We can calculate this final expression using the **TI BA II Plus**. The numerator is the value of an annuity-due of 1 minus the present value of n. We compute using the TVM keys as follows

1) Set mode to BGN

2) Enter numbers as indicated in Table 4.66.

TABLE 4.66

N	I/Y	PV	PMT	FV
n	i	$CPT = x$	-1	n

$$
(Ia)_{\overline{n}|,i} = \frac{x}{i}
$$

3) Set mode to END

We can use this same technique to compute the accumulated value of a general P - Q annuity. A little algebra shows that

$$
PV = Pa_{\overline{n}|,i} + Q(Ia)_{\overline{n-1}|,i} \cdot v
\tag{4.58}
$$

Rerurning to $P = Q = 1$ annuties, we obtain the following expression for the accumulated value:

$$
(Is)_{\overline{n}|,i} = \frac{s_{\overline{n+1}|,i} - (n+1)}{i}
\tag{4.59}
$$

Decreasing Annuity of term n with $P = n$, $Q = -1$

The present and accumulated values of such an annuity are computed in a similar fashion. The results are presented below in a summary table (Table 4.67).

Annuities 145

TABLE 4.67

Annuity Type	Present Value	Future Value
Increasing P, Q	$Pa_{\overline{n}|,i} + Q\frac{a_{\overline{n}|,i}-nv^n}{i}$	$Ps_{\overline{n}|,i} + Q\frac{s_{\overline{n}|,i}-n}{i}$
Increasing with rate of increase g	$P\frac{1-R^n}{i-g}$, $R=\frac{1+g}{1+i}$	$P\frac{1-R^n}{i-g}(1+i)^n$
Increasing $(P=Q=1)$	$(Ia)_{\overline{n}|,i} = \frac{\ddot{a}_{\overline{n}|,i}-nv^n}{i}$	$(Is)_{\overline{n}|,i} = \frac{s_{\overline{n+1}|,i}-(n+1)}{i}$
Decreasing $(P=n, Q=-1)$	$(Da)_{\overline{n}|,i} = \frac{n-a_{\overline{n}|,i}}{i}$	$(Ds)_{\overline{n}|,i} = \frac{n-a_{\overline{n}|,i}}{i}(1+i)^n$
Perpetuity P, Q	$\frac{P}{i} + \frac{Q}{i^2}$	Not possible
Perpetuity With rate of increase g	$P\frac{1}{i-g}$	Not Possible

Example 4.39 Find expressions for the present and future value of an annuity immediate for which payments start at 1, increase to 8 and then decrease to 1.

Solution: We compute the value of the two portions of the annuity and add.

For the increasing portion we have $(Ia)_{\overline{n}|,i}$. The decreasing portion is a bit tricky. Suppose $n=4$. The payments are then 1, 2, 3, 4, 3, 2, 1. The first four payments form an increasing annuity with four terms (1,2,3,4) while the decreasing portion consists of only three payments (3,2,1). It is also deferred by four periods and so must be discounted by v^4. In general the increasing portion has a present value of $(Ia)_{\overline{n}|,i}$ while the decreasing portion has a present value of $v^n(Da)_{\overline{n-1}|,i}$. We add these to get the present value and multiply by $(1+i)^n$ to get the future or accumulated value:

$$PV = (Ia)_{\overline{n}|,i} + v^n(Da)_{\overline{n-1}|,i} \tag{4.60}$$

$$FV = (1+i)^n((Ia)_{\overline{n}|,i} + v^n(Da)_{\overline{n-1}|,i}) \tag{4.61}$$

Example 4.40 Find an expression for the present value of an annuity in which payments start at 1, increase by 1 until reaching 10 and then remain level for an additional twenty payments.

Solution: There are several ways to approach this problem. We can consider this as a combination of an increasing annuity of ten payments followed by a deferred level payment annuity in which each payment is $10. The present value is then given by

$$PV = (Ia)_{\overline{10}|,i} + 10v^{10}a_{\overline{20}|,i}$$

We can also view it as an increasing annuity of thirty payments minus a deferred increasing annuity of twenty payments:

$$PV = (Ia)_{\overline{30}|,i} - v^{10}(Ia)_{\overline{20}|,i}$$

4.16 Varying Annuities Paid Less Frequently than Interest Is Convertible

We assume an annuity with $P = Q = 1$ for which payments are made every k interest conversion periods. We further assume that the term of the annuity (in interest conversion periods) is n. There will thus be $\frac{n}{k}$ payments, with the first payment deferred k periods. We go back to basics and then do some algebra:

$$PV = 1v^k + 2v^{2k} + 3v^{3k} + \cdots \left(\frac{n}{k} - 1\right)v^{n-k} + \frac{n}{k}v^n \qquad (4.62)$$

We divide both sides of Equation 4.62 by v^k which is the same thing as multiplying by $(1+i)^k$ to obtain

$$(1+i)^k PV = v + 2v^k + 3v^{2k} + \cdots \frac{n}{k}v^{n-k} \qquad (4.63)$$

Subtracting Equation 4.62 from Equation 4.63 and simplifying yields

$$PV = \frac{\frac{a_{\overline{n}|,i}}{a_{\overline{k}|,i}} - \frac{n}{k}v^n}{is_{\overline{k}|,i}} \qquad (4.64)$$

Example 4.41 An annuity consists of payments which start at $500 and increase by $500 every quarter for ten years. The nominal rate of annual interest is 6% converted monthly. Compute the present value of this annuity.

Solution We have $n = 120$, $k = 3$, $i = .5\%$

$$PV = 500 \cdot \frac{\frac{a_{\overline{n}|,i}}{a_{\overline{k}|,i}} - \frac{n}{k}v^n}{is_{\overline{k}|,i}} = \$276,611.93$$

4.17 Continuous Annuities

A continuous annuity is a theoretical entity which pays interest continuously over a period of time. If the total amount paid per conversion period is $1, the amount paid in time dt is $1\,dt$ and the total amount paid in the period from 0 to t is given by

$$\int_0^r 1\,dt = t \qquad (4.65)$$

To compute the present value of this (denoted by $\bar{a}_{\overline{n}|,i}$)annuity at time t we note that the present value of payment $1dt$ at time t is given by $1v^t dt$ so that the present value of the stream of payments at time n is given by

$$\bar{a}_{\overline{n}|i} = \int_0^n 1v^t dt = \frac{v^n}{\ln(v)} - \frac{v^0}{\ln v} = \frac{v^n - 1}{\ln(v)} \tag{4.66}$$

Recalling that $\ln(v) = -\delta$, we have

$$\bar{a}_{\overline{n}|,i} = \frac{1 - v^n}{\delta} \tag{4.67}$$

Since $v = e^{-n\delta}$, also have

$$\bar{a}_{\overline{n}|,i} = \frac{1 - e^{-n\delta}}{\delta}$$

Take note of the notation $\bar{a}_{\overline{n}|,i}$ for the present value of a continuous annuity. We can also compute the accumulated value of a continuous annuity as

$$\bar{s}_{\overline{n}|,i} = \int_0^n (1+i)^t dt = \frac{(1+i)^n - 1}{\delta} = \frac{e^{n\delta} - 1}{\delta} \tag{4.68}$$

Example 4.42 A continuous annuity pays $500 per quarter for three years. If the nominal annual interest rate is 12% per year compounded quarterly find the present and accumulated value of this annuity.

Solution: We have $i = .03, n = 12, \delta = \ln(1.03) = .0295588, n\delta = .354705627$

$$\bar{a}_{\overline{n}|,i} = \frac{1 - e^{-n\delta}}{\delta} = \frac{1 - e^{-.35472}}{.0295588} = \frac{.7580}{.029558802} = 10.10$$

$$\bar{s}_{\overline{n}|,i} = \frac{e^{n\delta} - 1}{\delta} = \frac{.425761}{.02955} = 14.4038$$

We multiply by $500 to obtain the present and accumulated values $5,051.29, $7,201.93.

Example 4.43 Find the force of interest at which a continous annuity with term 20 has an accumulated value which is exactly three times that of a continous annuity with a term of 10.

Solution: We want $\bar{s}_{\overline{20}|,i} = 3\bar{s}_{\overline{10}|,i}$ or

$$\frac{e^{20\delta}-1}{\delta} = 3\frac{e^{10\delta} - 1}{\delta}$$

We clear the fraction to obtain

$$e^{20\delta} - 1 = 3e^{10\delta} - 3$$
$$e^{20\delta} - 3e^{10\delta} + 2 = 0$$
$$(e^{10\delta} - 2)(e^{10\delta} - 1) = 0$$

Hence $e^{10\delta} = 1$ or 2 so that $\delta = 0$ or $\frac{\ln(2)}{10} = .06931 = 6.931\%$. Only the second value makes sense as $\delta = 0$ would imply an interest rate of 0.

Exercises For Chapter 4

1) Marge deposits $100,000 in an account paying 7% per annum convertible quarterly. She wishes to withdraw a fixed amount at the end of each quarter for twenty years. How much can she withdraw to use up the account exactly in twenty years?

2) A $50,000 loan is to be paid off in fifteen years. The effective annual rate of interest is 6.5%. Compute the amount of interest paid in each of the following scenarios and explain the difference.

   a) The entire loan is paid off, along with accumulated interest at the end of fifteen years.

   b) The interest due is paid at the end of each year and the principal is paid off in a lump sum at the end of the fifteen-year period.

   c) The loan is repaid with level payments at the end of each year.

   d) The loan is repaid with level payments at the end of each month. Hint: Compute the monthly interest rate $\frac{i^{(12)}}{12}$ and calculate in months.

3) Compute the present value of each annuity. The effective annual interest is 8%

   a) A series of ten level payments of $500 at the end of each year.

   b) A series of ten level payments of $500 at the start of each year.

4) Compute the present value of an annuity consisting of twenty level payments of $5,000 per quarter paid at the start of each quarter assuming

   a) Interest is paid at 5% compounded quarterly.

   b) Interest is paid at 6% compounded monthly.

   c) Interest is paid at 8% compounded annually.

5) Prove each identity and write a complete English sentence which explains why the identity is correct.

   a) $a_{\overline{n+am}|,i} = a_{\overline{n}|,i} + v^n a_{\overline{m}|,i} = a_{\overline{m}|,i} + v^m a_{\overline{n}|,i}$

   b) $s_{\overline{n+m}|,i} = s_{\overline{n}|,i} + (1+i)^n s_{\overline{m}|,i}$

6) Mark will retire in ten years and desires to purchase an annuity at that time which will provide thirty years of level annual payments of $10,000. He will save up for this purchase by making level payments at the end of each year into an account paying 5% effective per annum. Mark estimates that the price of the annuity will be based on an interest rate of 4% effective per annum. What is the amount of Mark's required annual payment?

7) Bob and Carol are, unfortunately, divorcing. They own an annuity which provides ten years of monthly payments of $5,000. They have owned the annuity for exactly six months. What is the value of this annuity at this time? Assume an interest rate of 2.5% effective per month.

8) Bob and Carol also own a perpetuity which will begin payments of $5,000 each year beginning ten years from now. Rather than fight over it, they decide to donate the perpetuity to good old NCC. If the IRS allows an interest rate of $i = 8\%$, what is the present value of this perpetuity?

9) A level payment annuity-immediate with $4n$ payments is to be shared by Bob and Carol and Ted and Alice. Bob will receive the first n payments, Carol, the second n payments, Ted the third n payments, and Alice the remaining payments. Ted's share is worth .49 of Bob's share. What is the ratio of Carol's share to Alice's share? (we are comparing present values at the present time).

10) Maria is buying a car with a negotiated price of $18,500. She will finance the car with level monthly payments (at the end of each month) over four years. Maria can afford a monthly payment of at most $250. How much must her down payment be if the interest rate is 2% per monthly convertible monthly?

11) A family is accumulating a college fund. The fund must total $100,000 at the end of twenty years. If they deposit $5,000 at the end of each year for the first ten years and earn 2% annual interest, how much must be deposited at the end of each of the last ten years to accumulate $100,000?

12) A professor will contribute $6,000 at the beginning of each year for the next twenty-five years. At that point she will retire and purchase a twenty-year annuity-due. If the deposits earn 5% effective per year and the annuity is priced based on $i = .07$ effective per year, what will be the amount of her annual payment from the annuity?

13) Verify each identity.

   a) $\ddot{a}_{\overline{n}|,i} = a_{\overline{n}|,i} + 1 - v^n$.

   b) $\ddot{s}_{\overline{n}|,i} = s_{\overline{n}|,i} - 1 + v^{-n}$.

   c) Interpret each result.

14) Bob buys an annuity for $X at a price equal to its present value plus a 5% service fee. The annuity pays $2,000 eat the end of each six-month period for the next ten years. If the nominal rate of interest used to compute the price is 6% compounded semiannually, find X.

15) An investment account is opened on January 1, 2009, with a deposit of $500 subsequent deposits of $500 are made at the start of every month

for twenty years (until December 1, 2028). After that payment, no further deposits are made. If the interest rate is a nominal 6% compounded monthly, what will the account balance be on December 31, 2031?

16) The nominal rate of interest, convertible quarterly is 8%. Today is January 1, 2008. An annuity immediate pays $1,000 at the end of each quarter. The first payment will be made on March 31, 2007. The last payment will be made on December 31, 2012. That is the present value of this annuity?

17) An investor is making deposits of $,1000 at the end of each month with the goal of accumulating $200,000. If the account earns a nominal annual rate of interest of 9% convertible monthly, how many months will it take to achieve the investment goal?

18) A thirty-payment annuity makes a payment of $500 at the end of the first quarter, with payments in each succeeding quarter increasing by $500. If the rate of interest is 2% per monthly compounded monthly what is the present value of this annuity?

19) A perpetuity-immediate has an initial payment of $5000 at the end of the first year. Payments increase by $500 each year. A level payment annuity-due provides twenty-five payments of $X per year. If the interest rate is 6.5% what is the value of X if these two annuities are of equal value to an investor?

20) Do the simplification to verify Equation 4.38.

21) You win a $4,000,000 Illinois Lottery prize and are offered the choice of twenty annual payments of $200,000 (the first payment to be made immediately) or a lump sum payment of $2,617,064.

a) What interest rate did the fine folk at the lottery use to compute your lump sum payment?

b) You can use your lump sum payment to purchase an annuity-due with twenty annual payments beginning one year from date of purchase. This annuity is priced based on 6% effective annual interest rate. What will your annual payments be? Should you elect the lump sum payment or the annual payments of $200,000?

c) What if the annuity in part b) is an annuity-immediate?

22) Marge will retire in twenty-five years and wishes to pre-purchase a twenty-year annual payment annuity which will pay her $75,000 each year for thirty years. Mega Investments LLC prices this class of annuity based on an interest rate of 7% and charges a .5% service fee. What will Marge pay for her annuity?

23) Mork[4] is using his TI **BA II Plus** to compute the present value of an annuity-immediate which consists of thirty annual payments of $5,000. Using an interest rate of 5%, Mork obtains a value of $80,705.36. Show Mork the error in his ways!

24) Aretha donates $2,000,000 to North Central College in the form of an irrevocable charitable trust. Aretha will receive annual payments (at the end of each year) until her death at which time NCC will begin receiving those payments. Assume an annual interest rate of 6%. If Aretha's life expectancy is thirty-five years, what is the present value of her donation to NCC?

25) An investor wishes to accumulate $50,000 with payments of $500 at the end of each month for as long as needed. The interest rate is nominal annual interest rate of 8.5% compounded quarterly. Compute the number of months of regular payments and the amount remaining after the final regular payment has been made.

26) An annuity-immediate pays $600 per month over a period of four years. Interest is year 1 is a nominal 8% convertible monthly. Interest in the second year is a nominal 5% convertible monthly. Interest in the third year is a nominal 7.55% convertible monthly. Interest in the final year is nominal 12% convertible monthly. Each of these rates applies during its respective year for all payments. What is the present value of this annuity at inception? Hint: don't try to write out all the terms. Take advantage of the fact that the interest rate is constant for each twelve-month period.

27) A twenty-year annuity consists of payments of $650 made at the end of every quarter Interest is a nominal 8.34% converted every month. Compute the present value of this annuity.

   a) By converting the interest rate to its equivalent in quarters.

   b) By using the method of Section 4.11.

28) At what interest rate does an fifteen-year annuity-immediate with payments of $650 have the same price as a twenty-year annuity-immediate with payments of $500 priced at 5%?

29) An annuity immediate pays $500 per month for the first three years. After that the annuity payments increase by $50 per month for five years and then remain level for an additional six years. At a nominal rate of annual interest of 12% convertible monthly what is the present value of this annuity?

---

[4]RIP Robin Williams.

30) A twenty-year annuity-immediate pays $5,000 the first year and increases by $x each year thereafter. The interest rate is 6.5% effective annual. If the present value of this annuity is $70,183.23, what is the value of x?

31) At what rate of effective annual interest rate will a twenty-year annuity-immediate paying $5,000 the first year and increasing by $500 each subsequent year have a present value of $80,000?

32) A perpetuity-immediate has an initial annual payment of $10 and increase by $5 in each subsequent year. At what effective annual interest rate is the present value of this annuity $2,000?

33) An annuity consists of payments made at the end of each six-month period for ten years. The initial payment is $600 and each subsequent payment is increased by $600. The nominal rate of interest is 8% converted every two months. Compute the present value of this annuity

a) Using the method of Section 4.16.

b) By converting the interest rate to a six-month rate and using the method of Section 4.14.

34) Find the present value and the accumulated amount of a continuous annuity which provides a total monthly payment of $650 for four years. Assume a nominal annual interest rate of 12% converted monthly.

35) Using the data from problem 34), by how much will the present and accumulated values of the continuous annuity exceed that of a simple annuity-immediate which pays $650 at the end of month for four years?

Chapter 5

Amortization Schedules and Sinking Funds

5.1 Introduction

In this chapter, we consider the calculation of the payments needed to pay off a loan with a series of level (equal) payments. There are two ways this is typically done: The amortization (or annuity) method and the sinking fund method.

In the amortization method, the borrower agrees to pay the lender a fixed amount each period –$500 per month for three years, for instance. This means that the loan is being paid off with an annuity. The amount of each periodic payment is computed so that the present value of the annuity is equal to the amount of the loan at the interest rate charged for the loan.

In the case of the sinking fund method, the borrower pays the lender only the interest due each period and accumulates a fund (called a sinking fund) sufficient to pay off the loan balance at the time it is due. Sinking funds are also used to accumulate the funds needed to purchase replacement equipment. As of March 2019, sinking fund problems are not part of the FM exam.

As we shall see, the amortization and sinking fund methods are equivalent to the lender and the borrower so long as the sinking fund earns interest at the same rate as the rate of interest for the loan. In either case, the total value of payments made is equal to the value of the loan at the time the agreement is entered into.

5.2 Amortization Method

If the borrower pays off the loan with a level payment annuity of term n, we can compute the payment needed by the basic equation of value.

Amount of Loan = Present Value of the Annuity

If the loan value is L and the payments are R, we have

$$L = Ra_{\overline{n}|,i}$$

$$R = \frac{L}{a_{\overline{n}|,i}} \tag{5.1}$$

These are the same equations we used to price an annuity. In return for a lump sum payment of L, the borrower is obligated to pay the lender \$R each period over the life of the loan. We can think of this transaction as the lender purchasing an annuity from the borrower.

The **TI BA II Plus** TVM keys use the PMT key for what we are calling R. It's easiest to enter $L = PV$ as a positive number and R as a negative number. Since the balance of the loan is \$0 at the end, we have $FV = 0$. For a loan being paid off with level payments we thus have (Table 5.1):

TABLE 5.1

N	I/Y	PV	PMT	FV
n	i	L	$-R$	0

We can enter any three of the four numbers N, I/Y, PV, PMT, and use the TVM keys to compute the fourth value. In addition, we could enter a value for FV and calculate, for example, how long it would take to reduce the value of the loan to a specified amount.

Example 5.1 A loan of \$50,000 is to be financed over thirty years with monthly payments at the end of each month. If the loan carries a nominal rate of annual interest of 6% compounded monthly what are the monthly payments? How much interest is paid on this loan?

Solution: We have $n = 12 \cdot 30 = 360$, $i = \frac{.06}{12} = .005$, and $L = 50,000$ so

$$R = \frac{50000}{a_{\overline{360}|,.005}} = \$299.78$$

The total payments are $299.77 \cdot 360 = \$107,917.20$, so the total interest is \$57,917.20.

TI BA II Plus solution (Table 5.2)

TABLE 5.2

N	I/Y	PV	PMT	FV
360	.5	50,000	$CPT = -299.78$	0

$$\pm \times 360 = 107,919.09 - 50,000 = \$57,919.09$$

The **TI BA II Plus** has a worksheet (AMORT) useful in this context. To use AMORT, we first enter the loan numbers into the TVM keys and compute the payment. We then press 2nd AMORT to enter the worksheet. The numbers from the TVM keys are loaded automatically.

P1 and P2 represent the starting and ending payment(s) we wish to use.

BAL represents the balance after payment P2. PRN represents the change (reduction) in principal as a result of payments P1 through P2.

INT represents the amount of payments P1 through P2 which went to interest.

To compute the total interest paid in our example we use $P1 = 1$, $P2 = 360$ (Table 5.3):

TABLE 5.3

Key	Value	Meaning	Then Press
P1	1		Enter ↓
P2	360		Enter ↓
BAL	0		↓
PRN	−50,000		↓
INT	−57,919.09		

The balance is $0 (as it should be), we reduced the principal by $50,000 and paid a total of $57,919.09 in interest. Press 2nd QUIT to exit the worksheet.

Example 5.2 An auto loan of $12,000 is being financed with monthly payments made at the end of each month. The nominal rate of annual interest of 12%.

a) Compute the monthly payments and total interest paid if the period of the loan is three years.

b) Compute the monthly payments and total interest paid if the period of the loan is five years.

c) Explain the difference in total interest paid.

Solution:

a) The monthly interest is 1%. We calculate the required payment and then use AMORT to compute the total interest paid (Table 5.4).

TABLE 5.4

N	I/Y	PV	PMT	FV
36	1	12,000	$CPT = -398.5717178$	0

Key	Value	Meaning	Then Press
P1	1		Enter ↓
P2	36		Enter ↓
BAL	0		↓
PRN	−12,000		↓
INT	−2,348.58		↓

b) The term of the loan is now 60 (Table 5.5).

TABLE 5.5

N	I/Y	PV	PMT	FV
60	1	12,000	$CPT = -266.93$	0

Key	Value	Meaning	Then Press
P1	1		Enter ↓
P2	60		Enter ↓
BAL	0		↓
PRN	−12,000		↓
INT	−4,016.00		↓

c) The term of the loan in case b) is longer and so more interest will be paid. This tactic is often used by car dealers who sometimes advertise only the payments required and neither the price of the car nor the term of the loan.

In each case the present value of all payments is $12,000 when evaluated using the interest rate in question.

5.3   Amortization of a Loan

Amortization of a loan consists in finding the answers to three questions:

1) What is the payment needed to pay off the loan given the interest rate and number of periodic payments?

2) What is the loan balance at any time during the life of the loan?

3) What portion of a given payment is interest and what portion goes toward reducing the principal?

In practice this is most often done by creating a table which displays these numbers for each payment in the life of the loan. Such a table is called an amortization table for the loan. The **TI BA II Plus** will create such a table, but Excel is the tool of choice for loans of substantial duration. For example, home loans typically run either fifteen or thirty years and are paid off monthly. That means the amortization table has either 180 or 360 rows! Not the sort of thing one wants to do on a calculator. That being said, there are FM problems which ask you to calculate specific entries, so you need to know the formulas involved.

5.4   Methods for Computing the Loan Balance

There are two standard ways to compute the loan balance (also called the outstanding balance, unpaid balance or outstanding principal). You will need to be able to use both methods.

Prospective Method

The prospective method computes the outstanding balance as the present value of the remaining payments. (The $a_{\overline{n}|,i}$ approach)

Retrospective Method

The retrospective method computes the outstanding balance as the accumulated value of the original loan minus the accumulated value of the payments made to date. (The $s_{\overline{n}|,i}$ approach)

We will denote the loan balance at time t by B_t. In the case where we want to specify the method used, we will use B_t^r for the retrospective method and B_t^p for the prospective method. Since it is always the case that $B_r^r = B_t^p$, we seldom need to keep track of which method we used.

We consider a loan of $L = B_0$ being paid off with equal payments of $R = \frac{L}{a_{\overline{n}|,i}}$ which are to be made at the end of each period. The two methods produce the following results:

Prospective and Retrospective Calculation of the Loan Balance

$$\text{Prospective Method: } B_t^p = R a_{\overline{n-t}|,i} = \frac{L}{a_{\overline{n}|,i}} a_{\overline{n-t}|,i}$$

$$\text{Retrospective Method: } B_t^r = L(1+i)^t - R s_{\overline{t}|,i} \tag{5.2}$$

Example 5.3 A loan of $500,000 is being financed with level monthly payments over thirty years. If the nominal rate of annual interest is 8% compounded monthly what is the balance of the loan at the end of ten years?

Solution: The monthly interest rate is $\frac{.08}{12} = .006666667$ and the period of the loan is 360 months. We are interested in the balance at time $t = 10 \cdot 12 = 120$. The payment is $\frac{5000000}{a_{\overline{360}|,.00666667}} = 3,668.82$

Prospective Method:

$$B_t^p = Ra_{\overline{n-t}|,i} = 3668.822824 \cdot a_{\overline{240}|,.00666667}$$
$$= \$438,623.52$$

TI BA II Plus solution. We first compute the payment R and then change N to 240 (the term **remaining**) to compute the PV (Table 5.6).

TABLE 5.6

N	I/Y	PV	PMT	FV
360	.66666666	500,000	$CPT = -3,668.82$	0

N	I/Y	PV	PMT	FV
240		$CPT = 438,623.52$		

Retrospective Method

$$B_t^r = L(1+i)^t - Rs_{\overline{t}|,i}$$
$$= 500000(1.00666)^{120} - 3668.82 \cdot s_{\overline{120}|,.006666}$$
$$= \$438,623.52$$

Example 5.4 A loan is being repaid with ten payments of $5,000 followed by fifteen payments of $6,000. All payments are made at the end of each half-year. If the nominal rate of interest convertible twice a year is 8%, find the loan balance immediately after the first seven payments have been made. Use both methods.

Solution. This problem is more difficult, since the original amount of the loan is not provided and the payments are not constant. Nonetheless, we can work it out using both methods.

Prospective Method

After seven payments we have three payments of $5,000 and fifteen payments of $6,000 still to make. The last fifteen payments are deferred three periods so the equation for the present value of the remaining payments is

$$B_t^p = 5000 \cdot a_{\overline{3}|,.04} + 6000 \cdot v^3 \cdot a_{\overline{15}|,.04} = \$73,180.69$$

Notice that we didn't need to know the loan amount to use this method.

Retrospective Method

This method requires us to compute the original amount of the loan. Since the given sequence of payments amortizes (pays off) the loan, the present value of all payments at the inception of the loan must be equal to the amount of the loan

$$L = 5000a_{\overline{a}|,.04} + 6000v^{10}a_{\overline{15}|,.04} = \$85,621.58392$$

The v^{10} accounts for the fact that the last fifteen payments commence ten periods after the inception of the loan.

We now use Equation 5.2 to compute the value of B_t^r

$$B_t^r = 85621.58(1.04)^7 - 5000s_{\overline{7}|,.04} = \$73,180.69$$

As expected, the two methods yield the same answer. In this case the Retrospective Method is easier as we do not need to compute the original loan amount.

TI BA II Plus

Prospective See Table 5.7. We compute the two terms in the formula and add.

TABLE 5.7

N	I/Y	PV	PMT	FV
3	4	$CPT = 13,875.45517$	$-5,000$	0

STO 1

N	I/Y	PV	PMT	FV
15	4	$CPT = 66,710.32459$	$-6,000$	0

N	I/Y	PV	PMT	FV
3		$CPT = 59,305.23565$	0	$-66,710.32459$

$$+RCL1 = 73,180.69$$

Retrospective See Tables 5.8 and 5.9. We compute the amount of the loan using

$$L = 5000a_{\overline{10}|,.04} + 6000v^{10}a_{\overline{15}|,.04}$$

TABLE 5.8

N	I/Y	PV	PMT	FV
10	4	$CPT = 40,554.48$	−5,000	0

STO 1

N	I/Y	PV	PMT	FV
15	4	$CPT = 66,710.32$	−6,000	0

N	I/Y	PV	PMT	FV
10		$CPT = 45,067.10499$	0	−66,710.32459

$$+RCL1 = 85,621.58$$

Now we compute the two terms in the equation below and subtract

$$B_t^r = 85621.58388(1.04)^7 - 5000s_{\overline{7}|,.04}$$

TABLE 5.9

N	I/Y	PV	PMT	FV
7		−85,621.58	0	$CPT = 112,672.1632$

STO 1

N	I/Y	PV	PMT	FV
7		0	5000	$CPT = -39,491.47240$

$$RCL1 = 73,180.69$$

Example 5.5 Suppose in Example 5.4 above that the borrower makes an extra payment of $20,000 along with the seventh payment and refinances the loan for ten years with monthly payments at a nominal interest rate of 6% convertible monthly. What is the new payment?

Solution.

The $20,000 payment reduces the loan balance to $53,180.69 which will be paid off with 120 monthly payments at a monthly rate of .005. We compute the payment needed using Equation 5.1

$$R = \frac{L}{a_{\overline{n}|,i}} = \frac{53,180.69}{a_{\overline{120}|,.005}} = \$590.41$$

TI BA II Plus (Table 5.10)

TABLE 5.10

N	I/Y	PV	PMT	FV
120	.5	53,180.69	$CPT = -590.41$	0

5.5 Allocation of Loan Payments between Principal and Interest

We now take up the second question posed at the start of this section: the division of each payment into principal and interest portions. Each payment goes both toward reducing the principal and toward paying the interest due on the principal during the prior period. The fundamental principle here is simple: interest is always paid before any reduction in premium is made. At each period, we compute the interest due and subtract this from the payment. The remaining amount goes to reducing the amount owed. The interest due is calculated using the balance at the start of the payment period. For example, the interest due at time $t = 6$, is computing using the loan balance at time $t = 5$.

We begin with an example and then state the general formulas. Suppose we borrow \$5,000 at 8% annual interest and will pay off the loan with five annual payments made at the end of each year. We first compute the amount of the payment

$$R = \frac{5000}{a_{\overline{5}|,.08}} = \$1,252.28$$

We now describe how to construct an **amortization table:** a table which allocates each payment between interest and principal reduction and shows the loan balance at each point in time that is of interest. At inception, the only entry is the amount of the original loan. At the end of the first period (one year in this case), there is interest due of \$5,000 $\cdot.08 = \$400$. This amount must be subtracted from the payment to obtain the amount which goes toward reducing the principal. Thus, $1252.28 - 400 = \$852.28$ will go towards reducing the principal. The amount still due after the first payment is \$5,000−\$852.28 = \$4,147.72. At the end of the second year, there is interest due of \$4,147.72 $\cdot$ .08 = \$331.82. Thus the second payment reduces the principal by \$1,252.28 − 331.82 = \$920.46. We continue this process to create each subsequent line in the table. We can automate this process and create a very nice looking table using Excel.

To make the maximal use of Excel, we first create a space where the user can enter the Principal amount, loan period and interest rate. Excel then calculates the required payment. Table 5.11 is a screen shot of such a space with the formulas displayed. The interest rate was formatted as a percent and the Excel PMT function was used to compute the payment.

TABLE 5.11

Enter the Loan Information	
Amount of Loan	$50,000.00
Number of Payments	5
Interest Rate Per Payment Period	5.60%
Computed Value	
Required Loan Payment	$11,740.95

Table 5.12 is the same file as Table 5.11 with the formulas displayed.

TABLE 5.12

Enter the Loan Information	
Amount of Loan	50000
Number of Payments	5
Interest Rate per Payment Period	0.056
Required Loan Payment	= −PMT(B4,B3,B2,0,0)

The next step is to set up the amortization table. The columns are: Year, Payment, Interest, Amount to Principal and Remaining Principal. We enter the years (or periods) as $0, 1, 2, \ldots, n$. We then set the entries in years 1 through n as indicated in Table 5.13. We enter just the values for year 1 and then copy them down. The completed table is displayed as Table 5.14.

TABLE 5.13

Period	Payment	Interest Due	Amount to Principal	Remaining Balance
0				=B2
=D2+1	=B$6	=H2*B$4	=E3-F3	=H2-G3
=D3+1	=B$6	=H3*B$4	=E4-F4	=H3-G4
=D4+1	=B$6	=H4*B$4	=E5-F5	=H4=G5
=D5+1	=B$6	=H5*B$4	=E6-F6	=H5-G6
=D6+1	=B$6	=H6*B$4	=E7-F7	=H6-G7

The formulas used are entered into Row 3 and then copied down. Table 5.14 is the completed amortization table.

TABLE 5.14

Period	Balance	Payment	Interest	To Principal
0	$50,000.00			
1	$41,059.05	$11,740.95	$2,800.00	$8,940.95
2	$31,617.41	$11,740.95	$2,299.31	$9,441.64
3	$21,647.04	$11,740.95	$1,770.57	$9,970.37
4	$11,118.32	$11,740.95	$1,212.23	$10,528.71
5	$ -	$11,740.95	$ 622.63	$11,118.32

If the period of the loan changes, we need to recopy the rows, erasing or adding rows as appropriate. In order to automate that process we will make use of Excel's Macro feature.

The **TI BA II Plus** AMORT Worksheet will compute all the entries in an amortization table line by line. We first compute the required payment for the loan using the TVM keys and then press 2ND AMORT to enter the AMORT Worksheet. Table 5.15 shows the set up for a loan with $N = 5$, $I/Y = 8\%$ and $PV = 5,000$.

TABLE 5.15

N	I/Y	PV	PMT	FV
5	8	5,000	$CPT = -1,252.28$	0

We press 2ND AMORT. The two values for $P1$ and $P2$ must be entered. While $P1$ is showing press 1 and then ENTER. Use the ↓ key to move through the worksheet. Here is what the complete table looks like on the **TI BA II Plus**

NOTE 1: When we get to year 2, we don't need to enter $P1 = 2$, $P2 = 2$. Pressing CPT causes the calculator to repeat the period between payments automatically. If we had started with $P1 = 1$, $P2 = 12$ pressing CPT would result in $P1 = 13$, $P2 = 24$. See Example 5.6 below.

NOTE 2: This is a worksheet so you need to press 2ND QUIT to exit. You might also want to press 2ND CLR WORK first to clear out the entries for the next problem.

This is not quite as quick as the Excel solution, but provides all the same numbers. The big advantage of the **TI BA II Plus AMORT** worksheet is that it can compute summary tables. See Table 5.16. Many loans are paid off with monthly payments. The tax deduction, if any, associated with the

TABLE 5.16

Key	Value	Meaning	Then Press
$P1$	$= 1$	Start with Period 1	↓
$P2$	$= 1$	End with Period 1	↓
BAL	$= 4,147,71$	Balance at end of Period 1	↓
PRN	$= -852.28$	Amount to Principal in P1	↓
INT	$= -400$	Amount to Interest in P1	↓
$P1$	$= 1$, Press CPT[13]	Amount to Interest in P1	↓
$P2$	$= 2$		↓
BAL	$= 3,227.252873$	Balance after two payments	↓
PRN	$= -920.4648547$	Amount of second payment to principal	↓
INT	$= -331.8174$	Amount of second payment to interest	↓
P1	$= 2$ Press CPT	$= 3$	↓ ↓
BAL	$= 2,233.150829$	Balance after three payments	↓
PRN	$= -994.1020430$	Amount of third payment to principal	
INT	$= -258.1802298$	Amount of third payment to interest	
P1	$= 3$ Press CPT	$= 4$	↓ ↓
BAL	$= 1,159.520623$	Balance after 4 payments	↓
PRN	$= -1,073.30206$	Amount of fourth payment to principal	↓
INT	$= -178.6520554$	Amount of fourth payment to interest	↓
P1	$= 4$ Press CPT	$= 5$	↓ ↓
BAL	$= 0.00000$	Balance after FV payments	↓
PRN	$= -1,159.52$	Amount of fifth payment to principal	↓
INT	$= -92.76164$	Amount of fifth payment to interest.	↓

interest is often computed on an annual basis. The AMORT worksheet makes that easy to do.

Example 5.6 A $10,000 is being paid of with monthly payments at the end of each month over five years. If the annual rate of interest is a nominal 11% compounded monthly, we would like to create a table containing the account balance, amount of payments to principal and amount of payments to interest for each **year**.

Solution: We begin by computing the monthly payment using the TVM keys (Table 5.17):

TABLE 5.17

N	I/Y	PV	PMT	FV
60	11/12	10,000	$CPT = -217.424$	0

---

[13]This is a clever feature of the AMORT worksheet. If you enter $P1 = 1$, $P2 = 12$, you are, presumably, interested in creating an annual amortization table. When you get to the second year you don't need to enter $P1 = 13$, $P2 = 24$, just press the CPT key and the **TI BA II Plus** fills in the needed numbers.

We then implement the AMORT worksheet as shown in Table 5.18.

TABLE 5.18

Key	Value	Meaning	Then Press
P1	= 1	Starting with Month 1	↓
P2	Press 12 ENTER	P2 = 12 Ending with Month 12	↓
BAL	= 8,412.45	Balance at end of year 1	↓
PRN	= −1,587.55	Amount of Principal in year 1	↓
INT	= −1,021.543246	Amount of interest paid in year 1	↓
P1	= 1 Press CPT = 13	Now starting with month 13	↓ ↓
BAL	= 6,641.195803	Balance at end of year 2	↓
PRN	= −1,771,2566	Amount of principal in year 2	↓
INT	= −837.8340946	Amount of interest paid in year 2	↓
P1	Press CPT = 25		↓ ↓
BAL	= 4,664.97	Balance at end of year 3	↓
PRN	= −1.976.22	Amount of principal in year 3	↓
INT	= −632.866	Amount of interest in year 3	↓
P1	= 25 Press CPT = 37		↓ ↓
BAL	2,460.06	Balance at end of yera 4	↓
PRN	= −2,204.91	Amount principal in year 4	↓
INT	= −404.18	Amount of interest paid in year 4.	↓
P1	= 37 Press CPT = 49		↓ ↓
BAL	0.00	Balance at end of year 5	↓
PRN	= −2,460.06	Amount to principal in year 5	↓
INT	= −149.03 Done? Press 2ND Quit	Amount of interest paid in year 5	↓

The entry in each INT row is the amount which could be deducted for income tax purchases if interest is a deductible expense. In any event, it needs to show up in the company audit as an expense for that year. We could have created a chart showing the PRN and INT amounts by quarters as well. In that case we would start with $P1 = 1$, $P2 = 3$ and use CPT to cycle through the quarters. See Table 5.18.

5.6 Formulas for the Balance, Amount to Interest, and Amount to Principal at any Time

We now compute general formulas for each entry in an amortization table. You will need to have these formulas memorized to succeed on the FM exam. As we have seen the required payment is computed as

$$R = \frac{L}{a_{\overline{n}|,i}} \qquad (5.3)$$

The amount of interest due at the end of the first year is simply $i \cdot L$. At the start of year 2, the loan balance is $B_t^p = Ra_{\overline{n-1}|,i}$ so the interest due is

$iRa_{\overline{n-1}|,i}$. In general, the interest due in year t is

$$i \cdot Ra_{\overline{n-t+1}|,i} = i \cdot R \cdot \frac{1 - v^{n-t+1}}{i} = R \cdot (1 - v^{n-t+1}) \qquad (5.4)$$

We can also express the interest due as a function of the original loan amount giving us two expressions for the interest portion of each payment.

Interest Portion of the nth Payment for an Annuity with n Payments

$$\frac{L}{a_{\overline{n}|,i}}(1 - v^{n-t+1}) = R \cdot (1 - v^{n-t+1}) \qquad (5.5)$$

Interest Portion (general version)
Interest Portion = Interest Rate × Previous Balance

The amount to principal is the payment minus the interest due

Principal Portion of the tth Payment

$$R - R(1 - v^{n-t+1}) = Rv^{n-t+1} = \frac{L}{a_{\overline{n}|,i}} \cdot v^{n-t+1} \qquad (5.6)$$

Finally, the outstanding loan balance (using the retrospective method) is

Loan Balance at Time t

$$B_t^r = R \cdot a_{\overline{n-t}|,i} = \frac{L}{a_{\overline{n}|,i}} a_{\overline{n-t}|,i} \qquad (5.7)$$

The total interest paid over the life of the loan is $R \cdot n - L$.

Summary of Mortgage Payment Formulas (Table 5.19)

TABLE 5.19

	Loan Balance	Payment to Principal	Payment to Interest
In terms of R	$Ra_{\overline{n-t}|,i}$	Rv_{n-t+1}	$R(1 - v^{n-t+1})$
In terms of L	$\frac{L}{a_{\overline{n}|,i}} a_{\overline{n-t}|,i}$	$\frac{L}{a_{\overline{n}|,i}} v^{n-t+1}$	$\frac{L}{a_{\overline{n}|,i}}(1 - v^{n-t+1})$

If we look at the values of the Payment to Principal they form a geometric sequence with common ratio $v^{-1} = 1 + i$: $\frac{L}{a_{\overline{n}|,i}}v^n$, $\frac{L}{a_{\overline{n}|,i}}v^{n-1}$, $\frac{L}{a_{\overline{n}|,i}}v^{n-2}$, and so forth. You will encounter problems where this fact is very useful. Here is an example.

Example 5.7 A loan is being paid off with annual payments. The amount of the third payment to principal is $333.25 and the amount to principal of the fifth payments is $367.05. What interest rate is being charged for this loan?

Solution: We have $Rv^{n-3+1} = R(1+i)^{2-n} = 333.25$ and $Rv^{n-5+1} = R(1+i)^{4-n} = 367.05$. Dividing we obtain $\frac{367.05}{333.25} = (1+i)^2$ so that $i = 4.95\%$. We can also solve this problem by simply noting that period 5 is two periods after period 2 and so we multiply by $(1+i)^2$ to obtain the payment in period 5 in terms of the payment in period 3.

Example 5.8 A loan of \$300,000 is being paid off over thirty years. Payments are made at the end of each month. The nominal annual interest rate is 12% compounded monthly.

a) What is the loan balance at the end of 15 years?

b) At what point is the loan balance \$150,000?

c) What is the balance after the 200th payment is made? How much of that payment went toward interest?

Solution. a)

The payment is $R = \frac{300000}{a_{\overline{360}|,.01}} = \$3085.84$ per month. At the end of fifteen years, 180 payments have been made and $t = 180$. The loan balance (Equation 5.2) is

$$B_t^r = 3085.84 a_{\overline{180}|,.01} = \$257,117.13$$

This is a rather shocking result – after fifteen years the principal has only been reduced by \$42,882.87.

b) To find out when the loan balance reaches \$150,000 we need to solve the equation

$$B_t^p = 150,000$$
$$= 3085.84 a_{\overline{360-t}|,.01}$$

We can use the **TI BA II Plus** to do both parts of this problem.

a)We first compute the payment and then the balance when there are $360 - 180 = 180$ payments remaining (Table 5.20):

TABLE 5.20

N	I/Y	PV	PMT		FV
360	1	30000	$CPT = -3,085.837791$		0

N	I/Y	PV		PMT	FV
180		$CPT = 257,117.1395$			

We can also use the AMORT worksheet as shown in Table 5.21

TABLE 5.21

Key	Value	Meaning	Then Press
P1	= 1		↓
P2	180 Enter P2 = 180		↓
BAL	257,117.14	Balance after 15 years	↓
PRN	= −42,882.86	Amount to principal in the first 15 years	↓
INT	= 512,567.94	Amount of interest paid in the first 15 years	↓

During the first fifteen years, the interest paid was $512,567.94 and the principal was reduced by only $42,882.86.

b) We set PV = 150,000 and compute N (Table 5.22).

TABLE 5.22

N	I/Y	PV	PMT	FV
$CPT = 66.90$		150,000		

This represents the **number of payments remaining**. Setting $N = 67$, yields a balance of $150,152 so we need to use $N = 66$ remaining payments which results in a balance of $148,568. To compute the number of payments we have made, we subtract this number from 360 obtaining 294. This means that we have to make 294 payments (81% of the total) before we bring the balance down to half of the original loan amount!

NOTE: When using the TVM keys to compute the remaining balance we enter the number of payments remaining and compute the PV. With AMORT we enter the number of the payment in question. Because varying the TVM registers interferes with the AMORT worksheet, it's probably better to use AMORT if you need to compute multiple balances/interest allocations.

c) We use the AMORT worksheet. Note that we must have the original loan numbers entered into the TVM keys to use AMORT. See Table 5.23.

TABLE 5.23

Key	Value	Meaning	Then Press
P1	= 200		↓
P2	= 200		↓
BAL	245,784.70		↓
PRN	621.77		↓
INT	2,464.065		

Example 5.9 A car is being financed with monthly payments over a five-year period. The original loan amount is $12,500 at a nominal annual rate of 8% compounded monthly. After two years the borrower refinances for an additional four years at a nominal annual rate of 8%. What is the new payment? What is the total amount of interest paid?

Solution:

We use $i = \frac{.08}{12}$ and compute in months. The original payment is

$$R = \frac{12,500}{a_{\overline{60}|,\frac{.08}{12}}} = \$253.45$$

At two years the balance is

$$B_t^r = 253.45 a_{\overline{60-24}|,\frac{.08}{12}} = 253.45 a_{\overline{36}|,\frac{.08}{12}} = \$8,088.24$$

We now compute the new payment (using a loan period of four years or forty-eight months at $i = \frac{.06}{12}$) as

$$R = \frac{8,088.24}{a_{\overline{48}|,\frac{.06}{12}}} = \$189.95$$

The payments total $253.45 \cdot 24 + 189.95 \cdot 48 = \$15,200.4$. The total interest paid is thus $15,2005.4 - 12,500 = \$2,700.40$

TI BA II Plus solution (Table 5.24)

TABLE 5.24

N	I/Y	PV	PMT	FV
60	.6666666	12500	$CPT = -253.45$	0

At the end of two years there are three years of payments remaining – a total of thirty-six payments. We change N to 36 and calculate the PV (Table 5.25)

TABLE 5.25

N	I/Y	PV	PMT	FV
36		$CPT = 8,088.24$		

To compute the new payment, we change N to 48 and I/Y to $\frac{6}{12} = .5$, (Table 5.26):

TABLE 5.26

N	I/Y	PV	PMT	FV
48	.5		$CPT = -189.95$	

As before, the payments total $253.45 \cdot 24 + 189.95 \cdot 48 = \$15,200.4$. The total interest paid is thus $15,200.4 - 12,500 = \$2,700.40$.

5.7 Examples Using the TI BA II Plus to Create Amortization Tables

Example 5.10 A car is being financed with monthly payments over a five-year period. The loan amount is \$12,500 at a nominal annual rate of 8% compounded monthly.

a) Construct the loan amortization table for the original loan using a period of one year. How much interest would be paid on this loan?

b) After two years the borrower refinances for an additional four years at a nominal annual rate of 6%. What is the new payment? What is the total amount of interest paid?

Solution:

a) We first compute the payment for the original loan (five years at 8/12% per month) (Table 5.27):

TABLE 5.27

N	I/Y	PV	PMT	FV
60	.66666	12,500	$CPT = -253.454$	0

Then we create the amortization table using AMORT. Since we want an annualized amortization table, we use P1 = 1, $P = 12$ (Table 5.28).

TABLE 5.28

Key	Value
P1	=1
P2	12 ENTER = 1
BAL	10,381.998
PRN	−2,118.00
INT	−923.458
P1	CPT = 13
BAL	8,088.20
PRN	−2,293.794
INT	−747.648
P1	CPT = 25

BAL	5,604.026
PRN	$-2,484.178$
INT	557.281
P1	CPT = 37
BAL	2,913.663
PRN	2,690.36
INT	351.095
P1	CPT = 49
BAL	-0.00
PRN	$-2,913.663$
INT	-127.797

Finally, we enter the results into the amortization table (Table 5.29).

TABLE 5.29

Year	Payment	Interest	Principal	New Principal
0			12,500	
1	253.454	-923.458	$-2,118.00$	10,381.998
2	253.454	-747.648	$-2,293.79$	8,088.204
3	253.454	-557.281	$-2,484.18$	5,604.026
4	253.454	-351.095	$-2,690.36$	2,913.663
5	253.454	-127.797	$-2,913.66$	-0.00

b) Since we will refinance after two years, we only compute the amortization table for the first two years. We set $P2 = 1$ and $P2 = 24$. We store the remaining balance and the interest paid over the first two years (Table 5.30).

TABLE 5.30

Key	Value	Meaning	Then Press
P1	=1		$\downarrow$
P2	=2		$\downarrow$
BAL	=8,088.204		$STO1\downarrow$
PRN	$-4,411.796$		$\downarrow$
INT	$-1,671.122$		$STO2\downarrow$

2ND QUIT

We now have the interest paid to date stored in register 2 and the loan balance stored in register 1. We exit the AMORT worksheet and use the TVM keys to compute the payments for the refinanced loan four years at $6/12 = .5\%$). See Table 5.31.

TABLE 5.31

N	I/Y	PV	PMT	FV
48	.5	RCL 1 ENTER = 8,088.20	CPT = −189.95	0

Now we use the AMORT worksheet (Table 5.32) to compute the total interest paid for this portion of the loan. That amount is added to the interest paid on the first portion which is stored in register 2

TABLE 5.32

Key	Value	Meaning	
P1	1 ENTER = 1		
P2	48 ENTER = 48		
BAL	−0.00		
PRN	−8,088.20		
INT	= −1,029.47	Total interest paid	+ RCL 2 = -2,700.6

Example 5.11 ACME Heavy Equipment is purchasing a road runner smashing machine for $5,456,900. They have obtained a four-year loan at a nominal 7% with monthly payments. They are required to report loan expenses each quarter.

a) Prepare the expense report for the first year of this loan.

b) What will the total interest expenses be?

c) The auditors request an estimate of the present value of the interest obligation for the first quarter. Compute an estimate based on a nominal interest rate of 5% convertible monthly.

Solution:
   a) We need a chart with the interest paid for each quarter. We use the TVM keys to set up the loan and then the AMORT worksheet with $P1 = 1$, $P2 = 3$ (three months per quarter) to create the loan expense chart (Table 5.33).

TABLE 5.33

N	I/Y	PV	PMT	FV
48	7/12	5,456,900	$CPT = -130,672.26$	0

Key	Value	Then Press
P1	= 1 ENTER	↓
P2	= 3 ENTER	↓
BAL	5,158,645.89	↓
PRN	$-298,254.10$	↓
INT	$-93,762.68$	STO 1 ↓
P1	CPT = 4	↓ ↓
BAL	4,855,141.83	↓
PRN	$-303,504.06$	↓
INT	$-88,512.73$	+ RCL 1 = STO 1 ↓
P1	CPT = 7	↓ ↓
BAL	4,546,295.41	↓
PRN	$-308,846.42$	↓
INT	$-83,170.36$	+RCL 1 = STO 1 ↓
P1	CPT = 10	↓ ↓
BAL	4,232,012.58	↓
PRN	$-314,282.83$	↓
INT	$-77,733.96$	+ RCL 1 = STO 1

The expense report for the year consists of the four quarterly interest payments (Table 5.34):

TABLE 5.34

Quarter	Interest Paid
One	$93,762.68
Two	88,512.728
Three	83,170.364
Four	77,733.96
Annual Total	$343,179.73

b) We use the AMORT worksheet with $P1 = 1$, $P2 = 48$ (Table 5.35)

TABLE 5.35

Key	Value
P1	1 ENTER
P2	48 ENTER
BAL	.000
PRN	$-5,456,900$
INT	$-815,368.6$

c) We are computing the present value of an annuity consisting of three payments. Since the amount to interest varies, the payments are not constant. We compute the amount of each payment and its present value and then add these up. See Table 5.36.

TABLE 5.36

Key	Value	Meaning	Then Press
P1	1 ENTER		
P2	48 ENTER		
INT	$-31,831.92$		STO 1
P1	CPT = 2		
INT	$-31,255.35$		STO 2
P1	CPT = 3		
INT	$-30,675.42$		STO 3

2 ND QUIT

Now we need to compute the present value of each of these payments. Payment 1 is delayed one month, payment 2 is delayed two months, and payment 3 is delayed three months. We use the TVM keys and store each computed present value. The interest rate is $5/12\%$ (Table 5.37).

TABLE 5.37

N	I/Y	PV	PMT	FV
1	5/12	CPT = 31,699.84 STO 4	0	$RCL1 = -31,831.92$

N	I/Y	PV	PMT	FV
2		CPT = 30,996.51 STO 5		$RCL2 = -31,255.35$

N	I/Y	PV	PMT	FV
3		CPT = 30,295.15 STO 6		RCL 3 = $-30,675.42$

We now compute the sum of the three present values = $92,991.50$. Note that it is smaller than the total interest paid of \$93,762.68. We will talk about how companies plan for such expenses in Chapter 8.

Example 5.12 A loan of \$450,123 is being financed over five years with monthly payments. The interest rate for the loan is a nominal 8.5% annually convertible quarterly. What is the total interest and the principal reduction paid in each indicated period?

a) The entire loan.
b) The third year.
c) The tenth quarter.

Solution

We first need to convert to monthly interest. The interest per quarter is $\frac{8.5}{12}$ (Table 5.38).

TABLE 5.38

N	I/Y	PV	PMT	FV
1/3	8.5/12	1	0	CPT $= -1.0023556$

N	I/Y		PV	PMT	FV
1	CPT $= .2355581$				

Now use the TVM keys to set up the loan (Table 5.39).

TABLE 5.39

N	I/Y	PV	PMT	FV
60		450,123	$CPT = -8,053.497$	0

Now we can compute the requested values using the AMORT worksheet.
a) We have $P1 = 1$, $P2 = 60$ (Table 5.40).

TABLE 5.40

Key	Value
P1	1 Enter
P2	60 Enter
BAL	0
PRN	450,123.00
INT	$-33,086.81$

b) We have $P1 = 25$, $P2 = 36$ (Table 5.41).

TABLE 5.41

Key	Value
P1	25 Enter $= 1$
P2	36 Enter $= 36$
BAL	187,707.14
PRN	$-89,952.88$
INT	$-6,689.08$

c) The first quarter ends at $P2 = 3$, the second at $P2 = 6$, etc. Hence, the tenth quarter ends at $P2 = 30$. Since there are three months per quarter, we have $P1 = 28$, $P2 = 30$ (Table 5.42).

TABLE 5.42

Key	Value
P1	28 Enter = 28
P2	30 Enter = 30
BAL	= 233,001.03
PRN	$-22,408.30$
INT	$-1,752.19$

5.8 Creating Annualized Amortization Tables Using Excel

Here is a screen shot of the first few rows of the Excel Amortization Table for **Example 5.10** in which a car loan of $12,500 is being paid off with monthly payments over five years. Interest is a nominal 8% per year converted monthly (Table 5.43).

TABLE 5.43

Period	Balance	Payment	Interest	To Principal
0	$12,500.00			
1	$12,329.88	$253.45	$83.33	$170.12
2	$12,158.62	$253.45	$82.20	$171.26
3	$11,986.23	$253.45	$81.06	$172.40
4	$11,812.68	$253.45	$79.91	$173.55
5	$11,637.97	$253.45	$78.75	$174.70
6	$11,462.11	$253.45	$77.59	$175.87
7	$11,285.07	$253.45	$76.41	$177.04
8	$11,106.84	$253.45	$75.23	$178.22

The periods here are months. Note that the Remaining Principal at the end of period 12 (one year) is in agreement with the value in our **TI BA II Plus** table. What we want to do is to total all the entries in the Excel table for each year. To do so, we use the Excel Subtotal Feature. We first need to add a column which indicates the year. The entries in the Interest column are computed as

$Ax = 1 + INT((Bx - 1)/12)$. For Example: $A11 = 1 + INT((B11 - 1)/12)$.

Table 5.44 shows the completed schedule.

TABLE 5.44

Year	Period	Balance	Payment	Interest	To Principal
0	0	$12,500.00			
1	1	$12,329.88	$253.45	$83.33	$170.12
1	2	$12,158.62	$253.45	$82.20	$171.26
1	3	$11,986.23	$253.45	$81.06	$172.40
1	4	$11,812.68	$253.45	$79.91	$173.55
1	5	$11,637.97	$253.45	$78.75	$174.70
1	6	$11,462.11	$253.45	$77.59	$175.87
1	7	$11,285.07	$253.45	$76.41	$177.04
1	8	$11,106.84	$253.45	$75.23	$178.22
1	9	$10,927.43	$253.45	$74.05	$179.41
1	10	$10,746.83	$253.45	$72.85	$180.61
1	11	$10,565.02	$253.45	$71.65	$181.81
1	12	$10,382.00	$253.45	$70.43	$183.02
2	13	$10,197.76	$253.45	$69.21	$184.24
2	14	$10,012.29	$253.45	$67.99	$185.47

We now highlight and copy the entire table to a new worksheet. We don't want to bring the formulas along, so we copy and then paste only the value in each cell. Here's the procedure

a) Choose Insert Worksheet to create a new worksheet for the summary.

b) Highlight and copy the entire table.

c) Go to cell A1 in the new worksheet and click on the down arrow just the below the paste button in the upper left hand corner of the Excel spreadsheet. The following dialog box shown in Figure 5.1 should open up:

FIGURE 5.1

Click on Paste Values. The table will look just like your other one, but the formulas are no longer there, just the numbers. Now highlight the entire table except for the title row (in the new worksheet) and choose Data Subtotals. Set Use Function to sum. Set Add subtotal to total Interest, Net to Principal, and Payment. Click OK.

Your spreadsheet should look like Table 5.45.

TABLE 5.45

Year	Period	Balance	Payment	Interest	To Principal
0	0	$12,500.00			
0 Total			$ -	$ -	$ -
	1	1 $12,329.88	$ 253.45	$ 83.33	$ 170.12
	1	2 $12,158.62	$ 253.45	$ 82.20	$ 171.26
	1	3 $11,986.23	$ 253.45	$ 81.06	$ 172.40
	1	4 $11,812.68	$ 253.45	$ 79.91	$ 173.55
	1	5 $11,637.97	$ 253.45	$ 78.75	$ 174.70
	1	6 $11,462.11	$ 253.45	$ 77.59	$ 175.87
	1	7 $11,285.07	$ 253.45	$ 76.41	$ 177.04
	1	8 $11,106.84	$ 253.45	$ 75.23	$ 178.22
	1	9 $10,927.43	$ 253.45	$ 74.05	$ 179.41
	1	10 $10,746.83	$ 253.45	$ 72.85	$ 180.61
	1	11 $10,565.02	$ 253.45	$ 71.65	$ 181.81
	1	12 $10,382.00	$ 253.45	$ 70.43	$ 183.02
1 Total			$ 3,041.46	$ 923.46	$ 2,118.00
	2	13 $10,197.76	$ 253.45	$ 69.21	$ 184.24
	2	14 $10,012.29	$ 253.45	$ 67.99	$ 185.47
	2	15 $ 9,825.58	$ 253.45	$ 66.75	$ 186.71
	2	16 $ 9,637.63	$ 253.45	$ 65.50	$ 187.95
	2	17 $ 9,448.43	$ 253.45	$ 64.25	$ 189.20
	2	18 $ 9,257.96	$ 253.45	$ 62.99	$ 190.47
	2	19 $ 9,066.23	$ 253.45	$ 61.72	$ 191.74
	2	20 $ 8,873.21	$ 253.45	$ 60.44	$ 193.01
	2	21 $ 8,678.91	$ 253.45	$ 59.15	$ 194.30
	2	22 $ 8,483.32	$ 253.45	$ 57.86	$ 195.60
	2	23 $ 8,286.42	$ 253.45	$ 56.56	$ 196.90
	2	24 $ 8,088.20	$ 253.45	$ 55.24	$ 198.21
2 Total			$ 3,041.46	$ 747.66	$ 2,293.79
	3	25 $ 7,888.67	$ 253.45	$ 53.92	$ 199.53

While this view allows us to read the totals for each year, it is much more useful to see just the yearly totals. To get that view, click on the 2 in the upper left-hand corner of the spreadsheet (the numbers 1, 2, and 3 appear in gray squares). Choosing option 2 results in a table with the subtotals and summary. Now it should look like Table 5.46.

To see the detail for any year, click on the + sign for that year. If you click the 1 in the upper right hand corner, only the Grand Total will display.

If you are going to include this in a report which others will view, you should change the labels in the Year column and add the Balance at each year.

TABLE 5.46

Year	Period	Balance	Payment	Interest	To Principal
0 Total			$ -	$ -	$ -
1 Total			$ 3,041.46	$ 923.46	$ 2,118.00
2 Total			$ 3,041.46	$ 747.66	$ 2,293.79
3 Total			$ 3,041.46	$ 557.28	$ 2,484.18
4 Total			$ 3,041.46	$ 351.10	$ 2,690.36
5 Total			$ 3,041.46	$ 127.80	$ 2,913.66
Grand Total			$ 15,207.30	$ 2,707.30	$ 12,500.00

5.9 Sinking Funds (Not on the FM exam as of October 2018, check the current syllabus to see the current status of this material.)

An alternate method of paying off a loan consists in paying only the interest due at the end of each period and then paying off the principal with a single payment at the end of the loan period. In this case the borrower will usually put aside enough money each period so as to accumulate to the value of the loan. Indeed, this may be a requirement of the loan. This "savings account" is called a sinking fund. Note that the amount of the loan remains constant in this case so the interest payments are always the same. The interest rate charged for the loan and the interest rate earned by the sinking fund deposits are usually not the same. In the case that these two interest rates are the same the amortization and sinking fund methods are exactly equivalent – that is, the payments made by the borrower and the interest earned by the lender are exactly the same. We will first compute the relevant numbers for the sinking fund method, and then look at the special case when the interest paid on the loan and the interest earned by the sinking fund are equal.

We suppose that the period of the loan is n, that the interest rate on the loan is i, and that the sinking fund earns an interest rate of j. Finally, suppose that the value of the loan is L. The buyer makes two payments each period:

Sinking Fund Payments

a) a payment of iL to pay the interest on the loan.

b) a payment of R into the sinking fund sufficient to accumulate to L in n payments.

The sinking fund is an annuity. We want the **accumulated** value of this annuity to equal the value of the loan. Since the accumulated value of an annuity with payments of R period is $R \cdot s_{\overline{n}|,i}$, the required value of the payment R is

$$R = \frac{L}{s_{\overline{n}|,j}} \tag{5.8}$$

This is not a typo, we are accumlating the amount L, not paying it off.

The total payment including both the interest and sinking fund payments is then

$$Li + \frac{L}{s_{\overline{n}|,j}} = L\left(\frac{1}{s_{\overline{n}|,j}} + i\right) \tag{5.9}$$

In period t, the interest paid is Li. But, beginning with the second period, the borrower will earn interest (at rate j) on the amount on deposit in the sinking fund. In order to account for this interest we compute what is called the **net interest in period** t as the difference between the interest paid and the interest earned by the sinking fund. Note that the interest earned by the sinking fund account for a given period is based on the amount on deposit at the end of the previous period.

$$\text{Net Interest Paid} = Li - \text{Interest Earned by Sinking Fund} \tag{5.10}$$

At the end of period $t - 1$, the sinking fund has accumulated to

$$Rs_{\overline{t-1}|,j} = L\frac{s_{\overline{t-1}|,j}}{s_{\overline{n}|,j}} \tag{5.11}$$

The interest earned on this amount (which will be credited in period t) is based on the sinking fund interest rate of j

$$\text{Sinking Fund Interest in period } t = j \cdot L\frac{s_{\overline{t-1}|,j}}{s_{\overline{n}|,j}} \tag{5.12}$$

The net interest is thus

$$Li - Lj\frac{s_{\overline{t-1}|,j}}{s_{\overline{n}|,j}} = L\left(i - j\frac{s_{\overline{t-1}|,j}}{s_{\overline{n}|,j}}\right) \tag{5.13}$$

We now compare the amortization and sinking fund methods for paying off a loan. For the amortization method, $R = \frac{L}{s_{\overline{n}|,i}}$ while for the sinking fund method $R = \frac{L}{s_{\overline{n}|,j}} + Li$. We now show that when $i = j$, the two methods are equivalent – the net interest paid using the sinking fund method is exactly the same as the interest paid using the amortization method.

In the case of the amortization method the interest portion of the payment at the end of period t is

$$R(1 - v^{n-t+1}) = \frac{L}{a_{\overline{n}|,i}}(1 - v^{n-t+1}) \tag{5.14}$$

We now compute the net interest paid using the sinking fund method under the assumption that $i = j$. The net interest for the sinking fund is computed (using Equation 5.13) as

$$L\left(i - i\frac{s_{\overline{t-1}|,i}}{s_{\overline{n}|,i}}\right) = Li\left(1 - \frac{s_{\overline{t-1}|,i}}{s_{\overline{n}|,i}}\right)$$

$$= Li \left(\frac{s_{\overline{n}|,i} - s_{\overline{t-1}|,i}}{s_{\overline{n}|,i}} \right)$$

$$= \frac{Li}{s_{\overline{n}|,i}} \left(\frac{(1-v^n)(1+i)^n}{i} - \frac{(1-v^{t-1})(1+i)^{t-1}}{i} \right)$$

$$= \frac{L}{s_{\overline{n}|,i}} \left((1+i)^n - 1 - ((1+i)^{t-1}) - 1) \right) \tag{5.15}$$

Still hanging in there? Here's the rest

$$\frac{L}{s_{\overline{n}|,i}} \left((1+i)^n - 1 - ((1+i)^{t-1}) - 1) \right)$$

$$= \frac{L}{s_{\overline{n}|,i}} \left((1+i)^n - (1+i)^{t-1} \right)$$

$$= \frac{L(1+i)^n}{s_{\overline{n}|,i}} (1 - (1+i)^{t-n-1}) \tag{5.16}$$

We almost have it! Since $s_{\overline{n}|,i} = a_{\overline{n}|,i}(1+i)^n$, we have $\frac{(1+i)^n}{s_{\overline{n}|,i}} = \frac{1}{a_{\overline{n}|,i}}$. Also, since $v = (1+i)^{-1}$, we have $(1+i)^{t-n-1} = v^{n-t+1}$. Putting these two results into the last line of Equation 5.16 we get the following expression for the net interest in the case of the sinking fund with $i = j$

$$\frac{L}{a_{\overline{n}|,i}} (1 - v^{n-t+1}) \tag{5.17}$$

This is exactly what we got for the interest in the amortization method (Equation 5.5). It's easier to see how this all works by looking at the loan amortization table versus a similar table for a sinking fund.

Example 5.13 Consider a $10,000 loan with interest at 5% per annum to be paid off in five years with annual payments. Construct the Amortization Schedule and the Sinking Fund Schedule assuming that the sinking fund deposits earn 5% as well.

Solution: For the amortization method, the annual payment is $\frac{10000}{a_{\overline{5}|,.05}} =$ $2,309.75. We compute the entries as we did in the previous section to obtain:

The sinking fund method requires two payments: a $500 interest payment plus a sinking fund payment sufficient to accumulate to $10,000 in five years. This payment is $\frac{10000}{s_{\overline{5}|,.05}} = $1,809.75. Adding these gives us exactly the same total payment as in the level payment method: $1,809.75+500 = $2,309.75. In the case of the level payment method, this amount is divided between interest and principal payments as indicated in the above Schedule. In the case of the sinking fund, we can compute a "net" payment by subtracting the interest earned by the sinking fund at each point. Here is the Sinking Fund Schedule:

There are three new columns in this table. Tables 5.47 and 5.48 display the differences between amortization and sinking fund schedules.

TABLE 5.47

AMORTIZATION SCHEDULE			
Payment	**Interest**	**Net to Principal**	**Remaining Principal**
			$ 10,000.00
$ 2,309.75	$ 500.00	$ 1,809.75	$ 8,190.25
$ 2,309.75	$ 409.51	$ 1,900.24	$ 6,290.02
$ 2,309.75	$ 314.50	$ 1,995.25	$ 4,294.77
$ 2,309.75	$ 214.74	$ 2,095.01	$ 2,199.76
$ 2,309.75	$ 109.99	$ 2,199.76	$ 0.00

TABLE 5.48

SINKING FUND SCHEDULE							
Period	**Interest Paid**	**Sinking Fund**	**Interest Earned**	**Amount in Sinking Fund**	**Net Amount of Loan**	**Net Interest**	**Net to Principal**
0	0	0	0	0	$ 10,000.00		
1	$ 500.00	$ 1,809.75	$ -	$ 1,809.75	$ 8,190.25	$ 500.00	$ 1,809.75
2	$ 500.00	$ 1,809.75	$ 90.49	$ 3,709.98	$ 6,290.02	$ 409.51	$ 1,900.24
3	$ 500.00	$ 1,809.75	$ 185.50	$ 5,705.23	$ 4,294.77	$ 314.50	$ 1,995.25
4	$ 500.00	$ 1,809.75	$ 285.26	$ 7,800.24	$ 2,199.76	$ 214.74	$ 2,095.01
5	$ 500.00	$ 1,809.75	$ 390.01	$ 10,000.00	$ -	$ 109.99	$ 2,199.76

Net Interest = Interest Paid − Sinking Fund Interest

The entries in this column are numerically equal to those in the Interest column in the amortization schedule since $i = j$ in our example.

Net to Principal = Sinking Fund Payment + Interest Earned on Sinking Fund

The entries in this column are numerically equal to those in the Amount to Principal Column in the amortization schedule since $i = j$ in our example.

Net Amount of Loan = Original Loan Amount − Amount in Sinking Fund

The entries in this column are numerically equal to those in the Remaining Principal column in the amortization schedule since $i = j$.

The same technique is used to create the Sinking Fund Schedule in the case the $i \neq j$.

Example 5.14: in Example 5.9, suppose that the sinking fund earns 6% interest while the interest on the loan remains at 5%. Create the sinking fund schedule for this situation.

Solution: The sinking fund payment is now $R = \frac{10000}{s_{\overline{5}|,.06}} = \$1,773.96$. The interest payment remains at \$500. The Sinking Fund Schedule now looks like Table 5.49.

TABLE 5.49

				SINKING FUND SCHEDULE				
Period	Interest Paid	Sinking Fund Deposit	Interest Earned on SF	Amount in SF	Net Amount of Loan	Net to Interest	Net to Principal
0	0	0	$ -	0	$ 10,000.00	0	0
1	$ 500.00	$ 1,773.96	$ -	$ 1,773.96	$ 8,226.04	$ 500.00	$ 1,773.96
2	$ 500.00	$ 1,773.96	$ 106.44	$ 3,654.37	$ 6,345.63	$ 393.56	$ 1,880.40
3	$ 500.00	$ 1,773.96	$ 219.26	$ 5,647.59	$ 4,352.41	$ 280.74	$ 1,993.23
4	$ 500.00	$ 1,773.96	$ 338.86	$ 7,760.41	$ 2,239.59	$ 161.14	$ 2,112.82
5	$ 500.00	$ 1,773.96	$ 465.62	$ 10,000.00	$ 0.00	$ 34.38	$ 2,239.59

5.9   How to Construct the Sinking Fund Schedule in Excel

It's easiest to create a small table which stores all the constants needed to fill in the Schedule. Table 5.50 is an example

TABLE 5.50

Loan Amount	$10,000.00
Term of Loan in Years	5
Number of Payments Per Year	2
Interest on Loan	6.00000%
Interest on Sinking Fund	4.50000%
Computed Values: do not change	
Loan Interest Per Period	$ 300.00
Number of Payments	10
Required Sinking Fund Payment	$902.88

We now fill in the table just as we did for a loan amortization table. Table 5.51 is the result for this example:

TABLE 5.51

SINKING FUND SCHEDULE							
Period	Interest Paid	Sinking Fund Deposit	Interest Earned	Amount in Sinking Fund	Net Amount of Loan	Net Interest	Net To Principal
0							
1	$ 300.00	$902.88	$ -	$ 902.88	$ 9,097.12	$ 300.00	$902.88
2	$ 300.00	$902.88	$ 20.31	$ 1,826.07	$ 8,173.93	$ 279.69	$923.19
3	$ 300.00	$902.88	$ 41.09	$ 2,770.03	$ 7,229.97	$ 258.91	$943.96
4	$ 300.00	$902.88	$ 62.33	$ 3,735.23	$ 6,264.77	$ 237.67	$965.20
5	$ 300.00	$902.88	$ 84.04	$ 4,722.15	$ 5,277.85	$ 215.96	$986.92
6	$ 300.00	$902.88	$ 106.25	$ 5,731.28	$ 4,268.72	$ 193.75	$1,009.13
7	$ 300.00	$902.88	$ 128.95	$ 6,763.11	$ 3,236.89	$ 171.05	$1,031.83
8	$ 300.00	$902.88	$ 152.17	$ 7,818.16	$ 2,181.84	$ 147.83	$1,055.05
9	$ 300.00	$902.88	$ 175.91	$ 8,896.94	$ 1,103.06	$ 124.09	$1,078.79
10	$ 300.00	$902.88	$ 200.18	$ 10,000.00	$ -	$ 99.82	$1,103.06

Example 5.15 A loan of $4,500 with annual interest at a nominal 6% compounded quarterly is to be paid off using a sinking fund which accumulates interest at a nominal 8% compounded monthly. Payments are to be made quarterly for five years.

a) What is the interest payment?
b) What is the sinking fund payment?
c) What is the net interest portion of the fifteenth payment?
d) What is the sinking fund balance at the end of three years?

Solution: We will compute in quarters since that is the period for the payments.

a) The quarterly interest rate is $\frac{.06}{4} = .015$. The interest due each quarter is thus $4500 \cdot .015 = \$67.50$

b) The sinking fund interest rate must be converted to a quarterly rate. The monthly rate is equal to $\frac{.08}{12} = .006667$. If j is the equivalent quarterly rate we have

$$1 + j = (1.006667)^3 = 1.02013363$$
$$j = .02013363$$

The required payment is then $\dfrac{4500}{s_{\overline{20}|,.02013361}} = \$184.96$

c) We can either construct a sinking fund schedule or use Equation 5.11

$$\text{Net Interest} = L\left(i - j\,\frac{s_{\overline{t-1}|,j}}{s_{\overline{m}|,j}}\right)$$

$$= 4500\left(.015 - .02013363 \cdot \frac{s_{\overline{14}|,.2013361}}{s_{\overline{20}|,.02013363}}\right) = \$7.96155172$$

d) We use Equation 5.10

$$\text{Sinking Fund} = L\,\frac{s_{\overline{t}|,j}}{s_{\overline{n}|,j}}$$

$$= 4500\,\frac{s_{\overline{3}|,.02013363}}{s_{\overline{20}|,.2013363}} = \$566.12$$

Table 5.52 is the sinking fund schedule for this problem:

TABLE 5.52

SINKING FUND SCHEDULE							
Period	Interest Paid	Sinking Fund Deposit	Interest Earned	Amount in Sinking Fund	Net Amount of Loan	Net Interest	Net To Principal
0					$ 4,500.00		
1	$67.50	$184.96	$ -	$ 184.96	$ 4,315.04	$ 67.50	$184.96
2	$67.50	$184.96	$ 3.72	$ 373.64	$ 4,126.36	$ 63.78	$188.68
3	$67.50	$184.96	$ 7.52	$ 566.12	$ 3,933.88	$ 59.98	$192.48
4	$67.50	$184.96	$ 11.40	$ 762.48	$ 3,737.52	$ 56.10	$196.36
5	$67.50	$184.96	$ 15.35	$ 962.79	$ 3,537.21	$ 52.15	$200.31
6	$67.50	$184.96	$ 19.38	$1,167.13	$ 3,332.87	$ 48.12	$204.34
7	$67.50	$184.96	$ 23.50	$1,375.59	$ 3,124.41	$ 44.00	$208.46
8	$67.50	$184.96	$ 27.70	$1,588.25	$ 2,911.75	$ 39.80	$212.65
9	$67.50	$184.96	$ 31.98	$1,805.18	$ 2,694.82	$ 35.52	$216.94
10	$67.50	$184.96	$ 36.34	$2,026.49	$ 2,473.51	$ 31.16	$221.30
11	$67.50	$184.96	$ 40.80	$2,252.25	$ 2,247.75	$ 26.70	$225.76
12	$67.50	$184.96	$ 45.35	$2,482.55	$ 2,017.45	$ 22.15	$230.30
13	$67.50	$184.96	$ 49.98	$2,717.49	$ 1,782.51	$ 17.52	$234.94
14	$67.50	$184.96	$ 54.71	$2,957.16	$ 1,542.84	$ 12.79	$239.67
15	$67.50	$184.96	$ 59.54	$3,201.66	$ 1,298.34	$ 7.96	$244.50
16	$67.50	$184.96	$ 64.46	$3,451.08	$ 1,048.92	$ 3.04	$249.42
17	$67.50	$184.96	$ 69.48	$3,705.52	$ 794.48	$ (1.98)	$254.44
18	$67.50	$184.96	$ 74.61	$3,965.09	$ 534.91	$ (7.11)	$259.56
19	$67.50	$184.96	$ 79.83	$4,229.88	$ 270.12	$ (12.33)	$264.79
20	$67.50	$184.96	$ 85.16	$4,500.00	$ -	$ (17.66)	$270.12

Example 5.16 Find i if the following two methods for paying of a loan of $10,000 over twenty years are equivalent.

a) A sinking fund in which the interest on the loan accrues at i and the sinking fund receives level payments at the end of each year earning interest at 10%.

b) Amortization by level payments with an effective annual interest rate of 12%.

Solution: To be equivalent, the two methods must require the same payment.

Method a): The total payment is given by equation 5.9

$$L\left(\frac{1}{s_{\overline{s}|,j}} + i\right) = 10000\left(\frac{1}{s_{\overline{20}|,.1}} + i\right) = 10000(0.0174596248 + i)$$

Method b): The payment is given by $\frac{L}{a_{\overline{20}|,.12}} = \frac{10000}{7.4694436230} = 1338.79$

We set these equal and solve for i : $1338.79 = 174.59 + 10000i$ and obtain $i = 11.64\%$

Example 5.17 A loan of $500,000 on which interest accrues at 8% compounded annually is repaid with twenty annual interest payments. A sinking fund earning 5% is used to accumulate the $500,000 needed to make the final payment. At what interest rate would the amortization method require the same payment?

Solution: The annual interest payment is $500,000 \cdot .08 = \$40,000$. The sinking fund deposit needed is

$$\frac{500000}{s_{\overline{20}|,.05}} = \frac{500000}{33.06595408} = 15,121.29$$

The payments thus total $40,000 + 15,121.29 = \$55,121.29$. If we use the amortization method a payment of $55,121.29 must pay off a loan of $500,000 over twenty years. We thus have $R = \$55,121.29$, $n = 20$, $L = \$500,000$ and want to find i.

We use the **TI BA II Plus** (Table 5.53) to compute the interest rate:

TABLE 5.53

N	I/Y	PV	PMT	FV
20	CPT = 9.089061119	500000	= -55,121.29	0

Thus, a sinking fund approach with loan interest of 8% and sinking fund interest of 5% is equivalent to an amortization of the same loan at a rate of 9.089%.

5.10 Amortization with Non-Standard Payments

Consider a loan of L which is being repaid with payments every k interest conversion periods for a total of n interest conversion periods. This means that there will be $\frac{n}{k}$ payments. We compute the payment, R, which is needed using Equation 5.1.

$$L = R\frac{a_{\overline{n}|,i}}{s_{\overline{k}|,i}}$$

$$R = L\frac{s_{\overline{k}|,i}}{a_{\overline{n}|,i}} \tag{5.18}$$

It's easier to work our way through the amortization schedule if we convert the interest to an equivalent rate at the period of the loan repayments. This rate is $(1+i)^k - 1$. We adopt the following notations

Effective Rate in Payment Periods: $j = (1+i)^k - 1$
Period of Loan in Payment Periods: $q = \frac{n}{k}$

We can then use Equations 5.5 through 5.7 to compute the amortization table.

Example 5.18 A loan of $50,000 accrues interest at a nominal 6% annual compounded monthly and will be paid off in sixty months. The loan is to be amortized by a series of level payments to be made quarterly. Construct the amortization table and write down the equations which compute the loan amount and the allocation of the payment between principal and interest at the end of each period. Use the AMORT worksheet to compute the amortization table using a period of one year.

Solution: We have

$$j = (1.005^3 - 1) = 0.015075125$$
$$q = \frac{60}{3} = 20$$

We can now compute the payment required using Equation 5.18

$$R = \frac{50,000}{a_{\overline{20}|,0.015075125}} = \$2,914.44$$

We now construct the amortization table (Table 5.54) exactly as we did in Section 5.1.

TABLE 5.54

Period	Payment	Interest Due	Amount to Principal	Principal
0	$0.00	$0.00	$0.00	$50,000.00
1	$2,914.44	$753.76	$2,160.68	$47,839.32
2	$2,914.44	$721.18	$2,193.26	$45,646.06
3	$2,914.44	$688.12	$2,226.32	$43,419.74
4	$2,914.44	$654.56	$2,259.88	$41,159.86
5	$2,914.44	$620.49	$2,293.95	$38,865.91
6	$2,914.44	$585.91	$2,328.53	$36,537.38
7	$2,914.44	$550.81	$2,363.63	$34,173.74
8	$2,914.44	$515.17	$2,399.27	$31,774.48
9	$2,914.44	$479.00	$2,435.44	$29,339.04
10	$2,914.44	$442.29	$2,472.15	$26,866.89
11	$2,914.44	$405.02	$2,509.42	$24,357.47
12	$2,914.44	$367.19	$2,547.25	$21,810.22
13	$2,914.44	$328.79	$2,585.65	$19,224.57
14	$2,914.44	$289.81	$2,624.63	$16,599.95
15	$2,914.44	$250.25	$2,664.19	$13,935.75
16	$2,914.44	$210.08	$2,704.36	$11,231.40
17	$2,914.44	$169.31	$2,745.13	$8,486.27
18	$2,914.44	$127.93	$2,786.51	$5,699.76
19	$2,914.44	$85.92	$2,828.52	$2,871.25
20	$2,914.44	$43.28	$2,871.16	$0.09

TI BA II Plus Solution: We use the TVM keys to compute the quarterly interest and to set up the loan. We then use the AMORT worksheet to create the table (Table 5.55).

TABLE 5.55

N	I/Y	PV	PMT	FV
3	.5	1	0	$CPT = -1.01507513$

N	I/Y	PV	PMT	FV
1	$CPT = 1.50751250$			

N	I/Y	PV	PMT	FV
20		50,000	$CPT = -2,914.444$	0

Key	Value	Key	Value
P1	1 Enter = 1	P1	CPT = 13
P2	4 Enter = 1	BAL	11,231.32
BAL	41,159.84	PRN	-10,578.84
PRN	-8,840.16	INT	-1,078.93
INT	-2,817.18	P1	CPT = 17
P1	CPT = 5	BAL	0.00
BAL	31,774.44	PRN	-11,231.32
PRN	-9,385.4	INT	-426.45
INT	-2,272.38		
P1	CPT = 9		
BAL	=21,810.17		
PRN	-9,964.27		
INT	-1,693.50		

Now we create the amortization table. See Table 5.56.

TABLE 5.56

Year	Payment	Interest	Principal	New Balance
0				$50,000
1	11,657.76	2,817.2	8,8410.2	41,159.84
2	11,657.76	2,272.38	9,385.4	31,774.4
3	11,657.76	1,693.5	9,964.3	21,810.2
4	11,657.76	1,078.9	10,578.8	11,231.3
5	11,657.76	426.5	11,231.3	0.0

5.11 Truth in Lending[1]

The 1968 Consumer Credit Protection Act requires that lenders provide "fair and accurate" disclosure as to the terms of consumer loans. The law requires that lenders provide consumers with two numbers:

1) The total finance charge.

2) The annual percentage rate (usually abbreviated as APR).

The finance charge includes all interest paid on the loan as well as some, but not necessarily all, of other charges which may incurred by the borrower. These include: points (sometimes called a loan origination fee), other loan fees, service charges, credit report fees, and premiums for credit insurance. The APR, which one might assume is the annual effective rate defined earlier, is often quoted as a nominal annual rate. Thus, a loan with monthly payments and a monthly interest rate of 1% would be reported as an APR of 12% even though the effective annual interest rate is 12.68%.

Credit card issuers need not report total finance charges, only the charge per loan period, usually one month. Often, a person who elects to pay only the minimum required amount on a credit card balance will never pay off the loan. The total possible finance charge is thus infinite! Additionally, the interest rate reported on almost all credit cards is the nominal rate of annual interest. Thus, a 1% monthly interest rate is reported as 12% APR.

Automobile and home loans are examples of closed end credit – the loan has a specific ending date at which the balance must be 0. In many cases, advertisements specify only a monthly payment and a number of years for the loan. We now discuss how to compute the reported APR and actual annual effective rate for such loans. We begin with a list of terms which will be used (Figure 5.2):

Truth In Lending Terms

L = original loan balance after any down payment

K = total finance charges (usually referred to as "the finance charge")

R = the installment payment

m = number of payments per year

n = total number of payments in the term of the loan

i = reportable annual percentage rate (the nominal annual rate)

j = rate of interest per payment period

I = the effective annual interest rate

FIGURE 5.2

---

[1]This topic is not part of the FM/2 Syllabus, but it is very useful knowledge to have the next time (or the first time) you make any major purchase and finance it: cars, houses, boats, appliances, etc.

As we know, we have $i = mj$ and $I = (1+j)^m - 1$. Thus, we need only compute the value of j to compute the others. Many short-term loans (cars, appliances) are stated in terms of a payment amount (R) and a number of payments (n). The total of all payments is nR which will equal the loan value, L, plus the finance charges. If we use K for the finance charge, we can compute the its value as

$$nR = L + K$$
$$K = nR - L \tag{5.19}$$

We can compute the value of j by using the equation for the payment required to amortize a loan and solving for the interest rate instead of the payment:

$$R = \frac{L}{a_{\overline{n}|,j}}$$
$$Ra_{\overline{n}|,j} = L \tag{5.20}$$

Example 5.19: Find the APR, the effective annual interest rate and the finance charge if a loan of \$1,000 is repaid by twelve monthly installments of \$90.

Solution. The finance charge is the total interest paid and is computed as:

$$K = 12 \cdot 90 - 1000 = \$80$$

We use the **TI BA II Plus** (Table 5.57) to compute the APR and the effective annual interest rate.

TABLE 5.57

N	I/Y	PV	PMT	FV
12	CPT = 1.20435678	1,000	-90	0

We now convert this to an effective annual rate as follows (Table 5.58):

TABLE 5.58

N	I/Y	PV	PMT	FV
		1	0	CPT = -1.154489

N	I/Y	PV	PMT	FV
1	CPT = 15.44893640			

Example 5.20 A car is advertised for "Sixty easy payments of $305 each month." Fine print at the bottom of the ad reveals that there is $1,300 due at signing. After considerable effort, the salesperson reveals that these payments are based on a purchase price of $13,300. What is the effective annual interest being charged in this case?

Solution: After subtracting the $1,300 initial payment, the amount financed is $12,000. This is being amortized by sixty monthly payments of $305.

TI BA II Plus Solution (Table 5.59)

We compute the effective annual interest in the same manner as Example 5.18

TABLE 5.59

N	I/Y	PV	PMT	FV
60	CPT = 1.503559069	12,000	-305	0

N	I/Y	PV	PMT	FV
12		1	0	CPT = -1.19612136

N	I/Y	PV	PMT	FV
1	CPT = 19.61213567			

The APR (what the dealer must report to you) is $1.503559069 \times 4 = 18.04\%$, with an effective annual rate of 19.6%. The interest paid is $60 \times 305 - 12,000 = \$6,300$ – over half the price of the car!

5.12   Real Estate Loans – Home Loans

Real estate mortgage loans represent, for most people, the largest single form of indebtedness. The loans typically have a term of either fifteen or thirty years and are paid off on a monthly basis. Payments are typically due on the first day of each month for the previous month. If the inception date (usually the day the buyer takes possession) does not fall on the first of the month, the buyer pays the interest due on the amount financed for the remainder of the month. This calculation generally is done using simple interest, and the actual/365 method of computing the number of days remaining in a month. This amount is paid at the time of the loan origination. A loan originating on

July 12 would incur an interest charge based on the twenty remaining days in July. The main loan would then commence on August 1 with first payment due on September 1 (annuity-immediate).

Example 5.21 A home is purchased for $150,000. There is a down payment of 20% and the loan has a nominal annual rate of 9.9% to be paid off monthly over thirty years. The loan is originated on July 12. What is the finance charge for the fractional portion of July?

Solution: The down payment is $.2 \cdot 150,000 = \$30,000$ so the initial amount of the loan is $120,000. There are twenty days remaining in July so the interest due (computed using simple interest and actual/365) is

$$.099 \cdot \frac{20}{365} 120,000 = \$650.96$$

The advertised annual rate of interest is used to compute the payments needed to amortize the entire loan. However any additional expenses paid at closing reduces the amount of money available to the borrower and thus increases the effective annual interest rate. For example, if a borrower must pay $2,000 at closing on a loan of $150,000, the loan payments will be based on a loan of $150,000 even though the borrower will only have $148,000. The annual percentage rate required by the truth in lending law must reflect this fact. Thus, a loan might be advertised as 8.7% (APR = 9.2%). We now develop the tools needed to compute the APR which is required by the truth in lending law.

We add a few letters to our long list (Figure 5.3):

Home Loan Terms

Q = finance charges incurred at inception which are required to be included in the truth in lending law

L^* = amount of loan for truth in lending purposes $(L^* = L - Q)$

j'' = monthly rate of interest on the loan (L)

i'' = a quoted annual rate of interest on the loan $(i'' = 12 \cdot j'')$

Some old variables now need to be defined more carefully

R = payment needed to amortize L at rate j'

j = computed rate of interest on L^* based on payments of R

i = quoted APR $(i = 12j)$

I = effective annual interest on L^*

FIGURE 5.3

The most common form of "extra charges" are called points. If a loan carries a charge of 2 points, the borrower must pay 2% of the total loan amount at closing. Note that this reduces the amount of money available to the lender. Points function just like a discount - they are paid up front. A loan with points is a mixture of a standard loan and a loan at discount. The loan payments are based on the original amount of the loan even though that much is never received by the borrower. There may be other loan charges as well. We will lump those under the catch all "other charges."

To compute the effective annual interest, I, and the finance charges required by the truth in lending law we proceed as follows.

Truth In Lending Law Formulas

The 6 Step Program

1) We compute the payment needed to amortize the loan using

$$R = \frac{L}{a_{\overline{n}|,j''}} \qquad (5.21)$$

2) The actual amount available to the borrower is

$$L^* = L - Q \qquad (5.22)$$

3) The finance charge for truth in lending purposes is given by

$$K = nR - L^* \qquad (5.23)$$

4) The monthly interest rate on L^* is computed by solving the following equation for j

$$R a_{\overline{n}|,j} = L^* \qquad (5.24)$$

5) The APR for truth in lending is

$$i = 12j \qquad (5.25)$$

6) The effective annual rate, I is

$$I = (1 + j)^{12} - 1 \qquad (5.26)$$

Example 5.22 (continuation of 5.12) A home is purchased for $150,000 with a down payment of $30,000, the remaining $120,000 to be financed with a thirty-year fixed-rate mortgage at 9.9%. Payments will be made monthly. The borrower is assessed 2 points as well as an additional $800 of "other costs" at closing. Compute all the numbers needed at settlement.

Solution: We already computed the interest charge for July of $650.96. We have $i'' = .099$ and so $j'' = \frac{.099}{12} = .00825$. The term of the loan is $n = 12 \cdot 30 = 360$. We now work our way through the six-step program above:

1) The monthly payments will be

$$R = \frac{120,000}{a_{\overline{360}|,.00825}} = \$1,044.23.$$

TI BA II Plus (Table 5.60)

TABLE 5.60

N	I/Y	PV	PMT	FV
360	$\frac{9.9}{12}$	120,000	CPT = 1,044.23	0

2) $Q = 120000 \cdot .02 + 800 = \$3,200$. Thus $L^* = 120,000 - 3,200 = \$116,800$

3) The finance charge is $K = 360 \cdot 1044.23 - 116,800 = \$259,122.8$

4) The effective monthly interest rate on L^* is computed by solving the equation below for j

$$1044.23 a_{\overline{360}|,j} = 116,800$$

TI BA II Plus: We need only change PV = 116,800 and compute the interest rate (Table 5.61):

TABLE 5.61

N	I/Y	PV	PMT	FV
	CPT = .851846008	116,800		

5) The APR is $12 \times .851846008 = 10.22\%$

6) See Table 5.62.

TABLE 5.62

N	I/Y	PV	PMT	FV
12		1	0	CPT = -1.10714939

N	I/Y	PV	PMT	FV
1	CPT = 10.71493876			

A typical disclosure statement (which usually appear as a footnote at the bottom of the ad) might look like this:

This loan is a fixed rate mortgage at 9.9% with 2 points and $800 in additional closing costs[2] (this footnote text is too large – in reality, they are usually barely readable!)

---

[2]APR = 10.22 with total finance charges of $259,123.

5.13 Balloon and Drop Payments

In some cases a loan payoff is described in terms of number of payments of a specified amount. These seldom result in an exact payoff. Here is an example

Example 5.23. A loan of $10,000 is to be paid off with annual payments (at the end of each year) of $1,000. If the effective annual interest is 5% how many payments will be needed?

Solution We use the TVM keys to compute the value of N (Table 5.63):

TABLE 5.63

N	I/Y	PV	PMT	FV
CPT = 14.2067	5	-10,000	1,000	0

Since we are dealing with annual payments the .2067 is an issue. There are two solutions.

1) Make thirteen payments of $1,000 and a fourteenth payment of slightly more than $1,000. This is called a balloon payment.

2) Make fourteen payments of $1,000 and a fifteenth payment to clear the balance. This is called a drop payment.

Method one. We compute the balance just after year 14 by changing n to 14 and computing (Table 5.64)

TABLE 5.64

N	I/Y	PV	PMT	FV
14				CPT = 200.68

We see that the fourteenth payment will be 1,200.68.

Method two: The loan balance after fourteen payments of $1,000 will be $200.68. We compute the accumulated value of this amount after one year to see what the final payment will be. To do that we convert the answer to Method one to PV, clear the payments and set $n = 1$. The answer is *FV* (Table 5.65)

TABLE 5.65

N	I/Y	PV	PMT	FV
1		200.68	0	CPT = 210.72

We make fourteen payments of $1,000 and a final drop payment of $210.72.

5.14   Problems Involving Loans

This section contains a small collection of problems illustrating some of the situations you might be called upon to deal with on the FM exam. The solutions have been shortened to better model the solutions you will find in the standard FM review texts.

Example 5.24 A loan of $5,000 is being financed at 5% annual interest. There will be five payments each made at the end of the year. The first three payments will be X and the final three payments will be $2X$ Find X.

Solution: The fundamental principle of loan amortization is simple: the present value of all payments must be equal to the amount of the loan at the stated interest rate. We thus have:

$$5000 = X a_{\overline{3}|,.05} + 2X v^3 a_{\overline{2}|,.05}$$

Solving for X yields $X = \$842.36$.

Example 5.25 Using the data from **Example 5.24**, what is the loan balance immediately after the second payment?

Solution: We can use either the retrospective or prospective method. In this case, since we know both the payments and the original loan amount, the retrospective method is easier:

$$B_2 = 5000 \cdot (1.05)^2 - 842.36 \cdot a_{\overline{2}|,.05} = \$3,946.21$$

Example 5.26 Mary takes out a $10,000 loan to be paid back in ten annual installments at the end of each year at an annual interest rate of 8%. Just after the fourth payment Mary renegotiates to pay off the loan in a total of eight payments. The new payments result in an overall yield rate for the bank of 6.5%.

  a) What is the amount of each of the four new payments?

  b) How much money does Mary save compared to the original loan?

Solution We first compute the original payment (Table 5.66):

TABLE 5.66

N	I/Y	PV	PMT	FV
10	8	10,000	CPT = -1,490.29	0

Under the original loan Mary would have a total of $14,902.9. With the new loan she will make four payments of $1,490.29 and an additional four payments of unknown value, say x For the bank to have a yield rate of 6.5% the present value of these eight payments must equal 10,000 when evaluated at 6.5%. We thus have

$$10000 = 1490.29a_{\overline{4}|,.065} + x \cdot v^4 \cdot a_{\overline{4}|,.065}$$

Solving, we find $x = 1,838.02$. The eight payments for the new loan total $13,313.27 compared to $14,902.0 for the original loan. Mary saves $1,589.68.

Example 5.27 A loan is being paid off with annual payments of $5,000 at an interest rate of 6% per year. The amount of interest in the fifth payment is $2,709.44. What is the amount of principal in the tenth payment?

Solution Note the switch from interest to principal. We are not told the amount of the loan or its length. That means we don't need this information to do the problem. Here's how: Since the interest in the fifth payment is $2,709.44, the amount to principal in this payment is $5000 - 2709.44 = 2290.55$. We now use the fact that the amount to principal is a geometric progression with common multiplier $1 + i = 1.06$. As a result the amount to principal in the tenth payment is $2290.44 \cdot 1.06^5 = 3,065.13$.

Example 5.28 A loan is being paid off with fifteen annual payments. The payments are $2x$ in the first five years, $3x$ in the next five years and x in the final five years. The interest repaid in year 6 is three times the interest repaid in year 11. Calculate the interest rate for this loan.

Solution: The interest due in any given payment is equal to the loan balance just prior to that payment. We can calculate the loan balance as the present value of the remaining payments and the multiply this by the interest rate to obtain the interest due. We need to recall that

$$a_{\overline{n}|,i} = \frac{1 - v^n}{i}$$

Year 6: There are five payments of $3x$ and five payments of $2x$ remaining so the interest due is

$$i \cdot (2xa_{\overline{5}|,i} + 3xv^5 a_{\overline{5}|,i})$$
$$= (3x(1 - v^5) + 2xv^5(1 - v^5))$$
$$= (1 - v^5)(2x + 3xv^5)$$

The interest due at year 11 is $x(1 - v^5)$. We then have

$$(1 - v^5)(2x + 3xv^5) = 3x(1 - v^5)$$

This gives us $3xv^5 = 1x$ so that $v^5 = \frac{1}{3}$ and $i = 24.57\%$.

Example 5.29 A loan is being paid off with 360 equal monthly payments beginning at the end of the first month. The effective annual rate of interest for the loan is 6%. Determine which payment will be the first in which amount of principal paid is more than two-thirds the amount of interest paid.

Solution It appears that we don't have enough information, but that is never really the case. Assume the payment is P. We then have

$$P_t = Pv^{360-t+1}$$
$$I_t = P(1 - v^{360-t+1})$$

We need to find t so that $P_t = \frac{2}{3}I_t$ which means that $v^{361-t} = \frac{2}{3}(1 - v^{361-t})$ which implies that $v^{361-t} = .4$. The value of i is the solution to $(1+i)^{12} = 1.06$ so $i = .00486755$. This results in $t = 172.29$. We need to round up to 173 to ensure that the principal portion is greater than the interest portion. Note: This is a problem you could do on the TVM keys. Make up a loan value, say 1,100 and set up the loan. Then go into AMORT and try each of the multiple choice answers to see which one works.

5.15   Important Concepts from this Chapter

There are a lot of formulas in this section which you will need to memorize. In order to make use of them you will need to be aware of the following basic concepts.

1) The yield rate of a loan is the interest rate at which the present value of all the payments equals the value of the loan.

2) The balance due at any integer time can be computed in one of two ways:

   a) As the present value of all remaining payments at the yield rate.

   b) As the accumulated value of the loan minus the accumulated value of the payments made to date.

3) The interest portion of any payment is equal to the loan balance just after the previous payment times the yield rate.

4) The payments to principal form a geometric sequence with common ratio $1 + i$ where i is the yield rate.

Exercises For Chapter 5

1) A loan of $5,000 is paid off at 5% effective converted annually over five years.

   a) What is the annual payment?

b) What is the total interest paid?

c) Construct the amortization table.

2) A loan of $5,000 will be paid off using a sinking fund. The sinking fund earns 5% while the interest on the loan is 5%. The loan is to be paid off in five years.

a) What is the annual sinking fund payment required?

b) Construct the amortization table.

c) Verify that the interest paid in this problem is the same as that paid in problem 1.

3) A loan of $50,000 is being financed with monthly payments over a period of ten years. The interest rate is a nominal 12% compounded monthly.

a) What is the required payment?

b) What is the total amount of interest paid?

4) The borrower in problem 3) obtains refinancing for eight years of monthly payments immediately after the twenty-first payment at an interest rate of $i^{(12)} = .08$

a) What is the required payment?

b) What is the total amount of interest paid?

5) A loan of $50,000 is being paid back with annual payments over twenty years. Interest is a nominal 6% annual compounded six times per year. After ten years the loan is refinanced at 6% compounded quarterly for twenty-five years with payments to be made every month.

a) What is the annual payment during the first ten years?

b) What is the loan balance at the time of re-financing?

c) How much interest was paid during the first ten years?

d) What is the monthly payment for the final twenty-five years?

e) How much interest is paid on the loan altogether?

6) A home is purchased on June 12 for $350,000. The home is to be financed with a thirty-year fixed rate 8.7% mortgage. The lender charges one and a half points plus $300 at inception. Compute all the numbers needed for truth in lending (our six-step program).

7) A car is advertised at a sale price of $25,456. The dealer says you can drive the car home for only forty-eight monthly payments of $641. When you declare that to be a bit too steep, he offers to sell you the car for sixty payments of $559. Compute all the truth in lending numbers for each scenario. Which is the better deal? Why might the interest rates be different in these two cases.

8) You opted for the second option in problem 7). After a year, you decide it's time for some new wheels. You return to the dealer and want to trade in your car on a brand-new model. The dealer offers you a trade-in value of $20,000. How much of a down payment will you be able to make after you have paid off the remainder of your loan on your first car?

9) Create a MAPLE plot which plots the time for a loan balance to decline to 50% of its original amount as a function of the interest. Start with $n = 100$, then vary the value of n see if this changes the shape of the plot.

10) A house is purchased for $350,450. A down payment of 15% is made and the remainder is financed with a thirty-year fixed loan with a nominal interest rate of 8% to be paid off in monthly installments at the end of each month.

     a) At what time does the balance reach 50% of the amount originally financed?

     b) At what time does the percentage of each payment to principal first exceed 50%?

     c) What is the total amount of interest paid?

     d) What is the loan balance just after payment number 145?

11) A loan is repaid with eighteen annual payments made at the end of each year. Half of the loan amount ($\frac{L}{2}$) is an amortized loan with effective annual interest of 8%. The other half is to be repaid by a sinking fund. The interest on the loan amount ($\frac{L}{2}$) is 7.5% while the sinking fund earns 3%. The total of the payments on the two halves of the loan is $1,200. What is the amount of the loan (L)?

12) Karl's Klassick Kars offers you a clean[3] '98 Buick for $7,500 cash or sixty monthly payments of $159 each to be made at the end of each month. What rate of interest is Klassick charging?

13) A home loan is signed on May 12. The purchase price is $450,000. The borrowers make a down payment of 15% of the purchase price. The bank charges 2 points, an $500 loan initiation fee and a nominal 7.5% for a thirty-year loan with monthly payments at the end of each month. Prepare a Truth in Lending report for this loan.

14) Maude has two possible ways to finance her home loan of $325,000.

     a) A thirty-year fixed rate loan at a nominal 5% with 2 points

     b) A thirty-year fixed rate loan at a nominal 4.7% with 1 point and other charges of C.

---

[3]The car was run through a $1 car wash within the past year.

Find the value of C if the APR based on truth in lending laws is the same in each case.

15) A \$75,000 loan, to be repaid over a five-year period, my be repaid by one of two methods

1) Level annual payments made at the beginning of each year.

2) Level semiannual payments made at the end of each six-month period.

If $d^{(4)} = .076225$. Find the absolute difference in the payments made each year under the two methods.

16) Bernard borrows \$100,000 on January 1, 2008, to be repaid in 360 monthly payments at a nominal annual interest rate of 9% convertible monthly. The first payment is due February 1, 2008. Bernard misses the first payment, but subsequently makes 359 monthly payments starting on March 1, 2008. Determine how much Bernard owes on the loan just after making payment 359.

17) Sally takes out a \$7,000 loan which requires sixty equal monthly payments starting at the end of the first month. The nominal annual interest rate is 8% convertible monthly. She arranges to make no payments for the first five months and then make the same payments she would have made plus an extra payment at the end of the sixtieth month. How much will this balloon payment be?

18) Kaneesha borrows X for ten years at an annual effective interest rate of 8%. If she pays the principal and interest in one lump sum at the end of ten years she will pay \$468.05 more than if she pays the loan off with level payments at the end of each year. Find X.

19) Smith borrows \$3000 and agrees to establish a sinking fund to repay the loan at the end of ten years. Interest on the loan at 8% is paid annually as it falls due. Level annual deposits to the sinking fund are made at the end of each year, with interest accumulating at an effective annual rate of 5% for the first four years and 3% thereafter. What is the size of the required annual sinking fund deposit?

20) A loan is repaid in ten equal annual installments with the first installment paid one year after the loan is made. The effective annual interest rate is 4%. The total amount of principal repaid in the fifth, sixth, and seventh payments combined is \$6,083. What is the total amount of interest repaid in the second, third, and fourth payments combined?

21) A loan of \$2,000 at an effective annual rate of 5% There are two options

a) Make level payments at the end of each year for five years.

  b) Make annual interest payments and accumulate the amount needed to pay off the principal in a sinking fund (payments made at the end of each year) which earns 4%. How much more does option b) cost than option a) in terms of total payments?

22) Joe has borrowed $50,000 on which he is paying interest an effective annual interest rate of 17.5% per year. He is contributing a constant amount P to a sinking fund at the end of each year. The sinking fund earns an annual effective rate of 9%. His combined payment for both the fund and the loan is $9,116.82 annually. Determine the year in which the sinking fund will be sufficient to pay off the loan.

23) A loan of $3,000 at 18% per year is being repaid by ten equal annual payments, the first payment due one year after the loan is made.

  a) Determine how much the interest portion of the sixth payment is.

  b) At the time of the fourth payment and additional payment of $100 is made and the loan is refinanced at 11% per year for the remainder of its life. What is the new payment?

24) You have just purchased a home with a $100,000 mortgage. The nominal rate of interest is 9% convertible monthly. A monthly payment of X is calculated based on a thirty-year amortization. If instead, you make a payment of one-half of x every two weeks, how long will it take to amortize the loan? (Assume fifty-two weeks per year, but not that a two-week period is one-half of a month).

25) Martha borrowed $1,000 to be repaid by level payments every two years with the first payment due two years after the loan starts. The effective annual interest rate is 9%. The amount of interest in the fourth payment is $177.72. Determine the amount of principal in the sixth installment.

26) Jack and Jill each borrow $5,000 for a term of five years at an effective annual interest rat of 7%. Jack repays his loan by making level annual payments at the end of each year. Jill accumulates the principal by making level deposits into a sinking fund at the end of each year. The sinking fund earns an effective annual interest rate of 4%. Jill makes annual interest payments to the creditor in addition to the sinking fund deposits. Determine the amount by which the sum of Jill's deposits and interest payments will exceed the sum of Jack's payments.

27) Sue borrows $10,000 for a term of ten years at an effective annual interest rate of 12%. At the end of each year she makes annual interest payments as well as deposits into two sinking funds. One-third of each deposit is made to a sinking fund earning 10% annual interest. Two-thirds of her

deposit is made into a sinking fund earning 8% annual interest. What is the sum of her three annual payments?

28) Georgia makes level annual year-end payments on a thirty-year, $100,000 mortgage loan. At the time of her fifteenth payment, she adds $25,000 to her regular payment and renegotiates the loan to reduce the remaining term from fifteen years to ten years. The effective annual interest rate is 8% throughout. What is the amount of the level payment required for the renegotiated loan?

29) George borrows $20,000 for a term of ten years at an effective annual interest rate of 8%. George repays the loan by making level annual payments at the end of each year. Immediately after making his first payment, George deposits an amount equal to the principal portion of the first payment into a fund earning an effective annual interest rate of 5%. Immediately after making his seventh payment, George applies the accumulated fund balance to the loan. George will then repay the remaining amount of the loan with three level payments of L. What is the value of L?

30) Iggy borrows $x for ten years at an effective annual interest rate of 6%. If he pays the principal and accumulated interest in one lump sum at the end of ten years, he would pay $356.54 more interest than if he repaid the loan with ten level payments at the end of each year. Find x.

31) The January 21, 2008, edition of the Chicago Tribune contained an ad for a recreational vehicle known as the Freedom Express. The Express was advertised as a real buy – its price had been reduced from $67,298 to $49,995. You could drive this beauty off the lot in return for a promise to make monthly payments of $355 for only 240 months! The ad provides an APR of 7.9%. In the fine print is a required down payment! Use the information provided to compute the value of the down payment. How much will you pay for this vehicle altogether?

32) Can't afford $355/month? Right next to the Freedom Express was a Freedom Spirit which was available for only $10,995. The financing offers was $95 per month over 160 months. Assuming the same percentage down payment as in problem 31), what is the APR for this loan?

33) A loan of $20,000 is to be paid off with annual payments of $1,500. The effective annual interest is 6%.

a) How many payments are required if a balloon payments is used? What is the value of the last payment?

b) How many payments are required if a sinking payment is used? What is the value of the last payment in this case?

Excel Projects for Chapter 5

Excel Project 1

Construct an Excel Spreadsheet which computes the level payment amortization schedule for an arbitrary loan. This will take some time, so get started on it well before the due date.

Required Features of your Spreadsheet

Create a space at the top for the user to enter the following information.

a) The loan amount

b) The number of payments per year

c) The term of the loan in years.

d) The nominal annual interest rate

Directly below this create a chart which contains the following information.

a) The required level payment

b) The number of payments required

c) The total interest which will be paid

d) The Effective Annual Interest for the loan

To the right (starting in row one) create the amortization schedule. You will need to use a macro to copy the relevant information. Your spreadsheet should look something like Table 5.67:

TABLE 5.67

Loan Amortization Calculator		Period	Payment	Interest	Amount to Principal	Principal
		0				
		1				
Enter Loan Amount		2				
Enter Number of Payments Per Year		3				
Enter Nominal Interest Rate Per Year						
Enter Period of the Loan in Years						
Calculated Values: do not change gray cells						
Required Payment						
Numer of Payments						
Total Interest Paid						
APR						

Excel Project 2

Modify Project 1 to allow the user to refinance the loan at any time during its lifetime. Create a space below the Loan Information area for the use to enter the following information:

a) The period in which the re-financing will start

b) The period in years of the re-financed loan

c) The number of payments per year

d) The nominal interest rate per year

e) The new APR

Create a Re-Financing Information Area which displays the following information

a) The number of payments required

b) The new payment

c) The total interest which will be paid (original payment plan plus the re-financing)

d) The savings/loss in total interest paid as a result of the refinancing

Write a macro which modifies the original amortization schedule beginning with the first period of the re-financing.

Excel Project 3

Create a Sinking Fund Calculator to pay off a loan using the Sinking Fund method. Remember that the sinking fund interest does not always equal the interest on the loan.

Your spreadsheet should allow the user to enter all the relevant information and then run a macro to create the table.

The Macro should also create the Subtotals worksheet which reports annual interest paid. No refinancing required for this one.

Excel Project 4

Create an Excel Spreadsheet which computes the numbers required for the truth in lending law. See Example 4.25 for the format required.

You will need to add entries for points and additional charges. The spreadsheet should fully implement our six-step program. Use a macro implemented by an action button.

Chapter 6

Yield Rates

6.1 Discounted Cash Flow Analysis

In this chapter we will generalize the ideas of Chapters 3, 4, and 5 to consider situations where money flows both into and out of an investment at various points in time. The deposits/withdrawals will not be constant, nor will the times at which they appear. For this reason the TVM keys won't be useful.

The technique used to solve these problems is referred to as **discounted cash flow analysis**. The sequence of payments and withdrawals is sometimes called an **investment scheme**. We can ask two distinct sorts of questions:

1. If the investment scheme earns a specified rate of interest throughout its life, what will the account balance be at the end of the scheme?

2. If the account shows a given balance at the end, what constant interest rate would be needed to produce that balance? This rate is called the internal rate of return (IRR) for the investment scheme.

In many cases the CF worksheet can be used to solve these problems. In the first case were are computing the NFV (net future value) given the value of the IRR (internal rate of return). In the second case we are computing IRR given the value of NFV. We will also consider two common estimates for the IRR of a scheme which often show up on the FM exam.

Notation

We consider a situation in which an investor makes deposits into an investment in amounts $C_0, C_1, C_2, \ldots, C_n$ at times $0, 1, 2, \ldots n$. If $C_t > 0$ there is a net cash flow into the investment; if $C_t < 0$ there is a net cash flow out of the investment. We assume $C_0 > 0$ and $C_n < 0$. This is a reasonable assumption – the investment scheme begins with a deposit and ends with a withdrawal. We can have $C_t = 0$ as well – no activity during that period. We will still need to account for that period in our calculations by entering $C_t = 0$.

We define a second sequence of terms representing withdrawals or returns on the investment. We will denote these as R_t and use the definition

$$R_t = -C_t \qquad (6.1)$$

If both a contribution and a withdrawal are made at the same time, we offset them and report the net contributions/withdrawal. For example if a contribution of $5,000 is accompanied by a withdrawal of $1,000 we have $C_t = \$4,000$ and $R_t = -\$4,000$.

Example 6.1 Consider a ten-year project in which the investor contributes $10,000 at inception and $5,000 in year 2. The project has expenses of $1,000 at the beginning of each remaining year of the project. The project is expected to generate income of $8,000 in year 6, $9,000 in year 7 and so on, ending with $12,000 in year 10. Table 6.1 is a chart illustrating this investment scheme:

TABLE 6.1

Year	Contibutions	Returns	Net Return
0	$ 10,000	$ -	$ (10,000)
1	$ 5,000	$ -	$ (5,000)
2	$ 1,000	$ -	$ (1,000)
3	$ 1,000	$ -	$ (1,000)
4	$ 1,000	$ -	$ (1,000)
5	$ 1,000	$ -	$ (1,000)
6	$ 1,000	$ 8,000	$ 7,000
7	$ 1,000	$ 9,000	$ 8,000
8	$ 1,000	$ 10,000	$ 9,000
9	$ 1,000	$ 11,000	$ 10,000
10	$ -	$ 12,000	$ 12,000
Total	$ 23,000	$ 50,000	$ 27,000

We now consider the question of computing the internal rate of return for such an investment scheme. We are not assuming that the account has earned a constant rate of interest. Rather, we are seeking to find a single rate, i, which would produce the same present value for the scheme.

If we assume a constant interest rate of i, we can compute the net present value of the investment as the present value of the net returns:

$$PV(i, \{R_t\}) = \sum_{t=0}^{n} v^t R_t \qquad (6.2)$$

Note that the net present value depends on both i and the values of R_t. We could also define net present value in terms of the C_t

$$PV(i, \{C_t\}) = -\sum_{t=0}^{n} v^t C_t \qquad (6.3)$$

The yield rate of an investment or internal rate of return (**IRR** in **TI BA II Plus** notation) is defined as the solution to $PV(i, \{R_t\}) = 0$. It is the rate at which the present value of returns from the investment is equal to the present value of the contributions into the investment.

Yield Rate of an Investment Scheme: The Value of i such that

$$\sum_{t=0}^{n} v^t R_t = -\sum_{t=0}^{n} v^t C_t = 0 \tag{6.4}$$

The idea is that we withdraw the final amount in the account so that the ending balance (FV) is 0. That number is R_n.

For **Example 6.1** Equation 6.4 becomes:

$$1000(-10 - 5v - v^2 - v^3 - v^4 - v^5 + 7v^6 + 8v^7 + 9v^8 + 10v^9 + 12v^{10}) = 0$$

We can use Excel to see how the net return $= PV(i)$ varies as i is changed. The Excel worsheet below computes the NPV based on a user-entered value of i. As the value of i increases, the net present value decreases. Table 6.2 displays the results for $i = 3\%$

TABLE 6.2

		i	v		
		3.00%	0.970873786		
Year	Contibutions	Returns	Net Return	Account Balance	NPV
0	$ 10,000	$ -	$ (10,000)	$ 10,000	$ (10,000)
1	$ 5,000	$ -	$ (5,000)	$ 15,300	$ (4,854)
2	$ 1,000	$ -	$ (1,000)	$ 16,759	$ (943)
3	$ 1,000	$ -	$ (1,000)	$ 18,262	$ (915)
4	$ 1,000	$ -	$ (1,000)	$ 19,810	$ (888)
5	$ 1,000	$ -	$ (1,000)	$ 21,404	$ (863)
6	$ 1,000	$ 8,000	$ 7,000	$ 15,046	$ 5,862
7	$ 1,000	$ 9,000	$ 8,000	$ 7,497	$ 6,505
8	$ 1,000	$ 10,000	$ 9,000	$ (1,278)	$ 7,105
9	$ 1,000	$ 11,000	$ 10,000	$ (11,316)	$ 7,664
10	$ -	$ 12,000	$ 12,000	$ (23,655)	$ 8,929
Total	$ 23,000	$ 50,000	$ 27,000		$ 17,602

Table 6.3 displays the formulas used to create Table 6.2.

TABLE 6.3

		0.03	=1/(1+C2)		
Year	Contibutions	Returns	Net Return	Account Balance	NPV
0	10000	0	=C4-B4	10000	=D4*v^A4
=A4+1	5000	0	=C5-B5	=(1+i)*E4-D5	=D5*v^A5
=A5+1	1000	0	=C6-B6	=(1+i)*E5-D6	=D6*v^A6
=A6+1	1000	0	=C7-B7	=(1+i)*E6-D7	=D7*v^A7
=A7+1	1000	0	=C8-B8	=(1+i)*E7-D8	=D8*v^A8
=A8+1	1000	0	=C9-B9	=(1+i)*E8-D9	=D9*v^A9
=A9+1	1000	8000	=C10-B10	=(1+i)*E9-D10	=D10*v^A10
=A10+1	1000	9000	=C11-B11	=(1+i)*E10-D11	=D11*v^A11
=A11+1	1000	10000	=C12-B12	=(1+i)*E11-D12	=D12*v^A12
=A12+1	1000	11000	=C13-B13	=(1+i)*E12-D13	=D13*v^A13
10	0	12000	=C14-B14	=(1+i)*E13-D14	=D14*v^A14
Total	23000	50000	=C15-B15		=SUM(F4:F14)

If $i = 13\%$, the net return is negative as shown in Table 6.4.

TABLE 6.4

		i	v		
		13.00%	0.884955752		
Year	Contibutions	Returns	Net Return	Account Balance	NPV
0	$ 10,000	$ -	$ (10,000)	$ 10,000	$ (10,000)
1	$ 5,000	$ -	$ (5,000)	$ 16,300	$ (4,425)
2	$ 1,000	$ -	$ (1,000)	$ 19,419	$ (783)
3	$ 1,000	$ -	$ (1,000)	$ 22,943	$ (693)
4	$ 1,000	$ -	$ (1,000)	$ 26,926	$ (613)
5	$ 1,000	$ -	$ (1,000)	$ 31,427	$ (543)
6	$ 1,000	$ 8,000	$ 7,000	$ 28,512	$ 3,362
7	$ 1,000	$ 9,000	$ 8,000	$ 24,219	$ 3,400
8	$ 1,000	$ 10,000	$ 9,000	$ 18,367	$ 3,385
9	$ 1,000	$ 11,000	$ 10,000	$ 10,755	$ 3,329
10	$ -	$ 12,000	$ 12,000	$ 153	$ 3,535
Total	$ 23,000	$ 50,000	$ 27,000		$ (45)

Since $PV(i)$ is a continuous function which is positive at $t = .03$ and negative at $i = .03$. The Mean Value Theorem ensures us that it must have at least one root in the interval $[.03, .13]$.

Using Goal Seek, we find that this investment scheme has an internal rate of return of $i = 12.96\%$. Table 6.5 is the table which resulted from using Goal Seek to set $NPV = 0$, by varying the value of i.

TABLE 6.5

Year	Contibutions	Returns	12.96% Net Return	0.885278642 Account Balance	NPV
0	$ 10,000	$ -	$ (10,000)	$ 10,000	$ (10,000)
1	$ 5,000	$ -	$ (5,000)	$ 16,296	$ (4,426)
2	$ 1,000	$ -	$ (1,000)	$ 19,408	$ (784)
3	$ 1,000	$ -	$ (1,000)	$ 22,923	$ (694)
4	$ 1,000	$ -	$ (1,000)	$ 26,893	$ (614)
5	$ 1,000	$ -	$ (1,000)	$ 31,378	$ (544)
6	$ 1,000	$ 8,000	$ 7,000	$ 28,444	$ 3,370
7	$ 1,000	$ 9,000	$ 8,000	$ 24,130	$ 3,409
8	$ 1,000	$ 10,000	$ 9,000	$ 18,257	$ 3,395
9	$ 1,000	$ 11,000	$ 10,000	$ 10,623	$ 3,340
10	$ -	$ 12,000	$ 12,000	$ 0	$ 3,548
Total	$ 23,000	$ 50,000	$ 27,000		$ (0)

6.1.1 Using the TI BA II Plus Cash Flow Worksheet to Compute Internal Rate of Return

Recall that **TI BA II Plus** Cash Flow worksheet is accessed by the CF key (it's not a 2nd function). You then enter in each cash flow (using plus signs for contributions and negative signs for withdrawals). When all the deposits/withdrawals have been entered, the keystrokes IRR, CPT will compute the internal rate of return. Make sure you enter 2nd QUIT when you are finished.

We compute the internal rate of return for Example 6.1. For convenience, the table of values is repeated below (Table 6.6).

TABLE 6.6

Year	Contibutions	Returns	Net Return
0	$ 10,000	$ -	$ (10,000)
1	$ 5,000	$ -	$ (5,000)
2	$ 1,000	$ -	$ (1,000)
3	$ 1,000	$ -	$ (1,000)
4	$ 1,000	$ -	$ (1,000)
5	$ 1,000	$ -	$ (1,000)
6	$ 1,000	$ 8,000	$ 7,000
7	$ 1,000	$ 9,000	$ 8,000
8	$ 1,000	$ 10,000	$ 9,000
9	$ 1,000	$ 11,000	$ 10,000
10	$ -	$ 12,000	$ 12,000
Total	$ 23,000	$ 50,000	$ 27,000

To begin enter CF, 2ND, CLR WORK to clear out any previous entries in the CF worksheet.

The display will show CFo = 0. CFo is the initial cash flow (at time $t = 0$, hence CFo). If there is no cash flow the first period, enter 0. We have 10,000 as CFo so press 10000 and then ENTER. Each entry in this worksheet must be confirmed by use of the ENTER key. The calculator will display CFo = 10,000.

Each subsequent contribution has both an amount (CO1, CO2, etc.) and a frequency (FO1, FO2, etc.). This saves time when the same amount is deposited/withdrawn for two or more consecutive periods. The frequency represents the number of periods its associated deposit/withdrawal is made.

These entries are accessed using the ↓ key. You can retrace your steps using the ↑ key. C01 represents the deposit/withdrawal at time $t = 1$. Pressing the ↓ once after entering CFo = 10,000 will result in a display of CO1 = 0.00000. In our example, this amount is 5,000, so we press 5,000 and then enter. The display now shows CO1 = 5,000. We use the ↓ to get to F01 = 1.00. This is the default setting and represents that fact that the deposit of $5,000 is only made at the end of period 1. That is correct for this example, so we don't need to change it.

We now use the ↓ key to enter each subsequent cash flow and its associated frequency. We must account for each year, so if there is no deposit or withdrawal in a period it is entered as 0. The deposit at the end of period 2 (CO2) is repeated in the next three periods. We thus enter $CO2 = 1,000$ and $FO2 = 4$.

Table 6.7 is a chart showing the values which are entered into the **TI BA II Plus** CF worksheet for this example. Note that each entry must be followed by pushing the ENTER key.

TABLE 6.7

Symbol	Amount	Frequency
$CF0$	$10,000	
$CO1$	$5,000	1
$CO2$	$1,000	4
$CO3$	$-$7,000	1
$CO4$	$-$8,000	1
$CO5$	$-$9,000	1
$CO6$	$-$10,000	1
$CO7$	$-$12,000	1

To find the IRR (value of i for which $PV(i) = 0$), we push the IRR key and then the CPT button. In this case a value of 12.96% is returned.

Note 1: We could have done this problem by entering the net returns. In that case the first six numbers are negative (net returns) and the last five are positive.

Note 2: If your IRR seems incorrect, press CF to return to the top of the worksheet and then use the ↓, ↑ keys to scroll through the numbers (and frequencies) you entered. You can correct them and then hit IRR, CPT at any point. It is recommended that you make a chart like the above so you can make sure you don't miss any entries. When using the CF worksheet you must account for each period of the scheme. The total of the entries in the Frequency column should equal the length of the scheme. In this case we have $1 + 4 + 1 + 1 + 1 + 1 + 1 = 10$.

6.2   Uniqueness of the Yield Rate

Unfortunately, the solution to $P(i) = 0$ need not be unique. If the investment lasts for n periods $P(i)$ is a polynomial in v of degree n. A polynomial of degree n can have as many as n distinct real roots. While it seems counterintuitive, it is possible for an investment scheme to yield a polynomial $P(i)$ which has multiple solutions, each of which yields a "reasonable" value for i. Here is an example of that situation.

Example 6.2 At what interest rate is the sum of the present value of payments of $100 at time 0 and $132 at time 2 equal to the present value of $230 at time 1?

Solution: Here we have

$$P(i) = 132v^2 - 230v + 100 = 0$$
$$v = \frac{10}{11} \text{ or } v = \frac{5}{6}$$

We used the quadratic formula to find v. Since $i = \frac{1}{v} - 1$, $i = \frac{1}{10}$ or $\frac{1}{5}$. That means the IRR for this scheme is either .1 or .2!

TI BA II Plus keystrokes. After this example, the ↓ *Enter* keystrokes won't be indicated (Table 6.8).

TABLE 6.8

Time	Deposit/Withdrawal	Amount	Symbol	Frequency
0	CFo	100 Enter ↓		
1	C01	−230 Enter ↓	F01	1 Enter ↓
2	C02	132 Enter ↓	F01	1 Enter
	IRRCPT = 10			

We observe that the **TI BA II Plus** only finds one solution (10%) and doesn't inform us that there may be other possible values.

Fortunately multiple solutions only occur in rare cases. Sometimes you can eliminate one or more values by observing that the interest rate should satisfy $-1 \le i$. As a percentage, that means the TI result should satisfy $-100 \le IRR$.

That doesn't help in Example 6.2 as both results are plausible. What went wrong? The answer is that this is an artificial example. The two values we found for $i = .1$ and $i = .2$. In the first case, the $100 deposit will be worth $110 at the end of year 1. At that point $230 is withdrawn, resulting in a negative balance ($110 - 230 = -\$120$). In the case $i = .2$ the balance at the end of year 1 is also negative: $-\$110$.

Negative balances in the middle of an investment schemem are rare. Eliminating that possibility also eliminates the possibility of multiple values for the internal rate of return. Suppose that B_t represents the balance of the account at the end of period t. Thus B_0 would be the amount in the account at inception, B_1 would be the amount in the account at the end of the first period, etc.

Theorem 6.1: If $B_t > 0$ for all t there is at most one solution to the equation $PV(i) = 0$ with $i > -1$.

Proof: We compute the value B_t be the outstanding balance at time t in terms of the values of C_t and i where i is a solution to

$$\sum_{t=0}^{n} C_t v^t = 0.$$

We have

$$B_0 = C_0$$
$$B_t = B_{t-1}(1 + i) + C_t \tag{6.5}$$

Assume we have a second solution, j with $j > i > -1$. Since $i > -1, 1 + i > 0$. We are solving

$$\sum_{t=0}^{n} C_t v^t = 0$$

Multiplying both sides of the equation by $(1 + i)^n$ gives us

$$\sum_{t=0}^{n} C_t (1 + i)^{n-t} = 0$$
$$C_0(1 + i)^n + C_1(1 + i)^{n-1} + \cdots + C_{n-1}(1 + i) + C_n = 0 \tag{6.6}$$

We now compute the values of B_t using the fact that $1 + i > 0$

$$B_0 = C_0 > 0$$

$$B_1 = B_0(1+i) + C_1 > 0$$
$$B_2 = B_1(1+i) + C_2 > 0$$

$$\vdots$$

$$B_{n-1} = B_{n-2}(1+i) + C_{n-1} > 0$$
$$B_n = B_{n-1}(1+i) + C_n > 0 \qquad (6.7)$$

We now compute a closed form solution for B_n by working our way down the list of equations above.

$$B_1 = B_0(1+i) + C_1 = C_0(1+i) + C_1$$
$$B_2 = B_1(1+i) + C_2$$
$$= (C_0(1+i) + C_1)(1+i) + C_2$$
$$= C_0(1+i)^2 + C_1(1+i) + C_2$$
$$B_3 = B_2(1+i) + C_3$$
$$= C_0(1+i)^3 + C_1(1+i)^2 + C_2(1+i) + C_3 \qquad (6.8)$$

Continuing in this manner we obtain a closed form expression for B_n which is, by assumption equal to 0.

$$B_n = 0 = \sum_{t=0}^{n} C_i(1+i)^{n-t} \qquad (6.9)$$

Suppose now that we have a second solution $j > 1$. We will denote the successive balances with an interest rate of j by B'_t. We must then have $B'_n = 0$. We must also have $1 + j > 1 + i$. We now show that $B'_t > B_t$ for all $t > 0$ using the relationships in Equation 6.6

$$B'_0 = C_0 = B_0$$
$$B'_1 = B'_0(1+j) + C_1 > B_0(1+i) + C_1 = B_1$$
$$B'_2 = B'_1(1+j) + C_2 > B_1(1+i) + C_2 = B_2$$

$$\vdots$$

$$B'_{n-1} = B'_{n-2}(1+j) + C_{n-1} > B_{n-1} = B_{n-2}(1+i) + C_{n-1}$$
$$B'_n = B'_{n-1}(1+j) + C_n > B_{n-1}(1+i) + C_n = B_n = 0 \qquad (6.10)$$

The last line tells us the $B'_n > 0$. However, by assumption, j is a yield rate and so $B'_n = 0$. That is our contradiction! We conclude that there is at most one possible yield rate if we assume that the account balance is positive throughout the life of the investment.

Example 6.3 Table 6.9 is an investment scheme with the account balance column added. The interest rate is 5%

TABLE 6.9

	i	v			
	5.00%	0.952380952			
Year	Contibutions	Returns	Net Return	Account Balance	NPV
0	$ 5,000	$ -	$ (5,000)	$ 5,000	$ (5,000)
1	$ 5,000	$ -	$ (5,000)	$ 10,250	$ (4,762)
2	$ 10,000	$ -	$ (10,000)	$ 20,763	$ (9,070)
3	$ 3,000	$ -	$ (3,000)	$ 24,801	$ (2,592)
4	$ 1,000	$ -	$ (1,000)	$ 27,041	$ (823)
5	$ 1,000	$ -	$ (1,000)	$ 29,393	$ (784)
6	$ 1,000	$ 9,000	$ 8,000	$ 22,862	$ 5,970
7	$ 1,000	$ 10,000	$ 9,000	$ 15,005	$ 6,396
8	$ 1,000	$ 10,000	$ 9,000	$ 6,756	$ 6,092
9	$ 1,000	$ 11,000	$ 10,000	$ (2,907)	$ 6,446
10	$ -	$ 12,000	$ 12,000	$ (15,052)	$ 7,367
Total	$ 29,000	$ 52,000	$ 23,000		$ 9,241

At 5% the investment shows a negative account balance in the last two years. This doesn't mean that we don't get a unique vale for i. We need to check the account balances when i is equal to the internal rate of return. We use the **TI BA II Plus** to find one value for the IRR as show in Table 6.10.

TABLE 6.10

Time	Deposit/Withdrawal	Amount	Symbol	Frequency
0	CFo	5,000		
1	C01	5,000	F01	1
2	C02	10,000	F02	1
3	C03	3,000	F03	1
4,5	C04	1,000	F04	2
6	C05	−8,000	F05	1
7,8	C06	−9,000	F06	2
9	C07	−10,000	F07	1
10	C08	−12,000	F08	1
	$IRRCPT = 10.655728562$			

Pushing IRR and CPT yields $IRR = 10.66$. We use Excel to compute the account balance at each period using this interest rate. As the chart below indicates, the account balance is positive throughout the life of the investment scheme; hence, the returned value of IRR is unique. See Table 6.11.

TABLE 6.11

		i	v			
		10.6573%	0.903691108			
Year	Contibutions	Returns	Net Return	Account Balance	NPV
0	$ 5,000	$ -	$ (5,000)	$ 5,000	$ (5,000)
1	$ 5,000	$ -	$ (5,000)	$ 10,533	$ (4,518)
2	$ 10,000	$ -	$ (10,000)	$ 21,655	$ (8,167)
3	$ 3,000	$ -	$ (3,000)	$ 26,963	$ (2,214)
4	$ 1,000	$ -	$ (1,000)	$ 30,837	$ (667)
5	$ 1,000	$ -	$ (1,000)	$ 35,123	$ (603)
6	$ 1,000	$ 9,000	$ 8,000	$ 30,866	$ 4,357
7	$ 1,000	$ 10,000	$ 9,000	$ 25,156	$ 4,430
8	$ 1,000	$ 10,000	$ 9,000	$ 18,837	$ 4,003
9	$ 1,000	$ 11,000	$ 10,000	$ 10,844	$ 4,020
10	$ -	$ 12,000	$ 12,000	$ (0)	$ 4,359
Total	$ 29,000	$ 52,000	$ 23,000		$ 0

The internal rate of return can be negative. That means the investor lost money on the scheme.

Example 6.4 An investment scenario with a negative yield rate is displayed in Table 6.12.

TABLE 6.12

Time	Symbol	Amount	Frequency
0	CFo	20,000	
1	C01	5,000	$FO1 = 1$
2	$C02$	10,000	$FO2 = 1$
3	$CO3$	3,000	$FO3 = 1$
4, 5	$CO4$	1,000	$FO4 = 2$
6	$CO5$	$-8,000$	$FO5 = 1$
7, 8	$CO6$	$-9,000$	$FO6 = 2$
9	$CO7$	$-4,000$	$FO7 = 1$
10	$CO8$	$-5,000$	$FO8 = 1$
	$IRR\,CPT = -1.99941966$		

The internal rate of return provides a way of comparing two investment schemes. The scheme with the higher internal rate of return did "better" than the other scheme.

6.3   Reinvestment

In many cases, an investor will reinvest some or all of the proceeds from an investment. The rate earned by the reinvested funds may or may not be the same as that earned by the original investment. For example, a bond might provide payments each month. Those payments could then be reinvested. The rate earned on the reinvested funds will generally not be the same as the rate earned by the bond.

Notation

We will use i for the rate earned by the original investment and j for the rate earned by reinvested funds.

Single Investment with Proceeds Reinvested

Suppose we make an initial investment of R at rate i and then reinvest all proceeds (as they are received) at a rate of j for n periods. Since we are re-investing the principal at each stage, its value does not change. As a result, the accumulated value of this investment is R plus the accumulated value of all reinvestments. Since we reinvest all proceeds, we are reinvesting Ri each time. The reinvestments thus comprise a level-term annuity at an interest rate of j and will accumulate to $(Ri)s_{\overline{n}|,j}$. The total accumulated value is then

$$R + (Ri)s_{\overline{n}|,j} \tag{6.11}$$

Example 6.5: \$5,000 is deposited in an account which pays interest at a nominal annual rate of 12% compounded quarterly. At the end of each quarter the accumulated interest is withdrawn and placed in a second account which pays interest at nominal annual rate of 15% compounded monthly. What is the accumulated value at the end of 10 years?

Solution: Each quarter the first account returns an interest payment .03 $\times$ 5000 = 150. This amount is then invested into the second account resulting in a ten-year level payment annuity with payments each quarter. Hence $N = 40$, $PMT = 150$. We want FV, the accumulated value. Since the interest rate for the second account is compounded monthly, we need to convert it to its equivalent quarterly rate;

$$\left(1 + \frac{.15}{12}\right)^3 - 1 = .037970703 \times 100 = 3.797070313\%$$

Now we use the TVM keys (Table 6.13).

The total value is $13,590.27 + 5,000 = \$18,590.27$.

If $i = j$ Equation 6.11 reduces to $R(1 + i)^n$ as we would expect.

TABLE 6.13

N	I/Y	PV	PMT	FV
40	3.797070313	0	150	$CPT = 13,590.27$

Equation 6.11 applies to a single investment of R at time $t = 0$. Now suppose that we invest R at the end of each of n periods, earning interest i and reinvest the proceeds at rate j. At the end of period 0 we will have R. That will earn Ri in interest which we reinvest at j and thus earn $R \cdot i \cdot j$ by the end of period 1. During the next period we will have $2Ri$ invested at j, then $3Ri$ invested at j and so forth. The reinvestment income is thus a P-Q annuity with $P = Ri$ and $Q = Ri$. Recall that P is the initial payment and Q is the increament in the size of the payment at each period.

Since the first deposit is made at the end of period 1, the first interest payment does not occur until the end of period 2. Hence the interest payments are delayed one period and there are only $(n - 1)$ increments, but a total of n terms.

Since the interest is withdrawn at each stage, the n payments of R do not accumulate value and so total nR at the end. This is added to the accumulated value of an increasing annuity with initial value Ri and a constant increment Ri. Recall that $(Is)_{\overline{n},j}$ is the symbol for the accumulated value of an annuity with initial payment of \$1 which increases by \$1 each period.

Accumulated Value of an investment of R at the end of each period for n periods earning interest at i with the interest reinvested at j.

$$nR + (Ri)(Is)_{\overline{n-1},j} = R\left(n + i\frac{s_{\overline{n},j} - n}{j}\right) \tag{6.12}$$

Since these calculations are usually done on the **TI BA II Plus** or the **TI30XS**, it is useful to write Equation 6.12 in a slightly different form:

Calculator Version of Equation 6.12

$$Rn\left(1 - \frac{i}{j}\right) + Rs_{\overline{n},j} \cdot \left(\frac{i}{j}\right) \tag{6.13}$$

If payments are made at the beginning of each period we obtain

Accumulated Value of an investment of R at the beginning of each period for n periods earning interest at i with the interest reinvested at j.

$$R\left(n + i\frac{s_{\overline{n+1}|,j} - (n+1)}{j}\right) \qquad (6.14)$$

Example 6.6 Payments of \$5,000 are made at the end of each month for ten years. Interest is paid at a nominal 6% compounded monthly. The earned interest is reinvested at a nominal 8% compounded monthly. What is the accumulated value of this fund?

Solution: We have $n = 120$, $i = \frac{.06}{12} = .005 = .5\%$, $j = \frac{.08}{12} = .0066666 = .6666\%$. $\frac{i}{j} = \frac{.06}{.08} = .75$, $1 - \frac{i}{j} = .25$

We compute $Ris_{\overline{n}|,j}$ using the TVM keys. We have $Ri = 25$. The result is displayed in Table 6.14.

TABLE 6.14

N	I/Y	PV	PMT	FV
120	.66666	0	−25	$CPT = 4573.65$

We multiply by $\frac{i}{j} = .75$: to obtain 3,430.24, **STO 1**. We now compute

$$Rn = 5000 * 120 = 600,000.$$

Total: \$603,430.24.

6.4   Interest Measurement of a Fund

In this section we consider the problem of computing the yield rate earned by an investment fund which has a term of one year. We will consider two methods of coming to a "reasonable" estimated value for the yield rate by finding an approximate solution to Equation 6.4.

We begin with the following terms for an investment fund over t periods in which the net deposits are made at times $0, t_1, t_2, t_3, \ldots, t_n = t = 1$ (Table 6.15).

TABLE 6.15

Symbol	Definition
A	The amount in the fund at inception. Hence: $A = B_0$
B	The amount in the fund at termination $B = B_{t_n} = B_1$
n	Number of times money is invested or withdrawn during the life of the investment fund
B_{t_i}	Net amount in the fund at time t_i. We then have: $B_0 = A$, $B_{t_n} = B_1$, $t_0 = 0$, $t_n = t$
I	The amount of interest earned by the fund over the life of the investment (not the rate, the $ amount!)
C_{t_i}	Net principal deposited at time t_i Hence, $B_0 = C_{t_0}$
C	The total net amount of principal contributed during the period $C = \sum_{i=1}^{n} C_{t_i}$
i	Effective rate of interest earned by the fund over the period

The amount in the fund at the end of the investment scheme will be the amount at inception plus the value of all deposits/withdrawals plus the total interest earned. That results in the following basic identity:

$$B = B_t = B_{t_n} = A + C + I$$

$$= A + \sum_{i=1}^{n} C_{t_i} + I \tag{6.15}$$

We now write down an equation which (in theory) allows us to compute the effective rate of interest (also called the rate of return: using either ROR or IRR). The amount (C_t) invested at time t_i is on deposit for a period of $t - t_i$ and thus accumulates to $C_{t_i}(1+i)^{t-t_i}$ by termination Recall that $B_0 = A$. Summing these values (some of which may well be negative) gives us

Formula to Compute B_t

$$B_t = B = \sum_{i=0}^{n} C_{t_i}(1+i)^{t-t_i} \tag{6.16}$$

If we know the values of B and C_{t_i} we can (in theory) solve for the value of i. We begin with an example where i is known.

Example 6.7 An account is opened with a deposit of $550 and earns interest $i = 3\%$ throughout its life. At the end of the first year, $100 is withdrawn, at the end of the second year $300 is withdrawn. At the end of the third year, a deposit of $50 is made. The account is closed (balance of $0) at the end of the fourth year. What is the amount of the closing withdrawal?

Solution. We have $B_0 = 555$, $C_1 = -100$, $C_2 = -300$, $C_3 = +50$ and want to calculate B_4

$$B_4 = 550(1.03)^4 - 100(1.03)^3 - 300(1.03)^2 + 50(1.03)$$
$$= \$242.99$$

In this section we are interested in the problem of finding i given the values of B_t. Note that the cash flow worksheet will provide an exact value in most cases. We will now consider two estimates for IRR which are sometimes used. In both cases we assume the the term of the scheme is exactly one year. These two estimates are highly likely to appear on the FM exam.

Our second example is a reminder of how the Cash Flow (CF) worksheet handles these types of problems.

Example 6.8 Table 6.16 is record of an investment fund over the course of one year with an initial balance of $10,000 and an ending balance of $10,176.22. The new symbols are attached to the numbers they represent.

TABLE 6.16

Date	Net Deposit	t_i	$1 - t_i$
January 1	$10,000 = A$	0	$1 - 0 = 1$
April 1	$\$1,500 = C_{\frac{3}{12}}$	$\frac{3}{12}$	$1 - \frac{3}{12} = \frac{9}{12}$
September 1	$-1,000 = C_{\frac{8}{12}}$	$\frac{8}{12}$	$1 - \frac{8}{12} = \frac{4}{12}$
January 1	$-10,176.22 = B$	1	$1 - 1 = 0$

Solution Using **Equation 6.16** we obtain:

$$10,176.22 = (1+i)^1 \cdot 10000 + (1+i)^{\frac{9}{12}} \cdot 1500 - (1+i)^{\frac{4}{12}} \cdot 1000$$

We can use the cash flow (CF) worksheet to compute the IRR with i convertible monthly as shown in Table 6.17.

TABLE 6.17

Time	Deposit/Withdrawal	Amount	Symbol	Frequency
0	CFo	10,000		
1, 2	C01	0	F01	2
3	C02	1,500	F02	1
4,5,6,7	C03	0	F03	4
8	C04	$-1,000$	F04	1
9,10,11	C05	0	F05	3
12	C06	$-10,176.22$	F06	1
	$IRRCPT = -.25350766$			

The scheme lost 2.53% per month. Note that the total of the frequencies is 12. We press the I/Y and exit the CF worksheet to compute the effecitive annual rate of interest (Table 6.18):

TABLE 6.18

N	I/Y	PV	PMT	FV
12	−.25350766	1	0	$CPT = -.96999967$

N	I/Y		PV	PMT	FV
1	$CPT = -3.00003270$				

6.4.1 Dollar-Weighted Estimate for the IRR

In the case of short investment cycles (almost always one year), it is possible to get a fairly reasonable estimate of the value of the IRR by assuming that the account earns simple interest rather than compound interest. In that case, we use $(1 + (t - t_i)i)$ rather than $(1+i)^{t-t_i}$ to compute the accumulated value of the contribution C_{t_i}. In the most common case $t = 1$ (so that the investment scheme has a duration of 1 period), resulting in an accumulation factor of $(1+(1-t_i)i)$. We are thus assuming that $(1+i)^{1-t_i} \approx (1+(1-t_i)i)$. The expression on the right is the first order Taylor Approximation to the function on the left. We will consider this approximation in more detail in a later chapter.

With this approximation Equation 6.16 becomes

$$B \approx \sum_{i=0}^{n} C_{t_i}(1 + (1 - t_i)i)$$

$$= \sum_{i=0}^{m} C_{t_i} + \sum_{i=0}^{n} C_{t_i} \cdot (1 - t_i) \cdot i \tag{6.17}$$

Solving this for i results in the Dollar-Weighted Estimate for the internal rate of return.

Approximation of IRR Assuming Simple Interest and a Period of 1 Dollar-Weighted Method

$$i \approx \frac{B - \sum_{i=0}^{n} C_{t_i}}{A + \sum_{i=1}^{n} C_{t_i} \cdot (1 - t_i)} \tag{6.18}$$

This is easier to remember in the form:

$$i \approx \frac{\text{Final Balance} - \text{Sum of Payments (withdrawals)}}{\text{Starting Balance} + \text{Estimated Interest Earned}} \tag{6.19}$$

The numerator of Equation 6.18 is the dollar value of the interest earned during the period in question and is ofter referred to as I. Using A for the initial balance and C for the net of all deposits and withdrawals $(\sum_{i=0}^{n} C_{t_i})$ we can write this in two other ways

$$i \approx \frac{B - A - C}{A + \sum_{i=1}^{n} C_{t_i} \cdot (1 - t_i)} \tag{6.20}$$

$$i \approx \frac{I}{A + \sum_{i=1}^{n} C_{t_i} \cdot (1 - t_i)} \tag{6.21}$$

Example 6.9 Use Equation 6.18 to obtain the dollar-weighted estimate of the rate of return for Example 6.8.

Solution: The chart for this scheme in displayed as Table 6.19.

TABLE 6.19

Date	Net Deposit	t_i	$1 - t_i$
January 1	$10,000 = A$	0	$1 - 0 = 1$
April 1	$\$1,500 = C_{\frac{3}{12}}$	$\frac{3}{12}$	$1 - \frac{3}{12} = \frac{9}{12}$
September 1	$-1,000 = C_{\frac{8}{12}}$	$\frac{8}{12}$	$1 - \frac{8}{12} = \frac{4}{12}$
January 1	$-10,176.22 = B$	1	$1 - 1 = 0$

We compute i using Equation 6.18

$$i \approx \frac{10176.22 - 10000 - 1500 + 1000}{10000 + 1500 \cdot \frac{9}{12} - 1000\frac{4}{12}} = -.03$$

Since the exact value we obtained earlier is -3.00003270, the dollar-weighted estimate is very accurate for this example.

The expression $\sum_{i=0}^{n} C_{t_i} \cdot (1 - t_i)$ is called the **exposure associated with** i.

Example 6.10 Compute the exact IRR and the dollar-weighted estimate of the IRR for the investment scheme below in Table 6.20.

TABLE 6.20

Date	Net Deposit	t_i	$1 - t_i$
January 1	$10,000 = A$	0	$1 - 0 = 1$
April 1	$8,000 = C_{\frac{3}{12}}$	$\frac{3}{12}$	$1 - \frac{3}{12} = \frac{9}{12}$
September 1	$-1,000 = C_{\frac{8}{12}}$	$\frac{8}{12}$	$1 - \frac{8}{12} = \frac{4}{12}$
January 1	$-19,000 = B$	1	$1 - 1 = 0$

Solution:

We use the cash flow (CF) worksheet to compute the IRR using months as the unit for t and hence computing effective monthly interest rate. We must then convert this to an effective annual rate or a nominal annual rate. See Table 6.21.

TABLE 6.21

Time	Deposit/Withdrawal	Amount	Symbol	Frequency
0	CFo	10,000		
1,2	C01	0	F01	2
3	C02	8,000	F02	1
4,5,6,7	C03	0	F03	4
8	C04	$-1,000$	F04	1
9,10,11	C05	0	F05	3
12	C06	$-19,000$	F06	1
	$IRRCPT = 1.01100448$			

Press I/Y and exit the CF worksheet to compute the effective annual rate of interest (Table 6.22):

TABLE 6.22

N	I/Y	PV	PMT	FV
12	1.01100448	1	0	$CPT = -1.1283$

The effective annual interest rate is 12.83%. See Table 6.23.

We now use Equation 6.18 to compute the dollar-weighted estimate:

$$i \approx \frac{19,000 - 10000 - 8000 + 1000}{10000 + 8000 \cdot \frac{9}{12} - 1000\frac{4}{12}} = .1276 = 12.76\%$$

In this case, the estimate is still fairly accurate, but not as close to the exact value as in Example 6.9.

TABLE 6.23

N	I/Y	PV	PMT	FV
1	$CPT = 12.83$			

We can get an even simpler formula if we assume that the total net contribution, C, occurs at $t = \frac{1}{2}$. In this case, equation 6.16 simplifies to

$$
\begin{aligned}
i &\approx \frac{B - C}{A + \frac{1}{2}C} \\
&= \frac{B - C}{\frac{2A+C}{2}} \\
&= \frac{2(B - C)}{2A + C}
\end{aligned}
\tag{6.22}
$$

Example 6.11 Compute the estimated value of the IRR for Example 6.9 and for Example 6.10 using formula 6.22

Solution:

Example 6.9: We have $B = 10,176.22$, $C = 10,000+1,500-1,000 = 10,500$, and $A = 10,000$. This gives us

$$
i \approx \frac{2(10,176 - 10,500)}{2 \cdot 10,000 + 10,500} \approx -.02123
$$

We obtained $i = -.0300003270$ for the exact value and $i \approx -.03$ for the dollar-weighted estimate and $i \approx -.02123$ for simple method. Clearly this is not a very good estimate.

Example 6.10: We have $B = 19,000$, $C = 10,000 + 8,000 - 1,000 = 17,000$ and $A = 10,000$. We obtain

$$
i \approx \frac{2(19,000 - 17,000)}{2 * 10,000 + 17,000} = .108
$$

The exact value was $i = 12.83\%$, the dollar-weighted estimate was $i \approx 12.76\%$ and our short method gives $i \approx 10.8\%$.

6.4.2   Time-Weighted Estimate for the IRR

We again assume a total time of $t = 1$. The **time-weighted rate of interest** is computed by solving for the interest earned over each period t_k using the formula

$$
(1 + i)^{t_k - t_{k-1}}(B'_{k-1} + C_{k-1}) = B'_k
\tag{6.23}
$$

In this equation, B'_k = Balance in the account just before the kth deposit/withdrawal is made and as before C_k = Amount of deposit/withdrawal in the kth period. If no interest were earned, we would have $B'_k = B'_{k-1} + C_{k-1}$. Equation 6.23 computes the interest earned as the ratio of the ending balance to the starting balance for this period. To see this, divide by $B'_{k-1} + C_{k-1}$ to obtain

$$(1+i)^{t_k - t_{k-1}} = \frac{B'_k}{B'_{k-1} + C_{k-1}} \tag{6.24}$$

The period in question runs from t_{k-1} to t_k.

We multiply these terms together to obtain:

$$1 + i = (1+i)^{t_1}(1+i)^{t_2 - t_1} \cdots (1+i)^{1-t_{n-1}}$$

$$= \frac{B'_1}{B_0 + C_0} \times \frac{B'_2}{B_1 + C_1} \times \cdots \frac{B}{B_{n-1} + C_{n-1}} \tag{6.25}$$

This is not as complicated as it seems. The first equation is true since the sum of all the exponents is 1 (the total period of the scheme is 1).

$$(t_1 + (t_2 - t_1) + (t_3 - t_2) + \cdots (1 - t_{n-1})) = 1$$

The numerator of each fraction in the second line of the equation is the end balance for each period, while the denominator is the beginning balance for that same period. Making a table of balances is the best way to solve these kinds of problems. In some cases a table will be provided, but it may or may not contain all the information you need. It may also contain uneeded information.

Example 6.12 Find the time-weighted rate of return for the investment described in Table 6.24

TABLE 6.24

Activity	Balance Before	Balance After
Deposit	0	50000
Withdraw	51000	46000
Deposit	46500	47500
Withdraw	50000	0

Solution: Using Equation 6.25 we obtain

$$1 + i = \frac{51000}{50000} \cdot \frac{46500}{46000} \cdot \frac{50000}{47500} \approx 1.085$$

Hence, $i \approx .085$. Notice that no dates are given – they aren't needed. In many cases dates will be provided as a distraction.

Example 6.13 On January 1, an investment has a value of $100,000. On May 1, the value has increased to $112,000 and $30,000 is deposited. On November 1, the value has declined to $125,000 and $42,000 is withdrawn. On January 1 a year later, the account is again worth $100,000. Find the yield rate using

1) The dollar-weighted and 2) the time-weighted method. Compute the exact IRR using the CF worksheet.

Solution: This is a typical exam problem. It's up to us to create the tables with the needed information. Its easier to make one table for each method.

   1) Dollar-Weighted Method.

   We first make a table of deposits and withdrawals by date as shown in Table 6.25:

TABLE 6.25

Date	Net Deposit	t_i	$1 - t_i$
January 1	$100,000 = A$	0	$1 - 0 = 1$
May 1	$30,000 = C_{\frac{4}{12}}$	$\frac{3}{12}$	$1 - \frac{4}{12} = \frac{2}{3}$
November 1	$-42,000 = C_{\frac{10}{12}}$	$\frac{8}{12}$	$1 - \frac{10}{12} = \frac{1}{6}$
January 1	$100,000 = B$	1	$1 - 1 = 0$

We find an approximate solution using Equation 6.17

$$i \approx \frac{100,000 - 100,000 - 30,000 + 42,000}{100,000 + \frac{2}{3}(30,000) - \frac{1}{6}(42,000)}$$

$$= \frac{12,000}{113,000} = .1062 = 10.6\%$$

   2) Time-Weighted Method. We first construct a chart like the one in Example 6.12. We omit the dates because they are not needed for the time-weighted method. See Table 6.26.

TABLE 6.26

Activity	Start		End
Open			100,000
Deposit	112,000	30,000	142,000
Withdrawal	125,000	-42000	83,000
Close Account	100,000		100,000

Now we can proceed just as in Example 6.12

$$1 + i \approx \frac{112,000}{100,000} \cdot \frac{125,000}{142,000} \cdot \frac{100,000}{83,000} = 1.1879$$

This gives us a value for i of 18.79%. The two methods result in dramatically different estimates of the IRR or yield rate. In this case the dollar weighted estimate is much more accurate as we can see by computing the exact value of the IRR using the **TI BA II Plus**. As usual, we first clear the CF registers: CF 2ND CLR WORK. Now we enter our data. See Table 6.27.

TABLE 6.27

Time	Deposit/Withdrawal	Amount	Symbol	Frequency
0	CFo	100,000		
1,2,3	C01	0	F01	3
4	C02	30,000	F02	1
5,6,7,8,9	C03	0	F03	5
10	C04	$-42,000$	F04	1
11	C05	0	F05	1
12	C06	$-100,000$	F06	1
	$IRRCPT = .84493341$			

Transfer this value to the TVM keys to compute the effective annual rate of interest. See Table 6.28.

TABLE 6.28

N	I/Y	PV	PMT	FV
12	.894493341	1	0	$CPT = -1.10623909$

N	I/Y	PV	PMT	FV
1	$CPT = 10.6239090950$			

Summary: The exact value is $i = 10.624\%$. The dollar-weighted estimate is $i \approx 10.6\%$. The time-weighted estimate is $i \approx 18.79\%$.

Example 6.14 A record of Funds A and B is displayed in Tables 6.29 and 6.30. On January 1 each fund was opened with \$5,000. The dollar-weighted rate of return of Fund A is equal to the time-weighted rate of return for Fund B. Find x.

Fund A

TABLE 6.29

Date	Balance	Deposit	Withdrawal
7/1	5,125		x
10/1	5,025	$2x$	
12/30	5,300		

Fund B

TABLE 6.30

Date	Balance	Deposit	Withdrawal
7/1	5,300		x
12/30	5,010		

Solution Note that the opening balance of \$5,000 is not in either table. For
Fund A we have (Table 6.29).

$$i = \frac{5300 + x - 2x - 5000}{5000 - x \cdot \frac{1}{2} + 2x \cdot \frac{1}{4}} = \frac{300 - x}{5000 - x \cdot \frac{1}{2} - 2x \cdot \frac{1}{4}} = \frac{300 - x}{5000}$$

For Fund B we have (Table 6.30)

$$1 + i = \frac{5300}{5000} \cdot \frac{5010}{5300 - x}$$

We thus have

$$\frac{5300}{5000} \cdot \frac{5010}{5300 - x} - 1 = \frac{300 - x}{5000}$$

Cross-multipling yields a quadratic in x. The positive solution is \$147.04. This
example serves as a reminder that you will need to be able to solve quadratic
equations to succeed on the FM exam.

Example 6.15 A record of two funds is displayed in Tables 6.31 and 6.32.
For Fund A, the dollar-weighted return was 0%. Calculate the time-weighted
return for Fund B

Fund A

TABLE 6.31

Date	Balance	Deposit	Withdrawal
1/1	5,000		
7/1	5,125		x
10/1	5,025	$2x$	
12/31	5,300		

Fund B

TABLE 6.32

Date	Balance	Deposit	Withdrawal
1/1	5,000		
7/1	5,300		x
12/31	5,100		

Solution Since the dollar-weighted return for Fund A is 0 the numerator in
Equation 6.18 must be 0. So we have $5300 - 5000 - 2x + x = 0$. Hence $x = 300$.
We now compute the time-weighted rate of return for Fund B using $x = 300$.

$$\frac{5300}{5000} \cdot \frac{5100}{5300 - 300} - 1 = .0812$$

The time-weighted rate of return for Fund B is 8.12%.

6.5   Selling Loans

Banks and other lending institutions often sell off loans. For accounting purposes, they need to know the yield rate for such transactions.

Example 6.16 A loan of $45,000 is being amortized with equal monthly payments made at the end of each year for ten years. Payments are computed at a nominal annual interest rate of 18%. The original lender sells the loan just after inception at a price which yields the purchaser 12%. What is the price?

Solution: We can compute the payment needed to amortize the loan (Table 6.33):

TABLE 6.33

N	I/Y	PV	PMT	FV
10	18	45,000	$CPT = -10,013.15886$	0

The original lender is thus selling an annuity which consists of ten payments of $10,013.16. The present value of this annuity will be the price to be charged. This, of course, depends on the interest rate chosen. In this case the original lender is selling it based on an interest rate of 12% per year. To compute the price we merely change the value of I/Y and then compute the new PV (price) as shown in Table 6.34.

TABLE 6.34

N	I/Y	PV	PMT	FV
	12	$CPT = 56,576.58$		

Sometimes loans are sold after several payments have been made. In this case, the yield rates for the seller and the purchaser need not be equal.

Example 6.17 A loan of $500,000 is being paid off with annual payments over twenty years. The payments are computed using an annual interest rate of 12%. Just after the tenth payment the original lender sells the loan to an investor at a price which yields an effective annual rate of 14% for the investor. What is the effective annual yield to the original lender for this transaction?

Solution: Let the effective annual yield rate be i. The original lender paid (loaned) $500,000 in return for ten payments under the original agreement plus the price the investor pays to purchase the loan just after the tenth payment.

We first compute the payment needed to amortize the loan and then compute the price paid by the investor (Table 6.35).

TABLE 6.35

N	I/Y	PV	PMT	FV
20	12	500,000	$CPT = -66,939.39$	0

The original lender will receive ten payments of \$66,939.39 plus the (future) selling price of the loan.

To compute the selling price to be paid by the investor, we change the term of the loan to $20 - 10 = 10$ and the interest rate to 14% and then compute the PV (Table 6.36).

TABLE 6.36

N	I/Y	PV	PMT	FV
10	14	$CPT = 349,163.60$		

At a yield rate of i, we want the original loan value of \$500,000 to equal the present value of the \$349,163.60 which is to be received at $t = 10$ plus the present value of the ten payments of \$66,939.39.

$$500,000 = 66939.39 a_{\overline{10}|,i} + 349163.599 \cdot v^{10}$$

This can be solved using the TVM keys. If you have just completed the last step the PMT is still stored and does not need to be entered again. Press FV to transfer the balance and change the sign (Table 6.37).

TABLE 6.37

N	I/Y	PV	PMT	FV
10	$CPT = 11.6385$	500,000		$-349,163.60$

The yield rate for the original lender is 11.64%. The yield rate for the investor is 14%.

Example 6.18 Bob borrows \$50,000 from Big Olde Bank and agrees to pay off the loan with 10 years of monthly payments based on an effective annual interest rate of 8%. Just after the twentieth payment Bob negotiates a lump sum payment with the bank of \$47,000 to clear the loan. What is the effective annual interest yield rate for the bank?

Solution: We first need to convert the 8% effective annual interest rate to a monthly rate as shown in Table 6.38.

TABLE 6.38

N	I/Y	PV	PMT	FV
12	$CPT = .6434$	1	0	-1.08

The loan payments are then computed using the TVM keys as shown in Table 6.39.

TABLE 6.39

N	I/Y	PV	PMT	FV
120	.643403	50000	$CPT = -599.29$	0

Big Olde Bank will recieve twenty payments of $599.29 plus the lum sum payment of $47,000. We compute the yield rate using the same technique as for the previous example. See Table 6.40.

TABLE 6.40

N	I/Y	PV	PMT	FV
20	$CPT = .92406$			$-47,000$

This is a monthly rate, so we must convert it to an effective annual interest rate. We can use the TVM keys as shown in Table 6.41.

TABLE 6.41

N	I/Y	PV	PMT	FV
12		1	0	$CPT = -1.116700599$

N	I/Y	PV	PMT	FV
1	$CPT = 11.67$			

6.6 Investment Year and Portfolio Methods

In some situations the interest earned by an account depends on the year in which the investment is first made (called the **new money rate** for that year). If the funds are left to accumulate, they earn a different interest rate each year for a specified number of years. At that point the **portfolio rate** is paid. If funds are withdrawn and redeposited they earn the new money rate each year. There are thus three possible scenarios:

P: Investment Year Method – interest rates depend on the first year an investment is made.

The rate in the first year is called the new money rate.

We assume no new investment or reinvestment of these funds.

Q: The portfolio method.

All investors earn the same rate of interest (the portfolio rate), which can also depend on date of first investment.

After a specified number of years all investments earn the portfolio rate.

R: Investment Year Method with Reinvestment.

The interest is based on the new money rate for each year.

Table 6.42 displays is an example. All accounts are transferred to the portfolio rate five years after the initial investment. The first column (i_1^y) is called the new-money rate for year y.

TABLE 6.42

Year	Investment Year Rate					Portfolio Rate	Portfolio Year
y	i_1^y	i_2^y	i_3^y	i_4^y	i_5^y	i^{y+5}	$y + 5$
2005	4.5	6.6	5.4	5.4	7.1	7.23	2010
2006	5.1	6.5	3.6	6.3	6.5	6.2	2011
2007	9.0	6.4	5.2	7.0	6.2	5.35	2012
2008	8.0	6.3	5.4	5.2	7.1	4.5	2013
2009	6.3	5.6	6.1	4.8	8.1	6.1	2014
2010	6.5	4.2	6.7	4.9	6.8		
2011	7.5	8.5	6.2	5.2			
2012	8.2	8.1	6.4				
2013	4.5	7.3					
2014	6.2						

The symbol i_k^y represents the interest rate paid to an investment made in year y during the kth period. If $y = 2008$, the entries in the 2008 row represent the rates earned in years 2008, 2009, etc., assuming the money is left in the account during that entire period. Starting in 2013 investments made in 2008 are transferred to the portfolio account and earn 4.5% and 6.1% in years 2013 and 2014. The table is et up to display interest rates through 2014 only. If a deposit is made in 2008 and then withdrawn and redeposited in 2010, it will earn interest of 8.0% and 6.3% in years 2008 and 2009. In year 2010 it will earn 6.5% (the new money rate for 2010).

Example 6.19 Suppose $1,000 is deposited in year 2010. Compute the accumulated value after three years using each method.

Solution:

P: We use the rates displayed in the 2010 row

$$1000(1.065)(1.042)(1.067) = 1,184.08$$

Q: We use the portfolio rates for years 2010, 2011, and 2012 which are in the next to the last column:

$$1000(1.073)(1.062)(1.035) = 1,179.41$$

R: We go down the second (new money) column starting with 2010:

$$1000(1.065)(1.075)(1.082) = 1,238.75$$

Example 6.20 Using the data in Table 6.42 compute the accumulated value of a \$4,500 investment made in year 2005 which is left to accumulate for seven years. Assume the investment year rates are paid for five years after which the account is switched to the applicable portfolio rates.

Solution: Investment year rates are paid in years 2005, 2006, 2007, 2008, 2009. For the final two years the portfolio rate applies. We go across the 2005 row and then down the portfolio column starting with the 2010 rate.

$$4500(1.045)(1.066)(1.054)^2(1.071)(1.0733)(1.062) = 6,792$$

Example 6.21 What is the yield rate for the transaction in Example 6.17?

Solution: As shown in Table 6.43 \$4,500 grows to \$6,792 in seven years:

TABLE 6.43

N	I/Y	PV	PMT	FV
7	$CPT = 6.057$	4,500	0	$-6,792$

The yield rate is 6,057%.

Summary of Formulas
Reinvestment Formulas
Accumulated Value of an investment of R at the end of each period for n periods
earning interest and i with the interest reinvested at j.

$$FV = Rn\left(1 - \frac{i}{j}\right) + Rs_{\overline{n}|,j} \cdot \left(\frac{i}{j}\right) \tag{6.26}$$

Accumulated Value of an investment of R at the beginning of each period for n periods earning interest at i with the interest reinvested at j.

$$R\left(n + i\frac{s_{\overline{n+1}|,j} - (n+1)}{j}\right) \tag{6.27}$$

Approximations to IRR $(t = 1)$
Dollar Weighted Method

$$i \approx \frac{B - \sum_{i=0}^{n} C_{t_i}}{A + \sum_{i=1}^{n} C_{t_i} \cdot (1 - t_i)} \tag{6.28}$$

Time-Weighted Method

$$1 + i = \frac{B'}{B_0 + C_0} \times \frac{B'_2}{B_1 + C_1} \times \cdots \frac{B}{B_{n-1} + C_{n-1}} \tag{6.29}$$

Exercises for Chapter 6

1. Use the **TI BA II Plus** CF worksheet to compute the IRR for the investment scheme shown in Table 6.44.

TABLE 6.44

Year	Contributions	Returns
0	$20,000	$0
1	$5,000	$0
2	$10,000	$0
3	$3,000	$0
4	$1,000	$0
5	$1,000	$0
6	$1,000	$9,000
7	$1,000	$10,000
8	$1,000	$15,000
9	$1,000	$20,000
10	$0	$5,000
	$44,000	$59,000

2. Find the internal rate of return for an investment scheme with $C_o = 3000$, $C_1 = 1000$, $R_1 = 2000$, and $R_2 = 4000$.

3. An investor deposits $5,000 to open an account. Two years later $1,000 is withdrawn. A year later $2,000 is deposited. After five years the account is closed with a withdrawal of $7,000. What is the internal rate of return for this scheme?

4. WaMu Mutual wishes to pay off a $10,000 loan which is being amortized with twenty annual payments at 12% effective annual interest rate. In order to do so, they must offer a yield rate of 15%. What is the sale price?

5. A loan of $10,000 is being paid off with twenty annual payments of $1,000 at the end of each year. The payments are immediately reinvested at 5% effective annual interest rate. What is the effective annual interest rate earned over the twenty-year period?

6. Marge agrees to contribute $7,000 immediately and additional $1,000 at the end of two years in return for $4,000 at the end of one year and $5,500 at the end of three years. Find:

   a) $P(.09)$

   b) $P(.10)$

   c) Make an argument that a solution to $P(i) = 0$ exists with $.09 \le i \le .1$. Find it. Is it unique?

7. $1,000 is deposited at the beginning of each year for five years in a fund which earns 5%. The interest from this fund can be reinvested, but earns only 4%. Show that the accumulated value of the account at the end of ten years (not payments made after five years) is:

$$1250 \cdot (s_{\overline{11}|,.04} - s_{\overline{6}|,.04} - 1)$$

Hint: Compute the value as if the deposits had continued and subtract the value of the phantom deposits.

8. A fund earning 4% effective annual interest rate has a balance of $1,000 at the beginning of the year. $200 is added to the fund at the end of three months and $300 is withdrawn at the end of nine months. Find the ending balance (at the end of one year) using the simple interest approximation.

9. On 1/1/2007, an account had a balance of $10,210. Activity for this account consisted of a deposit of $4,000 on March 1, 2007, a withdrawal of $3,000 on June 1, 2007, and a deposit of $1,000 on December 1, 2007. At the end of the year the balance was $12,982.

   a) Compute the annual effective yield using the approximate dollar-weighted method.

   b) Find the equation of value for this problem.

   c) Solve the equation in b) and compare your answer to that in a).

10. A stock portfolio had the following activity.

    1/1/2004   Value = $100,000

    1/1/2005   Value = $115,000 immediately preceding a deposit of $18,000

    1/1/2006   Value = $145,000 immediately preceding a deposit of $23,000

    1/1/2007   Value = $185,000 immediately preceding a withdrawal of $x

    1/1/2008   Value = $100,000

    The portfolio had a time-weighted rate of return for the period of 10%. What is the value of x?

11. On 1/1/2006, Anthony deposits $90 into an investment account. On 4/1/2006 when the acount in the account is $x, a withdrawal of $W is made. The dollar-weighted rate of return using Equation 6.16 is 20%. The time-weighted rate of return is 16%. Find W and x.

12. Tables 6.45 and 6.46 display two funds. The time-weighted return for Fund 2 is the same as the dollar-weighted return for Fund 1. Find the value of F.

Fund 1

TABLE 6.45

Date	Deposits	Withdrawals	Value of fund before deposits/withdrawals
1/1/06			50,000
3/3/07			55,000
5/1/07	24,000		50000
11/1/07		36,000	77,310
12/31/07			43,100

Fund 2

TABLE 6.46

Date	Deposits	Withdrawals	Value of fund before deposits/withdrawals
1/1/07			100,000
7/1/07		15,000	105,000
12/31/07			F

13. $1,000 is deposited into a fund on 1/1/06. An additional deposit of $x is made on 7/1/06. On 1/1/07 the balance in the fund is $2,000. The time-weighted estimate for the yields is 10% and the dollar-weighted estimate of the yield is 9%. Find the effective annual interest rate for the first six months.

Hint. Find x using the dollar-weighted yield.

14. On 1/1/06 an investment account is worth $300,000. M months later, the value has increased to $315,000 and $15,000 is withdrawn. 2M months prior to the end of the year, the account is again worth $315,000 and an additional $15,000 is withdrawn. On 12/31/06 the account is worth $315,000. The effective annual yield is 16% using the approximate dollar-weighted method. What is M?

15. Bert invests $1,000 at the end of each year for five years at an effective annual interest rate of 9% and reinvests the interest at an effective annual interest rate of 9%. Call the value of Bert's account at the end of five years X. Ernie invests $1,000 at the end of each year for five years into an account paying interest at 10% effective annual interest rate. He reinvests the interest in an account which pays 8% effective annual interest rate. Call the value of Ernie's account Y.

Find $Y - X$.

16. The present value of $100 today, $200 in a year, and $100 in two years is $375. What is the effective annual interest rate of interest?

17. Joe's retirement scheme at work pays $500 at the end of each month. Joe puts his money in an account which earns a nominal 12% converted monthly, the interest is reinvested at a nominal 4% converted monthly. Carol's account also pays $500 at the end of each month, but she earns a nominal 12% convertible monthly (principal and interest both earn 12%). After twenty years Joe and Carol retire. How much more money will Carol have?

18. Same Scenario as in 17). How long will it be before Carol's account exceeds Joe's' by $1,000,000?

19. Bob loses a bundle in Vegas and sells his retirement account to the casino. His account pays $5,000 at the end of each year for twenty years. Bob would have received the accumulated value of the account at the end of twenty years. The money is re-deposited at 5% with the interest earning 4%. The kindly folk at the casino pay Bob $26,233 for this account. What was the rate of return for the casino?

20. An investment has a value of $110,000 at the start of the year. On 6/30 of that year the value of the fund has declined to $95,000 and a deposit of $55,000 is made. At the end of the year the account is worth $180,000.

a) Compute the time-weighted estimate for the IRR of this investment.

b) Compute the dollar-weighted estimate for the IRR of this investment.

21. A $20,000 loan is to be repaid with three annual payments at the end of each year of $7,500. What is the yield to the investor?

22. Marge finances her house through Big Olde Bank. She borrows $450,000 and will repay the loan with monthly payments at the end of each month for 15 years. Payments will be based on a nominal annual interest rate of 8% per year compounded monthly. Just after Marge makes the final payment for year 4, the bank sells the mortgage to Mortgage Grabbers, LLC at a price which yields Mortgage Grabbers a nominal annual rate of 6%. What is the effective annual interest rate for Big Old Bank for this transaction?

Use the table of interest rates in Table 6.47 for problems 23 and 24.

TABLE 6.47

Year	Investment Year Rate					Portfolio Rate	Portfolio Year
y	i_1^y	i_2^y	i_3^y	i_4^y	i_5^y	i^{y+5}	$y+5$
2005	4.5	6.6	5.4	5.4	7.1	7.23	2010
2006	5.1	6.5	3.6	6.3	6.5	6.2	2011
2007	9.0	6.4	5.2	7.0	6.2	5.35	2012
2008	8.0	6.3	5.4	5.2	7.1	4.5	2013
2009	6.3	5.6	6.1	4.8	8.1	6.1	2014
2010	6.5	4.2	6.7	4.9	6.8		
2011	7.5	8.5	6.2	5.2			
2012	8.2	8.1	6.4				
2013	4.5	7.3					
2014	6.2						

23. Find the accumulated amount and the yield rate for an investment of $1,200 which is made in 2010 and allowed to accumulate for three years.

a) Using the investment year method.

b) Using the portfolio method.

c) Assuming the account and accumulated interest is withdrawn and reinvested at the end of each year.

24. A account is opened with $5,500 in 2008 and remains on deposit for six years. Find the accumulated amount and the yield rate.

Chapter 7

Bonds

7.1 Introduction

Bonds, preferred stock, and common stock are all types of intruments which are known as securities. There are three questions which will be addressed.

a) Given a desired yield rate, what price should be paid for a given security?

b) Given a purchase price, what is the yield rate for a given security?

c) What is the value of a given security at any date after its purchase?

In the case of stocks, these numbers can't be calculated directly. Only in the case of bonds it is possible to do so without recourse to stochastic[1] techniques. We will limit our discussion to the valuation of bonds.

We begin with some definitions:

A bond is an interest-bearing security which promises to pay a stated amount at one or more times in the future. There are two basic types of bonds:

TYPES OF BONDS

Zero-Coupon Bonds:

These bonds make no intermediate payments. They are redeemable for a specified amount at the end of a specified period. Zero-coupon bonds are sometimes called accumulation bonds.

Coupon Bonds:

These bonds provide a sequence of payments (just like an annuity) and can be redeemed for a specified amount at the end of a specified period.

Most bonds have a term. The bond is redeemable for an agreed-upon amount at the end of its term. The amount to be paid is called the redemption value (typically referred to as C) or the maturity value. The end date is called the **maturity date**. In some cases, the bond may be redeemed by the buyer or the seller at one or more dates prior to its maturity date. In others, it may not be possible to redeem the bond earlier or there may be a penalty if it is

---

[1]Using probability to estimate the value of the stock at some future time.

redeemed prior to its maturity date. Any earlier dates at which the bond may be redeemed by the buyer are called **redemption dates**. Thus, a five-year bond might be redeemable at the ends of years 2, 3, and 5. Years 2 and 3 are redemption dates, while year 5 is the maturity date. Bonds which the seller may redeem prior to the maturity date are called **callable bonds**.

A bond provides no additional payments after its maturity date. It is sometimes possible to renew the bond – this process is called a rollover. Rollovers are common in the case of retirement accounts, since they can, in some cases, delay tax payments. IRA accounts are a good example of using a rollover to defer taxes.

Example 7.1   A bond might sell for $1,000 and provide payments of $40 per year for ten years. At the end of the ten-year term of the bond, it can be redeemed for $1,000 or, perhaps, rolled over. Describe this bond in terms of an annuity.

Solution: The bond provides nine payments of $40 and a final payment (year 10) of $1,040 which consists of the redemption value plus the final coupon payment. It is thus a varying payment annuity priced at $1,000.

Coupon bonds pay a specified amount (the coupon payment) at regular intervals. Historically, the owner would clip off and turn in an actual coupon at the specified date and in turn be paid an agreed-upon amount. The amount of each coupon payment is usually a specified percentage of a number called the **par value of the bond**. The term **face value** is also used in this context. A bond with a par (face) value of $1,000 and a 4% annual coupon rate convertible annually would provide coupon payments of $40 at the end of each year. In some cases the coupon payments are stated directly as a fixed (or variable) amount.

We can view a coupon bond as a sort of loan being paid off like a sinking fund. The purchaser pays the seller an amount P and receives an amount C at some time in the future. The coupon payments represent the interest the seller pays the purchaser for the "loan" of P. If the coupon payment matches the prevailing interest rate the redemption value will equal the purchase price. If the payments are less than prevailing interest rate the seller will pay the buyer more than the purchase price when the bond is redeemed. In this second case we say the bond was **purchased at a discount**. Conversely, if the payments are greater that the prevailing interest rates, the seller will owe less than the purchase price at redemption time. In this case we say that the bond was **purchased at a premium**. A bond can thus be amortized – its value will change over time in a precise way. As we will see, the **TI BA II Plus** AMORT worksheet is one way of doing this.

There are three "prices" associated with a bond: its **selling price**, its **par value** (used only for determining the coupon payments), and its **redemption value**. The par value and redemption value are usually constant. In many cases the par value and the redemption value are equal. Bonds for which the par value equals the redemption value are called par value bonds. If only the par value is given in a problem, you are to assume that the bond is a par value bond and hence that the redemption value is equal to the par value.

In some cases, the redemption value and all accumulated interest are paid at maturity. Bonds of this type are referred to as **accumulation bonds** or **zero coupon bonds**. The purchaser pays a price, P and, at the maturity date receives a (presumably larger) amount C. If the bond has a period of n we can infer an interest rate by the formula:

Interest Rate Implied by a Zero Coupon Bond

$$i = \left(\frac{C}{P}\right)^{\frac{1}{n}} - 1 \tag{7.1}$$

We can also express this relationship as $(1+i)^n = \frac{C}{P}$ or

$$\frac{P}{C} = v^n \tag{7.2}$$

The value of i in Equation 7.1 is referred to as the yield rate for the bond. The expression $\frac{P}{C}$ represents the ratio of the price of a zero-coupon bond to its redemption value. As Equation 7.2 shows it can also be used to calculate the value of v^n. The values of $\frac{P}{C} = v^n$ are often quoted in financial publications.

Example 7.2 A zero-coupon bond which matures in three years at $5,000 is currently selling for $4,400. What is the implied three-year interest rate?

Solution: Using Equation 7.2 we have

$$\frac{4400}{5500} = v^3$$

Solving for i yields $i = 7.72\%$.

The most common zero coupon bonds are U.S. Government T-Bills. This rate establishes the lowest (no-risk) interest which an investor would expect to earn or a borrower would expect to pay. T-Bills will be discussed in Chapter 10. Companies with poor credit would pay substantially more to borrow funds and/or have to offer a higher interest rate on bonds they issue.

We can think of a bond as a loan made to the seller by the purchaser of the bond. The seller of a bond takes out a loan equal to the purchase price of the bond and agrees to pay off the loan via a sequence of coupon payments and one final (balloon) payment equal to the sum of the final coupon payment and the redemption value of the bond.

A coupon bond can also be viewed as a form of annuity. In return for the purchase price the purchaser is entitled to a sequence of payments (called coupon payments in this context) at specified intervals as well as final balloon payment consisting of the last coupon payment and the **redemption value** of the bond. Coupon bonds are sold as retirement instruments just like annuities.

Unless otherwise stated coupon bonds are assumed to make **two payments per year**. In addition, if the annual coupon rate is r, we use $\frac{r}{2}$ to compute the coupon payment to be made every six months. If the yield rate is given as i then $\frac{i}{2}$ is assumed to be the six-month rate. This means that we are treating the annual coupon rate and the yield rate as nominal annual rates. In some cases you might be asked to compute the effective annual yield rate. To do this compute the nominal rate and then use ICONV or the TVM keys as we have done in earlier chapters.

Example 7.3 A bond with a par value of $40,000 has a coupon rate of 6%. What is the coupon payment?

Solution: The payment frequency is not provided so by convention we assume that coupons are paid semiannually. Likewise, we assume the semiannual coupon rate is 3%. Hence the semiannual coupon payment is $1,200.

We can analyze bonds using cash flow analysis or the TVM keys. A bond requires an initial cash inflow (CF0 in **TI BA II Plus** terms) and provides a sequence of cash outflows (the coupon payments) and one final outflow equal to the redemption value of the bond plus the final coupon payment. The **TI BA II Plus** also has a BOND worksheet which can be used in some cases as well.

7.2   Basic Bond Terminology

For now we assume the following:

1. All obligations (coupon payments and final redemption value payment) will be paid on the dates specified on the bond.

2. The bond has a fixed maturity date. It can only be redeemed/called on that date.

3. The price of the bond will be computed immediately after a particular coupon payment has been made.

We begin with a long list of symbols.

Bond Terminology

P = price paid for the bond – the **purchase price**.

F = the **par value** or **face amount** of the bond. Coupon payments are computed as a percentage of this amount.

C = the **redemption value** of the bond. The amount to be paid at the redemption date of the bond.

r = the **coupon rate** of the bond. This is usually stated as a nominal annual rate. In the usual case of semiannual payments $\frac{r}{2}$ is multiplied by the par value to compute the coupon payment.

Fr = the value of each coupon payment.

n = the number of coupon payment periods from the current date to the maturity date. At issue date this would be the term of the bond.

i = the yield rate of the bond. This is interest rate at which the present value of the coupon payments added to the present value of C is equal to the price, P.

If $F = C$, the bond is called a **par value bond**. We will assume a bond is a par value bond unless different values for C for F are provided.

These values are entered into the TVM keys as indicated below. As usual, we can calculate the value of any one of these five numbers given the values of the other four. For bonds it is essential that PMT and FV be entered as negative numbers and that P be entered as a positive number. See Table 7.1 for the proper way to enter bond variables into the TVM keys.

TABLE 7.1

N	I/Y	PV	PMT	FV
n	i	P	$-Fr$	$-C$

Example 7.4: A $1,000 par value bond with a 4% annual coupon rate, annual coupon payments, a term of ten years, and a $900 redemption value sells for $950. What are the values of the variables defined above?

Solution:

$P = \$950$

$F = \$1000$

$C = \$900$ (this is entered as -900 when using the TVM keys)

$r = 4\%$

$Fr = \$40$ (this is entered as -40 when using the TVM keys)

$n = 10$

If the purchaser had put the purchase price in a savings account earning 4%, the annual interest in the first year would have been $.04 \times 950 = \$38$. Since the bond pays $40, the purchaser is earning more than 4%. That results in a redemption value of $900 which is less than the purchase price. We say that this bond was purchased at a premium of $50.

In general, we have the following two types of bonds:

Premium and Discount Bonds

If $P > C$, the bond was purchased/sold at a premium. Its book value will decline from P to C over its lifetime.

If $P < C$ the bond was purchased/sold at a discount. Its book value will increase from P to C over its lifetime.

If $P = C$ the bond is referred to as a par value bond. Its calculated value will remain constant. Note that if the prevailing interest rates change, the bond's value may also change. We will not discuss that situation.

More Bond Terminology

g = the **modified coupon rate** of the bond.

$$g = r \cdot \frac{F}{C} \tag{7.3}$$

The modified coupon rate represents the interest paid per unit of redemption value.

We also have

$$gC = Fr \tag{7.4}$$

The interest rate i is called the **yield rate** of the bond. It is also referred to as the **yield to maturity** for the bond. This is the number we called the internal rate of return (IRR) which we discussed earlier in the context of cash flow (CF) analysis. If we are using the TVM keys it is I/Y.

K = the present value, computed at the yield rate (i), of the redemption value.

$$K = Cv^n \tag{7.5}$$

G = the **base amount** of the bond.

$$G = \frac{Fr}{i} \tag{7.6}$$

Note that F, C, g, r, and n are fixed by the conditions of the bond. P and i often vary throughout the life of the bond in an inverse relationship. As the price goes up, the yield rate goes down and as the price goes down, the yield rate goes up.

Example 7.5 (previous example continued) Compute the value of g, K, and G. What is the internal rate of return? This number is also referred to as the yield rate of the bond.

Solution: $g = r \cdot \frac{F}{C} = 04 \cdot \frac{1000}{900} = 4.44\%$. To compute G and K we need the interest rate.

To compute i we can use the TVM keys or the cash flow analysis (CF) keys.

Using the CF keys: We pay \$950 now and receive nine payments of \$40, plus a final payment of \$940 (the redemption value, plus the final coupon payment). See Table 7.2.

TABLE 7.2

Time	Deposit/Withdrawal	Amount	Symbol	Frequency
0	CFo	950		
1	C01	−40	F01	9
	C02	−940	F02	1

We inter IRR, CPT, $IRR = i = 3.767402732$. The yield rate is 3.77%.

Using the TVM keys: We have an initial deposit of \$950 and receive ten payments of \$40 each. Just after the last payment, the account is worth \$900, which we enter as the FV. See Table 7.3.

Note: Both PMT and FV are entered as negative quantities. Note also that the value of FV is −900 and not −940 as the final \$40 payment is included in the PMT column.

TABLE 7.3

N	I/Y	PV	PMT	FV
10	$CPT = 3.767402732$	950	−40	−900

Now we can compute K and G:

$K = 900v^{10} = \$621.7746972$. The present value of the redemption amount of this bond is \$621.77.

$G = \frac{1000 \cdot (.04)}{.03767402732} = \$1,061.74$. This is the amount we would have to invest at the yield rate to earn the coupon payment. That is: $Gi = Fr$.

7.3 Pricing a Bond

Suppose we are given F, r,c C, i, and n. What price should be paid for this bond? The simplest method is to equate the price with the present value of all coupon payments plus the present value of the redemption value of the bond. This treats the pricing problem as a cash flow problem. Equating the present value (P) to the present value of all future payments results in a formula known as the **Cash Flow Valuation Formula**. There are three other formulas to compute the price of a bond, each of which is a modification of the cash flow formula. Here are the four ways the price of a bond can be computed. While these formulas are often useful you also need to keep in mind the most general relation between price, coupon payments, and redemption value:

Basic Bond Pricing

Price = Present Value of the Coupon Payments + Present Value of the Redemption Value

Formulas for the Valuation of a Bond

Given its Yield to Maturity, Face, and Redemption Values

1: Cash Flow Valuation Formula (the most common method used).

$$P = Fra_{\overline{n}|,i} + Cv^n = Fra_{\overline{n}|,i} + K \qquad (7.7)$$

Price = Present Value of the Coupon Payments + Present Value of the Redemption Value

2: Premium/Discount Valuation Formula

$$P = Fra_{\overline{a}|,i} + Cv^n$$
$$= Fra_{\overline{n}|,i} + C(1 - ia_{\overline{n}|,i})$$
$$= C + (Fr - Ci)a_{\overline{n}|,i} = C(1 + (g - i)a_{\overline{n}|,i}) \qquad (7.8)$$

Price = Redemption Value $\pm$ The Premium/Discount. Equation 7.8 takes this into account, but you might also be given a value of the premium/discount and need to compute the price directly.

3: Base Amount Valuation Formula

$$P = Cv^n + Fra_{\overline{n}|,i}$$
$$= Cv^n + Cg\left(\frac{1 - v^n}{i}\right)$$
$$= G(1 - v^n) + Cv^n$$
$$= G + (C - G)v^n \qquad (7.9)$$

4: Makeham Valuation Formula

$$P = Cv^n + Fra_{\overline{n}|,i}$$
$$= Cv^n + Cg\left(\frac{1 - v^n}{i}\right)$$
$$= Cv^n + \frac{g}{i}(C - Cv^n) \qquad (7.10)$$

Example 7.6 Find the price of a ten-year bond with a par value of $1,000 with coupons at 8.4% convertible semiannually and redeemable at $1,050. The bond is priced to produce a yield rate of a nominal rate of annual interest of 10% convertible semiannually.

Solution: We will compute the price using each of the four standard formulas. We first compute all the numbers we will need.

$$F = 100$$

$$C = 1050$$

$$i = \frac{.1}{2} = .05$$

$$r = \frac{.084}{2} = .042$$

$$g = \frac{1000}{1050}(.042) = .04$$

$$n = 2 \cdot 10 = 20$$

$$K = 1050 \left(\frac{1}{1.05}\right)^{20} = 395.7340$$

$$G = \frac{.042}{.05} 1000 = 840$$

Method One: Cash Flow Analysis

$$P = F \cdot r \cdot a_{\overline{n}|,i} + K$$

$$= 42 \cdot a_{\overline{20}|,.05} + 395.7340$$

$$= \$919.15$$

TI BA II Plus (Table 7.4)

TABLE 7.4

N	I/Y	PV	PMT	FV
20	5	$CPT = 919.15$	−42	−1,050

Method Two: Premium/Discount Formula

$$P = C + (Fr - Ci) \cdot a_{\overline{n}|,i}$$

$$= 1050 + (42 - 52.40) \cdot a_{\overline{20}|,.05} = \$919.15$$

Method Three: Base Amount Formula

$$G + (C - G) \cdot v^n$$

$$840 + (1050 - 840) \cdot (1.05)^{-20} = \$919.15$$

Method Four: Makeham Formula

$$P = K + \frac{g}{i}(C - K)$$

$$= 395.7340 + \frac{.04}{.05} \cdot (1050 - 395.7340)$$

$$= \$919.15$$

This bond was sold at a discount of $\$130.85 = C - P$.

Note: You can use the CF worksheet to calculate bond prices. Enter $CFo = 0$ and then the coupon payments and final redemption value plus the last coupon payment. Use NPV and enter the yield rate. See Table 7.5. Note that only 19 payments are entered. The final payment is received at the time the bond is redeemed.

TABLE 7.5

Time	Symbol	Amount	Frequency
0	$CF0$	0	
1	$C01$	-42	19
2	$C02$	-1092	1

$NPV, I = 5, CPT = 919.15$

Note that we only entered nineteen coupon payments. The final coupon payment is paid at the same time as the redemption value.

Example 7.7 An eleven-year \$9,000 8% bond is purchased for \$9,500. If the effective yield to maturity of this bond is 6.5%, what is the redemption value of the bond? Use the Cash Flow Valuation Formula and the **TI BA II Plus TVM** worksheet.

Solution. Unless otherwise stated, interest rates are assumed to be expressed as nominal annual rates. In addition, we assume payments are made twice a year unless stated otherwise in the problem.

We have $i = \frac{.065}{2} = .0325$ and $r = \frac{.08}{2} = .04$. We also have $F = 9000$, $P = \$9,500$, and $n = 22$. The semiannual coupon payment is then $9000 \cdot .04 = 360$. In eleven years we will have twenty-two semiannual payments. We are to compute the value of C, the redemption (or future) value of the bond. Our Cash Flow equation of value is:

$$9500 = 360 \cdot a_{\overline{22}|,.0325} + Cv^{22}$$

We solve for C to obtain

$$C = \frac{9500 - 360 \cdot a_{\overline{22}|,.0325}}{v^{22}} = \$8,071.75$$

TI BA II Plus TVM worksheet (Table 7.6):

TABLE 7.6

N	I/Y	PV	PMT	FV
22	3.25	9,500	-360	$CPT = -7,889.831$

This bond sold at a premium of \$1,610.15.

Using the CF worksheet as displayed in Table 7.7.

TABLE 7.7

Time	Symbol	Amount	Frequency
0	CF0	9,590	
1	C01	−360	22

NPV, $I = 3.25$, $NFVCPT = 8,071.75$. We put in all twenty-two coupon payments and compute the NFV which is the redemption value in this case.

Example 7.8 A bond with a par value of \$50,000 has 8.4% semiannual coupons and a redemption value of \$47,500. If the bond has a ten-year period and a yield rate of 7.5, compute the price.

Solution: We use the TVM keys (Table 7.8):

TABLE 7.8

N	I/Y	PV	PMT	FV
20	3.75	$CPT = 51,929.41$	−2,100	−47,500

This bond sold at a premium of \$4,429.42.

Example 7.9 A ten-year \$10,000 face value bond with annual coupons is priced at \$9,086.38. The present value of the redemption value is \$6,142.34 and the ratio of the coupon rate to the yield rate is 2/3. Find the yield rate and coupon rate.

Solution: We use the version of the Cash Flow Valuation formula which includes the present value of C.

$$P = Fra_{\overline{n}|,i} + Cv^n = Fra_{\overline{n}|,i} + K$$

We know $n = 10$, $P = 9,086.38$, $K = 6,142.34$ and $\frac{r}{i} = 2/3$. Since we don't know the values of r or i, we need to recall the definition of $a_{\overline{n}|,i} = \frac{1-v^n}{i}$. This results in the following modified version of Equation 7.7:

$$P = Fr\frac{1 - v^n}{i} + K = F\frac{r}{i}(1 - v^n) + K$$

For the numbers provided we obtain

$$9086.38 = 10000\frac{2}{3}(1 - v^{10}) + 6142.34$$

Solving for v yields $v = .94339$ and hence $i = .06$, $r = .04$. The yield rate is 6% and the coupon rate is 4%.

The modified version of Equation 7.7 is probably worth remembering on its own:

Alternate Bond Pricing Formula

$$P = F\frac{r}{i}(1 - v^n) + K \tag{7.11}$$

Example 7.10 A five-year bond has a face value of 10,000. The present value of its redemption value is \$6,000 and the ratio of the coupon rate to the yield rate is 1.2. Its price is \$9,554. Calculate the coupon and yield rates.

Solution: We have

$$9554 = 10000 \cdot 1.2 \cdot (1 - v^5) + 6000$$

Solving yields $i = 7.28\%$, $r = 8.73\%$

7.4   Premium and Discount

As we have seen, the purchase price, P, and the redemption value, C, need not be equal. Recall that if $P > C$ the bond is said to have **sold at a premium**. The **premium** is $P - C$. If $P < C$ the bond was **sold at a discount**. The **discount** is $C - P$. We can derive expressions for the premium and discount using Equation 7.8 and recalling that $Fr = Cg$.

Premium

$$P - C = (Fr - Ci)a_{\overline{n}|,i} = C(g - i)a_{\overline{n}|,i} \tag{7.12}$$

Discount

$$C - P = C(i - g)a_{\overline{n}|,i} \tag{7.13}$$

Note that in either case we have

$$P = C \cdot \left(1 + (g - i) \cdot a_{\overline{n}|,i}\right) \tag{7.14}$$

In the case of a premium we have $g > i$, while in the case of a discount, $g < i$. If a bond is purchased at a premium the value of the bond will decrease from its purchase price of P to its redemption value of C over its lifetime. If the bond is purchased at a discount, the bond's value will increase from its purchase price to its redemption value.

We can construct a **bond amortization schedule** in much the same way as we constructed an amortization schedule for loans. The purpose of a bond amortization schedule is to compute the value of the bond just after each of its coupon payments. The value computed after a given payment is called the **book value** of the bond. Note that this number need not represent the selling price of the bond at the point in time it is computed. Book value is used as an accounting tool. If the book value declines, the holder may be able to report that as a loss to defray taxes.

We assume a bond with redemption value C, price P, and yield rate i. If P were invested at i it would earn iP after one period. The coupon value is $gC = Fr$. The book value calculation allocates the amount gC between earned interest for a given period and an adjustment to the book value at the end of that period. Using Equation 7.8 we see that the interest in period 1, which we will designate as I_1 is

$$I_1 = iP = iC[1 + (g - i)a_{\overline{n}|,i}] \tag{7.15}$$

We subtract this from the total interest of gC to obtain the adjustment in the book value:

$$\begin{aligned} P_1 &= gC - iC[1 + (g - i)a_{\overline{n}|,i}] \\ &= C(g - i)(1 - ia_{\overline{n}|,i}) \\ &= C(g - i)v^n \end{aligned} \tag{7.16}$$

The new book value is then given by $P - P_1$ or

$$\begin{aligned} P - P_1 &= C(1 + (g - i)a_{\overline{n}|,i}) - C(g - i)v^n \\ &= C(1 + (g - i)a_{\overline{n-1}|,i}) \end{aligned} \tag{7.17}$$

We then continue this process in the same way as we did for loan amortization tables. Table 7.9 displays the formulas used to compute bond variables.

Bond Amortization Table

TABLE 7.9

Period	Coupon	Interest Earned (at i)	Principal Adjustment	Book Value
0				$C(1 + (g - i)a_{\overline{n}|,i})$
1	Cg	$iC[1 + (g - i)a_{\overline{n}|,i}]$	$C(g - i)v^n$	$C(1 + (g - i)a_{\overline{n-1}|,i})$
2	Cg	$iC[1 + (g - i)a_{\overline{n-1}|,i}]$	$C(g - i)v^{n-1}$	$C(1 + (g - i)a_{\overline{n-2}|,i})$
t	Cg	$I_t = iC[1 + (g - i)a_{\overline{n-t+1}|,i}]$	$P_t = C(g - i)v^{n-t+1}$	$C(1 + (g - i)a_{\overline{n-t}|,i})$
n	Cg	$iC[1 + (g - i)v]$	$C(g - i)v$	C

Bond Price Adjustment Formulas

$$I_t = iC \cdot (1 + (g - i)a_{\overline{a-t+1}|,i})$$
$$P_t = C(g - i)v^{n-t+1} \tag{7.18}$$

As with loan payments, the amount to principal forms a geometric sequence with common ratio $(1 + i)$. For a par value bond $F = C$ and so $g = r\frac{F}{C} = r$. In this case $(g - i)$ is the difference between the yield rate and the coupon rate. If F is not equal to C we have the following alternate expression for P_t

$$P_t = (Fr - Ci)v^{n-t+1} \tag{7.19}$$

We begin with a sampling of problems which require using Equation 7.18 or 7.19 in addition to the bond pricing formulas.

Example 7.11 You are given two twenty-year annual coupon par value bonds.

Bond A: Has a face value of $1,000, coupon rate of 10% and was purchased to yield 8%

Bond B: Has a coupon rate of 6% and was purchased to yield 8% as well.

In year 10 the absolute value of the change in principal is the same for each bond.

Find the price of Bond B.

Solution: Using Equation 7.18 the change in principal for Bond A is

$$1000(.1 - .08)v^{20-10+1} = 20v^{11}$$

If the face value of Bond B is F the change in principal for Bond B is

$$F(.08 - .06)v^{20-10+1} = .02Fv^{11}$$

Equating these two expressions yields $F = 1,000$. We now price the second bond using the TVM keys we obtain the result shown in Table 7.10.

TABLE 7.10

N	I/Y	PV	PMT	FV
20	8	CPT = 803.64	−60	−1000

Example 7.12 A fifteen-year bond with annual coupon payments was purchased at a premium. The principal adjustment in the second coupon was $150 and in the fourth coupon it was $160. What was the premium paid for this bond?

Solution: Since the amount to principal forms a geometric sequence with common ratio $1 + i$, we have

$$160 = 150 \cdot (1 + i)^2$$

Solving yields so $i = 3.27\%$. The total of all principal payments is equal to the premium which was paid. We are given that the adjustment in the second period was $150. Since these adjustments form a geometric sequence, the first adjustment was $\frac{150}{1+i}$. The sequence of price adjustments is then:

$$\frac{150}{1+i}, 150, 150(1+i)\cdots+\cdots 150(1+i)^{13}$$

This is a geometric sequence with $a = \frac{150}{1+i}$, $r = (1+i)$. There are fifteen terms so our sum is

$$\frac{150}{1+i}\cdot\frac{1-(1+i)^{15}}{1-(1+i)} = 2,757.28$$

The bond sold at a premium of $2,757.28. Note that you don't need to know the value of the last adjustment to compute the sum.

Example 7.13 A twenty-year annual coupon with a par value of $1,500 is purchased to yield 5%. The amount to principal in the tenth year is $30. Find the book value (price) of the bond at the end of year 11.

Solution We assume that this is a par value bond so that $C = \$1,500$. We use Equation 7.19

$$30 = (Fr - Ci)v^{20-10+1} = (Fr - 75)v^{11}$$

Solving for Fr we obtain

$$Fr = 30(1.05)^{11} + 75 = 126.31$$

At the end of year 11 the bond has nine years remaining and is priced as a nine-year bond with a yield rate of 5% and $Fr = 126.31$ as shown in Table 7.11.

TABLE 7.11

N	I/Y	PV	PMT	FV
9	5	$CPT = 1,864.70$	-126.31	$-1,500$

The price of the bond at the end of year 11 is $1,864.70.

Bond Amortization Tables

We now consider the problem of constructing a bond amortization table. This process is virtually identical to that for amoritzing a loan.

Example 7.14 Construct the amortization table for a five-year annual coupon bond with face value and redemption value of $20,000. The bond has an annual coupon rate of 8% and is purchased at a price which yields 4%.

Solution: Since $F = C$ we have $g = r = .08$.

The price paid is the computed using the **TI BA II Plus** (Table 7.12):

TABLE 7.12

N	I/Y	PV	PMT	FV
5	4	$CPT = \$23,561.46$	$-1,600$	$-20,000$

Note: Since the bond had annual coupons, we used $N = 5$ and $I/Y = 4$.

This bond was purchased at a premium of $3,561.46. See Table 7.13 for all the starting numbers. The amortization table represents a way to compute the changing value of the bond during its lifetime. It will decline in value from its original purchase price of $23,561.46 to its redemption value of $20,000 over its five-year life span. These intermediate values are referred to as the book value of the bond. To compute them, we compare the dollar value of the interest earned at the yield rate to the amount of the coupon payment. At 4%, $23,562.46 would yield interest in the first period of

$$iP = .04 \cdot 23,561.46 = \$942.46$$

The coupon pays $Fr = g \cdot C = .08 \cdot 20,000 = \$1,600$. This means that the interest (at the yield rate) does not cover the required coupon payment. The difference of 657.542 is subtracted from the purchase price, P, to obtain $P_1 = \$22,903.92$. This process continues with each subsequent payment. Table 7.14 is an Excel Spreadsheet with the complete solution to this problem.

TABLE 7.13

Length of Bond	5
Face (Par) Value	**$20,000.00**
Redemption Value	**$20,000.00**
Coupon Rate	8.00%
Yield Rate	4.00%
Price to yield i	$23,561.46
Premium/Discount	$3,561.46

BOND AMORTIZATION TABLE

TABLE 7.14

Year	Coupon Payment	Interest at Yield Rate	Change in Price	Balance
0			$23,561.46	
1	$1,600.00	$942.46	$(657.54)	$22,903.92
2	$1,600.00	$916.16	$(683.84)	$22,220.07
3	$1,600.00	$888.80	$(711.20)	$21,508.88
4	$1,600.00	$860.36	$(739.64)	$20,769.23
5	$1,600.00	$830.77	$(769.23)	$20,000.00

TI BA II Plus solution: We invoke the AMORT worksheet after computing the price using the TVM keys. We want to set P1 = 1 and P2 = 1 and then step through the table one line at a time. See Table 7.15.

TABLE 7.15

Key	Value	Key	Value
$P1$	1	$P1$	$CPT = 4\downarrow\downarrow$
$P2$	1	BAL	20,769.23
BAL	22,903.92	PRN	−739.645
PRN	−657.542	INT	−860.356
INT	−942.458	$P1$	$CPT = 5\downarrow\downarrow$
$P1$	$CPT = 2\downarrow\downarrow$	BAL	20,000
BAL	22,220.07283	PRN	−769.231
PRN	−683.84	INT	−830.769
INT	−916.16		
$P1$	$CPT = 3\downarrow\downarrow$		
BAL	21,508.88		
PRN	−711.20		
INT	−888.803		

Example 7.15 Construct the amortization table for the same bond as in Example 7.14 if the coupon rate remains 8%, but the bond is purchased at a price which yields 12%.

Solution: In this case, the bond will be purchased at a discount. See Table 7.16 for the starting numbers. The calculations are all the same, except that the entries in the Principal column are now all positive. Table 7.17 displays the Excel Spreadsheet. You should verify at least a few of these entries.

TABLE 7.16

Length of Bond	5
Face (Par) Value	**$20,000.00**
Redemption Value	**$20,000.00**
Coupon Rate	8.00%
Yield Rate	12.00%
Price to yield i	$17,116.18
Premium/Discount	$(2,883.82)

BOND AMORTIZATION TABLE

TABLE 7.17

Year	Coupon Payment	Interest at Yield Rate	Change in Price	Balance
0			$17,116.18	
1	$1,600.00	$2,053.94	$453.94	$17,570.12
2	$1,600.00	$2,108.41	$508.41	$18,078.53
3	$1,600.00	$2,169.42	$569.42	$18,647.96
4	$1,600.00	$2,237.76	$637.76	$19,285.71
5	$1,600.00	$2,314.29	$714.29	$20,000.00

TI BA II Plus solution:
We first price the bond as shown in Table 7.18.

TABLE 7.18

N	I/Y	PV	PMT	FV
5	12	$CPT = 17,116.17904$	$-1,600$	-20000

We can now amortize the bond using the AMORT worksheet by stepping through each period in the same way as Example 7.14.

Most bonds pay interest every six months (semi-annual coupons). If the interest is compounded at periods other than annually, appropriate adjustments must be made.

Example 7.16 A bond with a par value of $50,000 and a redemption value of $55,000 has a coupon rate of 6% convertible semiannually and is purchased to yield 8% converted semiannually over its eight-year life span. Construct the bond amortization table.

Solution:
We convert to six-month periods. We are allowed to assume that both interest rates are nominal annual rates. The life of the bond is sixteen the coupon rate is .03 and the yield rate is .04. We price the bond using the TVM keys as displayed in Table 7.19.

TABLE 7.19

N	I/Y	PV	PMT	FV
16	4	$CPT = 46,846.39307$	$-1,500$	$-55,000$

Table 7.20 is the amortization table for this bond.

BOND AMORTIZATION TABLE

TABLE 7.20

Year	Coupon Payment	Interest at Yield Rate	Change in Price	Balance
0			$46,843.39	
1	$1,500.00	$1,873.74	$373.74	$47,217.13
2	$1,500.00	$1,888.69	$388.69	$47,605.81
3	$1,500.00	$1,904.23	$404.23	$48,010.05
4	$ 1,500.00	$1,920.40	$420.40	$48,430.45
5	$ 1,500.00	$1,937.22	$437.22	$48,867.67
6	$1,500.00	$1,954.71	$454.71	$49,322.37
7	$1,500.00	$1,972.89	$472.89	$49,795.27
8	$1,500.00	$1,991.81	$491.81	$50,287.08
9	$1,500.00	$ 2,011.48	$511.48	$50,798.56
10	$1,500.00	$2,031.94	$531.94	$51,330.50
11	$1,500.00	$2,053.22	$553.22	$51,883.72
12	$1,500.00	$2,075.35	$575.35	$52,459.07
13	$1,500.00	$2 ,098.36	$598.36	$53,057.44
14	$1,500.00	$2 ,122.30	$622.30	$53,679.73
15	$1,500.00	$2 ,147.19	$647.19	$54,326.92
16	$1,500.00	$2 ,173.08	$673.08	$55,000.00

Example 7.17 (optional – requires sinking funds) Find the price of a $1,000 par value two-year 8% bond with semiannual coupons which is purchased at a price which will yield 6% convertible semiannually if the investor can replace the premiums by means of a sinking fund which earns 5% convertible semiannually.

Solution: Since this is a par value bond and $r < i$ it will sell at a premium. The coupon payment will be $\frac{.08}{2} \cdot 1000 = 40$, while the interest earned will be $\frac{.06}{2}P = .03P$.

Thus $40 - .03P$ is available to go into the sinking fund. This must accumulate to the premium paid for the bond, which is $P - 1000$. The sinking fund has a term of $2 \cdot 2 = 4$ six-month periods and an interest rate per period of $\frac{.05}{2} = .025$. We then have

$$(40 - .03P) \cdot s_{\overline{4}|,.025} = P - 1000$$

Solving for P gives us

$$P = \frac{1,000 + 40s_{\overline{4}|,.025}}{1 + .03s_{\overline{4}|,.025}} = \$1,036.93$$

The **TI BA II Plus** won't solve this problem directly, but it will compute the value of $s_{\overline{4}|,.025}$ as shown in Table 7.21:

TABLE 7.21

N	I/Y	PV	PMT	FV
4	2.5	0	-1	$CPT = 4.152515625$

We then compute P using the formula above.

7.5   Determination of Yield Rates

So far, we have assumed a known yield rate and computed the price. If one is contemplating purchasing a bond, the price is usually given, and the yield rate (Internal Rate of Return or IRR on the **TI BA II Plus**) must be determined. It is this number which is used to compare the relative worth of the bond in question versus other possible investments. If we had the choice of investing in the bond or putting our money in a savings account paying an effective annual interest rate of 5% we would choose the bond only if its yield rate was 5% or greater.

Computing the IRR involves solving one of Equations 7.4, 7.5, 7.6, or 7.7 for i. We will consider Equation 7.5, the premium/discount formula:

$$P = C + (Fr - Ci)a_{\overline{n}|,i}$$
$$= C + C(g - i)a_{\overline{n}|,i} \tag{7.20}$$

If we let $k = \frac{P-C}{C}$ we obtain

$$k = \frac{P - C}{C} = (g - i)a_{\overline{n}|,i}$$
$$i = g - \frac{k}{a_{\overline{n}|,i}} \tag{7.21}$$

Example 7.18 A three-year bond with face value of $10,000 and redemption value of $10,000 and an annual coupon rate of .04 is purchased for $9,500. What is the yield rate? Since we are not informed otherwise, we can assume that this is a par value bond.

Solution: Since $F = C$, we have $g = r = .04$. Also, $k = \frac{9500-10000}{10000} = -\frac{1}{20} = -.05$. We need to solve

$$i = .04 - \frac{-.05}{a_{\overline{3}|,i}}$$

TI BA II Plus solution using the TVM keys. We enter N, PV, PMT, and FV and CPT I/Y. See Table 7.22.

TABLE 7.22

N	I/Y	PV	PMT	FV
3	$CPT = 5.865910278$	9,500	−400	−10,000

TI BA II Plus solution using the CF keys:

The only tricky part is that the last coupon payment is redeemed at the same time as the bond is redeemed. Hence, in this case, there are only two regular coupon payments plus a final payment (at the end of year 3) of 10,000 + 400 = 10,400. See Table 7.23.

TABLE 7.23

Time	Deposit/Withdrawal	Amount	Symbol	Frequency
0	CFo	9,500		
1	C01	−400	F01	2
3	C02	−10,400	F02	1

IRR, CPT, IRR = 5.87

Example 7.19 In the same situation as Example 7.18, what is the yield rate if the term is increased to thirty years?

Solution:

TI BA II Plus.

Using the CF worksheet, we need only change the number of coupon payments from two to twenty-nine (again, the final payment is part of the balloon payment which consists of the coupon payment plus the redemption value). Then IRR *CPT* IRR = 4.29977.

Using the TVM keys we change N to 30 and compute the new value. See Table 7.24.

TABLE 7.24

N	I/Y	PV	PMT	FV
30	$CPT = 4.29977$	9,500	−400	−10,000

Example 7.20 In Example 7.18, what is the yield rate if the coupon rate on the bond is changed to 10% and the term of the bond is fifteen years?

Solution: This changes the annual payment from \$400 to \$1,000 and the frequency to fourteen. We must also adjust the final payment from 10,400 to 11,000 (Table 7.25).

TABLE 7.25

Time	Deposit/Withdrawal	Amount	Frequency
0	CFo	9,500	
1	C01	$-1,000$	$FO1 = 14$
	C02	$-11,000$	$FO2 = 1$

$IRR, CPT, IRR = 10.68\%.$

7.6   The Term Structure of Interest

Equation 7.5 allows us to compute any bond variable given the values of the other variables. For convenience, we repeat Equation 7.5 below:

$$P = Fr \cdot a_{\overline{n}|,i} + C \cdot v(i)^n \qquad\qquad ((7.5)7.22)$$

The **TI BA II Plus** TVM keys are used as below. Enter any four of the five and CPT the fifth. Table 7.26 displays the entry protocol for bonds.

TABLE 7.26

N	I/Y	PV	PMT	FV
n	i	P	$-Fr$	$-C$

When we compute the yield rate[2] (denoted by the letter i) we are assuming that interest rates are constant throughout the life of the bond. The yield rate is thus an average rate. In this section we consider the relationship between the length of time to maturity of a security and its yield rate. These rates are published on a regular basis. A graph of yield rate versus time to maturity is known as the **yield curve**. As a general rule yield rates increases with the length of time to maturity. For example, a five-year CD usually pays greater interest than a two-year CD. This is known as a normal yield curve. If the yield rate decreases over time the curve is referred to as inverted. Finally, if yield-to-maturity is constant the yield curve is referred to as flat.

---

[2]In this context, i is also called the yield-to-maturity.

> **Definition:** The yield-to-maturity of a zero-coupon bond purchased today which matures in a given number (n) of years is referred to as the **spot rate for year** n. Spot rates always refer to transactions which occur at time $t = 0$.

The yield curve is thus a plot of spot rates versus length of time to maturity.

A plot of the spot rates results in what is called the yield curve. Consider the set of spot rates in the Table 7.27:

TABLE 7.27

Years to Maturity	Spot Rate
1	2.3%
2	1.4%
3	3.5%
4	2.5%
5	4%

Based on the definition of a spot rate we can see (for example) that a zero-coupon bond purchased today with a term of three years would have a yield rate of 3.5%.

The yield curve for this situation is plotted as Figure 7.1.

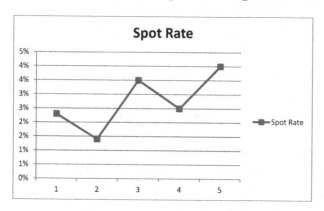

FIGURE 7.1

As a general rule spot rates increase with time to maturity. A "normal yield curve" refers to that situation. See Figure 7.2. We will discuss this further in Chapter 10.

We can view a coupon bond as a set of zero-coupon bonds which consist of the coupon payments and the redemption value. Hence spot rates can be used to price coupon bonds as well.

Example 7.21: Suppose that a bond provides coupon payments of $50 at the end of years 1, 2, and 3 and is redeemable at the end of year 3 for $550. We

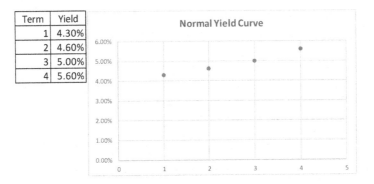

Term	Yield
1	4.30%
2	4.60%
3	5.00%
4	5.60%

FIGURE 7.2

can view this as purchasing four zero-coupon bonds with maturity dates one, two, and three years in the future. Suppose further that the spot rates (yield curve) for the next three years are as given in Table 7.28.

TABLE 7.28

Payment	Time to Maturity	Spot Rate
50	1	2.3%
50	2	1.4%
50	3	3.5%
550	3	3.5%

We compute the present value of the bond by computing the present value of each payment and the redemption value using the appropriate spot rates. Using the yield curve data provided we then obtain

$$PV = \frac{50}{1.023} + \frac{50}{1.014^2} + \frac{50+550}{1.035^3} = \$638.67$$

We can then compute the yield rate for this bond by using the TVM keys (Table 7.29):

TABLE 7.29

N	I/Y	PV	PMT	FV
3	$CPT = 3.353$	638.67	−50	−550

It is possible to buy some or all of the coupon payments for a bond. A bond which is sold in this way is referred to as a **stripped** bond. In our example the bond in question has been stripped into four zero-coupon bonds which consist of the three coupon payments plus the redemption value of the bond.

In the general case we assume each spot rate applies to the coupon payments made in that period. Hence, if a payment of B is made at the end of period t at spot rate s_t, its value at time $t = 0$ will be computed as:

$$\frac{B}{(1 + s_t)^t} = B \cdot (1 + s_t)^{-t} \tag{7.22}$$

Suppose that we have a sequence $\{s_t\}$ of spot rates, how can we adjust our cash flow bond price formula to account for this? Recall our basic pricing rubric

Price = Present Value of Payments + Present Value of the Redemption Value

Since the interest rates vary each payment must be valued using the spot rate for the period in which it was made. For the coupon paid in period k, the value at $t = 0$ is given by

$$Fr(1 + s_k)^{-k}$$

The redemption value of the bond has present value equal to $C(1 + s_n)^{-n}$. We add the values of the coupon payments to the redemption value to obtain our revised equation relating the bond variables.

$$P = \sum_{k=1}^{n} Fr(1 + s_k)^{-k} + C(1 + s_n)^{-n} \tag{7.23}$$

As we will see later in this chapter the expression $(1 + s_k)^{-k}$ is often referred to as P_k or $P(0, k)$.

The yield to maturity, i, is a single interest rate which would result in the same present value. It is thus a sort of average rate and can be computed using the TVM keys once we have computed the price of the bond. We first use Equation 7.23 to compute the price and then compute I/Y using the TVM keys.

Example 7.22 Consider the following **spot rate table** shown in Table 7.30 (also called the **Yield Curve**)

TABLE 7.30

Two – Year Yield Curve

Year	Spot Rate
1	.07
2	.09

Find the price and yield to maturity of each bond:
Bond A: A two-year bond with $F = C = 900$ and 10% annual coupons.
Bond B: A two-year bond with $F = C = 900$ and 5% annual coupons.

Solution:

Bond A: The coupon payment is \$90, so we have

$$P_A = \frac{90}{1.07} + \frac{90}{1.09^2} + \frac{900}{1.09^2} = \$917.38$$

To find the yield to maturity, we use the **TI BA II Plus** TVM keys. See Table 7.31.

TABLE 7.31

N	I/Y	PV	PMT	FV
2	$CPT = 8.9039$	917.375	−90	−900

Bond B: The coupon payment is \$45, so we have

$$P_B = \frac{45}{1.07} + \frac{45}{1.09^2} + \frac{900}{1.09^2} = \$837.44$$

The yield to maturity is 8.95%. See Table 7.32.

TABLE 7.32

N	I/Y	PV	PMT	FV
2	$CPT = 8.94874$	837.44	−45	−900

For marketing purposes, bond issues often desire that $P = F = C$. In most circumstances, the values of s_k are known (or assumed). We then set $F = C$, to find the coupon rate, r, which will result in $P = F = C$ by solving Equation 7.23 for r:

$$F = \sum_{k=1}^{n} Fr(1 + s_k)^{-k} + F(1 + s_n)^{-n}$$

$$= Fr \sum_{k=1}^{n}(1 + s_k)^{-k} + F(1 + s_n)^{-n}$$

$$\Rightarrow$$

$$r = \frac{1 - (1 + s_n)^{-n}}{\sum_{k=1}^{n}(1 + s_k)^{-n}} \tag{7.24}$$

The number we have just computed is a coupon rate, but it is also referred to as the at-par yield rate.

At-Par Yield Rate For A Bond With $P = F = C$

$$r = \frac{1 - (1 + s_n)^{-n}}{\sum_{k=1}^{n}(1 + s_k)^{-k}} \tag{7.25}$$

There is a nice interpretation of the terms in 7.26. Suppose we price a $1 zero-coupon bond with term n. We obtain $P = 1 \cdot (1 + s_n)^{-n}$. Hence the expression $(1 + s_n)^{-n}$ represents the price as a fraction of the redemption value. These numbers appear in various financial publications. The notation is: $(1 + s_k)^{-k} = P_k = P(0, k)$. Using that notation the at-par yield rate can be expressed as

$$r = \frac{1 - P_n}{\sum_{k=1}^{n} P_k} \tag{7.26}$$

Example 7.23 Compute the at-par yield rate for an annual coupon bond given the set of spot rates, as shown in Table 7.33.

TABLE 7.33
Yield Curve

Year	Spot Rate
1	.06
2	.04
3	.05

Solution: We use Equation 7.26. Note that we don't need the value of $F = C = P$.

$$r = \frac{1 - (1.05)^{-3}}{(1.06)^{-1} + (1.04)^{-2} + (1.05)^{-3}} = .0498$$

Example 7.24: Compute the at-par yield rate for the bonds in Example 7.22.

Solution: The at-par rate depends only on the yield curve

$$r = \frac{1 - 1.09^{-2}}{1.07^{-1} + 1.09^{-2}} = 8.91311279\%$$

In some cases, you will be asked to compute the spot rates given a set of bond prices.

Example 7.25 Find the spot rates implied by the prices for bonds with $F = C = 1000$ and 10% annual coupons, as shown in Table 7.34.

TABLE 7.34

Bond	Term	Price
1	1 year	1028
2	2 years	1035
3	3 years	1030

Solution: We are looking for the values of s_1, s_2, and s_3.

We begin with the one-year bond, since its price involves only s_1. We have

$$1028 = \frac{100}{1+s_1} + \frac{1000}{1+s_1} = \frac{1100}{1+s_1}$$

Hence:

$$1028(1+s_1) = 1100$$

$$1+s_1 = \frac{1100}{1028} = 1.0700389$$

$$s_1 = .0700389$$

We could have computed this value directly from the TVM keys as well as shown in Table 7.35.

TABLE 7.35

N	I/Y	PV	PMT	FV
1	$CPT = 7.004$	1,028	-100	$-1,000$

We can now use this value to find the value of s_2 using the price of the two-year bond

$$1035 = \frac{100}{1.0700389} + \frac{1100}{(1+s_2)^2}$$

Solving yields $s_2 = .0808755$. In this case we can't use the TVM keys since the interest rates vary.

Finally, we find the value of s_3 using the price of the three-year bond

$$1030 = \frac{100}{1.0700389} + \frac{100}{1.0808755607^2} + \frac{1100}{(1+s_3)^3}$$

Solving yields $s_3 = .089338$

Example 7.26 Compute that at-par yield rates for the bonds in Example 7.25.

Solution: We use Equation 7.23

One-Year Bond

$$r = \frac{1 - 1.0700389^{-1}}{1.0700389^{-1}} = .0700389 = 7\%$$

Two-Year Bond

$$r = \frac{1 - 1.0808755607^{-2}}{1.0700389^{-1} + 1.0808755607^{-2}} = .08045 = 8.045\%$$

Three-Year Bond

$$r = \frac{1 - 1.0893383939^{-3}}{1.0700389^{-1} + 1.0800755607^{-2} + 1.0893383939^{-3}} = .088256 = 8.83\%$$

The result for Bond One is not surprising. If the spot rate is constant, the coupon rate must equal the spot rate (or yield rate) in order for the bond to sell at par. Remember that we assumed $F = C$.

Example 7.27. Compute that at par yield rate for a three-year bond given the prices for zero-coupon bonds as given in Table 7.36.

TABLE 7.36

Years to Maturity (k)	$P_k = (1 + s_k)^{-k}$
1	.902
2	.897
3	.877

Solution. We are given the values $P_k = (1 + s_k)^{-k}$ and just need to insert into Equation 7.26

$$r = \frac{1 - .877}{.902 + .897 + .877} = .04596 = 4.6\%$$

NOTE: In some cases the table looks like Table 7.37.

TABLE 7.37

Years to Maturity	P_k
1	902
2	897
3	877

In this case, we are to assume that 900 represents the price of a $\$1,000$ one-year bond, 897 represents the price of a $\$1,000$ two-year bond, etc. You can divide all entries by 1,000 and then use Equation 7.26.

Value of an Annuity

We can compute the value of an annuity under the assumption of a yield curve as well. Recall that an annuity provides a sequence of payments at the start (or end) of each of a specified number of periods. If we have a set of spot rates $\{s_t\}$ and a set of payments $\{R_t\}$ made at the end of each period (an annuity-immediate) we have the following expression for the present value (sometimes called the net present value of NPV) of these payments:

$$NPV(\{s_t\}) = \sum_{t=1}^{n}(1+s_t)^{-t} \cdot R_t \qquad (7.27)$$

If the payments are constant (a level payment annuity), we can compute an average interest rate (the internal rate of return - IRR) for the annuity using the **TI BA II Plus** TVM keys as shown in Table 7.38.

TABLE 7.38

N	I/Y	PV	PMT	FV
n	CPT	NPV	−R	0

Example 7.28: An annuity provides payments of $100, $200, $400, and $500 at the end of each year. Compute the NPV of this annuity for the given yield curve. Find the value of $R for a level-payment annuity with payments of $R which has the same NPV as this annuity. See Table 7.39.

TABLE 7.39

Year	Spot Rate
1	.07
2	.04
3	.05
4	.07

Solution: To compute the NPV of the varying annuity, we use Equation 7.23

$$NPV = \frac{100}{1.07} + \frac{200}{1.04^2} + \frac{400}{1.05^3} + \frac{500}{1.07^4}$$
$$= \$1,005.3518322$$

To compute the required payment for a level-payment annuity of equal NPV solve:

$$1005.3518322 = R\sum_{i=1}^{4}(1+s_t)^{-t}$$

$$= R\left(\frac{1}{1.07} + \frac{1}{1.04^2} + \frac{1}{1.05^3} + \frac{1}{1.07^4}\right) = R \cdot 3.4858684620$$

$$R = \$288.611$$

7.7 Forward Rates

Definition: An n-year one-year forward rate, f_n is the rate agreed upon today for a one-year loan to be made n years in the future. For example, f_4 would represent the interest rate for a one-year loan to be taken out at the end of year 4 and paid off at the end of year 5. Put another way, the loan is taken out at $t = 4$ and repaid at $t = 5$.

The forward rates depend on the spot rates and vice versa. Forward rates are usually determined using the current set of spot rates. As we shall see, it is also possible to compute the spot rates given the forward rates.

Computing Forward Rates From a Yield Curve:

Suppose we have a set of spot rates $\{s_k\}$ and a set of forward rates $\{f_k\}$. The number f_k is the interest rate to be charged for a one-year loan made in period k. To compute the values of f_k implied by the set of spot rates we equate the accumulated value of \$1 under two scenarios:

1. The amount is accumulated for $k + 1$ periods at s_{k+1}

$$(1 + s_{k+1})^{k+1}$$

2. The amount is accumulated for k periods at s_k and for the last period at f_k

$$(1 + s_k)^k(1 + f_k)$$

For these rates to be consistent we must have

$$(1 + s_{k+1})^{k+1} = (1 + s_k)^k(1 + f_k)$$

Solving for f_k results in the following expression for the kth forward rate in terms of given spot rates:

Computing the Forward Rates from the Spot Rates

$$f_k = \frac{(1 + s_{k+1})^{k+1}}{(1 + s_k)^k} - 1 \qquad (7.28)$$

Note that $f_0 = s_0$ since a loan of one year made right now will pay the spot for year 1. Note also that we need s_k and s_{k+1} to compute f_k

Example 7.29 Compute the forward rates implied by the spot rates in Table 7.40.

TABLE 7.40

Year	Spot Rate
1	.05
2	.08
3	.065
4	.09

Solution: With the information provided we can compute f_1, f_2, and f_3. To compute f_4 we would need s_5.

$$f_1 = \frac{1.08^2}{1.05} - 1 = .110857 = 11.0857\%$$

$$f_2 = \frac{1.065^3}{1.08^2} - 1 = .035622106 = 3.5622\%$$

$$f_3 = \frac{1.09^4}{1.065^3} - 1 = .168576554 = 16.858\%$$

Multi-Year Forward Rates

The symbol $_m f_t$ is used to indicate the predicted (forward) interest rate for an m-year investment beginning at time t. We can compute the value of $_m f_t$ using either the one-year spot rates or the one-year forward rates. Another notation is $i_{t,t+m}$ which represents the m-year forward rate starting in year t. We will use this last notation most of the time. You will also encounter the phrase "m-year forward rate delayed (or deferred) t years," which means the same thing.

Method One: Using the Spot Rates (the yield curve)

We compute $_m f_t$, we compute the future value of $1 at time $t+m$ in two ways:

a) Using the value of s_{t+m} - the $t+m$ - spot rate
b) Using the value of s_t for the first t periods and $_m f_t$ for the final m periods

We then equate the results and solve for $_m f_t$

$$(1 + s_{t+m})^{t+m} = (1 + s_t)^t (1 +_m f_t)^m$$

$$(1 +_m f_t) = \frac{(1 + s_{t+m})^{\frac{t}{m}+1}}{(1 + s_t)^{\frac{t}{m}}}$$

$$_m f_t = \frac{(1 + s_{t+m})^{\frac{t}{m}+1}}{(1 + s_t)^{\frac{t}{m}}} - 1 = i_{t,t+m} \tag{7.29}$$

Method Two: Using the Forward Rates

The accumulated value of an investment at $_m f_t$ should equal that of an investment with forward rates (starting at time t) of $f_t, f_{t+1}, f_{t+2}, \ldots, f_{t+m-1}$:

$$(1 +_m f_t)^m = (1 + f_t)(1 + f_{t+1})(1 + f_{t+2}) \ldots (1 + f_{t+m-1})$$

$$1 +_m f_t = ((1 + f_t)(1 + f_{t+1})(1 + f_{t+2}) \ldots (1 + f_{t+m-1}))^{\frac{1}{m}}$$

$$_m f_t = i_{t,t+m} = [((1 + f_t)(1 + f_{t+1})(1 + f_{t+2}) \ldots (1 + f_{t+m-1}))]^{\frac{1}{m}} - 1$$

$$(7.30)$$

Example 7.30 For the yield curve of Example 7.29 compute $_2 f_2 = i_{2,4}$. This rate is also referred to as the two-year forward rate deferred two years.

Solution:

Method One: We have $m = t = 2$ and use Equation 7.29

$$_2 f_2 = i_{2,4} = \frac{(1 + s_4)^{\frac{2}{2}+1}}{(1 + s_2)^{\frac{2}{2}}} - 1 = 10.00925\%$$

Method Two:

$$_2 f_2 = i_{2,4} = ((1 + f_2)(1 + f_3))^{\frac{1}{2}} - 1$$

$$= (1.035622106 \cdot 1.168576554)^{\frac{1}{2}} - 1$$

$$= .1000925 = 10.00925\%$$

7.8 Converting from Forward Rates to Spot Rates

We can also convert from forward rates to spot rates.
We have:

$$(1 + s_{k+1})^{k+1} = (1 + f_k)(1 + s_k)^k \qquad (7.31)$$

Now we can apply this same result again to expand $(1 + s_k)^k$

$$(1 + s_{k+1})^{k+1} = (1 + f_k)(1 + s_k)^k = (1 + f_k)(1 + f_{k-1})(1 + s_{k-1})^{k-1} \quad (7.32)$$

Continuing in this fashion and noting that $f_0 = s_1$ we obtain the following formula

Computing Spot Rates Given Forward Rates

$$(1 + s_{k+1})^{k+1} = (1 + f_0)(1 + f_1) + \cdots (1 + f_k)$$

$$s_{k+1} = ((1 + f_0)(1 + f_1) + \cdots (1 + f_k))^{\frac{1}{k+1}} - 1 \qquad (7.33)$$

Example 7.31 Compute the spot rates corresponding to the table of forward rates given in Table 7.41.

TABLE 7.41

t	Forward Rate
0	$f_0 = s_1 = .03$
1	$f_1 = .04$
2	$f_2 = .056$
3	$f_3 = .071$

Solution:

$$s_2 = (1.03 \cdot 1.04)^{\frac{1}{2}} - 1 = .034987923$$

$$s_3 = (1.03 \cdot 1.04 \cdot 1.056)^{\frac{1}{3}} - 1 = .041945078$$

$$s_4 = (1.03 \cdot 1.04 \cdot 1.056 \cdot 1.071)^{\frac{1}{4}} - 1 = .049134064$$

$s_1 = 3.5\%$, $s_2 = 4.2\%$, $s_3 = 4.9\%$

As a check on our calculations, we compute f_3 using the computed values of s_k;

$$f_3 = \frac{(1 + s_4)^4}{(1 + s_3)^3} - 1 = \frac{1.049134064^4}{1.041945078^3} - 1 = .071$$

Alternative Notation and Calculations for Spots and Forwards

Spot and forward rates are based on published estimates from major banks. One of these is the LIBOR rate. LIBOR stands for London Inter Bank Offered Rate. These are benchmark interest rates for many adjustable rate mortgages, business loans, and financial instruments traded on global financial markets. LIBOR tables are based on the zero-risk interest rate which is defined as the yield rate of a zero-coupon bond with no default risk. These are typically U.S. Treasury Bonds. Thus

$$s_n = \text{Yield Rate of zero-risk, zero-coupon n-year bond}$$

As we saw, we can denote the price of a zero-coupon bond by $P(0, n)$ as a fraction of its redemption value to obtain

$$P(0, n) = \frac{1}{(1 + s_n)^n} \qquad (7.34)$$

NOTE: The notation $P_n = P(0, n)$ is also used.

Our basic equation for computing forward rates from a spot rate table looks like this:

$$(1 + i_{t_1,t_2})^{t_2-t_1}(1 + s_{t_1})^{t_1} = (1 + s_{t_2})^{t_2} \tag{7.35}$$

Using Equation 7.34 this simplifies to

$$(1 + i_{t_1,t_2})^{t_2-t_1} = \frac{P(0, t_1)}{P(0, t_2)} \tag{7.36}$$

For one-year forward rates, the formula is even simpler:

$$f_t = i_{t,t+1} = \frac{P(0, t)}{P(0, t + 1)} - 1 \tag{7.37}$$

Using the alternate notation $P(0, n) = P_n$ we have

$$i_{t,t+1} = \frac{P_t}{P_{t+1}} - 1 \tag{7.38}$$

Example 7.32 Given the prices shown in Table 7.42 for zero-coupon bonds[3] of the indicated durations compute $i_{3,4}$ and $i_{2,4}$.

TABLE 7.42

Duration	Price
1	95
2	93.2
3	90.7
4	88.7

Solution: We have $i_{3,4} = \frac{P_3}{P_4} - 1 = .0225 = 2.25\%$.
We compute $i_{2,4}$ using

$$(1 + i_{2,4})^2 = \frac{P_2}{P_4} = 1.0507$$

Solving yields $i_{2,4} = 2.5\%$

NOTE: When forward rates are computed based on a price table for zero-coupon bonds, the results are often referred to as the **implied forward rates**

---

[3]Note: Unless specified otherwise, the par value is assumed to be 100. In some cases prices are given as decimals. A "price" of .987 would indicate that the price is .987 · (Par Value). The **TI BA II Plus** bond worksheet uses this convention, expressing prices in terms of the par value.

for the given spot rate table. Example 7.32 could have been phrased: Find the implied forward rates $i_{3,4}$ and $i_{2,4}$ for the table below. You might also be asked to find the "implied" spot rates for a set of forward rates.

Finally, rates are sometimes reported in terms of the "Par Coupon." The Par Coupon is the coupon rate that would result in an $F = P = C$ bond. Note that for a zero-coupon band we always have $P < C$. The Par Coupon is the payment required to make these two numbers equal. In terms of the price structure $P(0,t)$ for each period we have:

$$c = \frac{1 - P(0,n)}{\sum_{t=1}^{n} P(0,t)} = \frac{1 - P_n}{\sum_{t=1}^{n} P_t} \tag{7.39}$$

Finally, the equivalent continuously compounded yield is equivalent rate of interest computed as being continuously compounded. We solve $e^r = (1 + f_t)$ and obtain

$$r = \ln(1 + f_n) \tag{7.40}$$

Example 7.33 Fill in the remaining entries in the Table 7.43.

TABLE 7.43

Years to Maturity	0-Coupon Yield	0-Coupon Price	Implied Forward	Par Coupon	Continuous Rate
1	7%	$P(0,1)$	$f_0 = i_{0,1}$	c_1	$r_{0,1}$
2	5%	$P(0,2)$	$f_1 = i_{1,2}$	c_2	$r_{0,2}$
3	6.5%	$P(0,3)$	$f_2 = i_{2,3}$	c_3	$r_{0,3}$

We have

$$P(0,1) = P_1 = \frac{1}{1.07} = .9346$$

$$P(0,2) = P_2 = \frac{1}{(1.05)^2} = .907$$

$$P(0,3) = P_3 = \frac{1}{(1.065)^3} = .828$$

We now have Table 7.44 to complete.

TABLE 7.44

Years to Maturity	0-Coupon Yield	0-Coupon Price	Implied Forward	Par Coupon	Continuous Rate
1	7%	.9346	$f_0 = i_{0,1}$	c_1	$r_{0,1}$
2	5%	.9070	$f_1 = i_{1,2}$	c_2	$r_{0,2}$
3	6.5%	.8278	$f_2 = i_{2,3}$	c_3	$r_{0,3}$

The implied forwards are computed using the prices we just computed

$$f_0 = s_1 = 7\%$$

$$f_1 = \frac{P_1}{P_2} - 1 = 3.04\%$$

$$f_2 = \frac{P_2}{P_3} - 1 = 9.56\%$$

The Par Coupon rates are computed as

$$c_1 = \frac{1 - P_1}{P_1} = 7\%$$

$$c_2 = \frac{1 - P_2}{P_1 + P_2} = 5.05\%$$

$$c_3 = \frac{1 - P_3}{P_1 + P_2 + P_3} = 6.45\%$$

Finally, we have the following values for continously compounded interest

$$r_{0,1} = \ln(1.07) = 6.77\%$$

$$r_{0,2} = \ln(1.05) = 4.9\%$$

$$r_{0,3} = \ln(1.065) = 6.3\%$$

Table 7.45 shows the final table

TABLE 7.45

Years to Maturity	0-Coupon Yield	0-Coupon Price	Implied Forward	Par Coupon	Equivalent Continuous Interest
1	7%	.9346	7%	7%	6.7659%
2	5%	.9070	3.037%	5.048%	4.879%
3	6.5%	.8278	9.565	6.449%	6.2975%

Summary of Formulas for Sport and Forward Rates
Price of a Bond

$$P = \sum_{k=1}^{n} Fr(1 + s_k)^{-k} + C(1 + s_n)^{-n} \tag{7.41}$$

At Par Yield Rate $(F = P = C)$

$$r = \frac{1 - (1 + s_n)^{-n}}{\sum_{k=1}^{n}(1 + s_k)^{-k}} = \frac{1 - P_n}{\sum_{k=1}^{n} P_k} \tag{7.42}$$

Computing a Forward Rate from Spot Rates

$$i_{k,k+1} = \frac{(1+s_{k+1})^{k+1}}{(1+s_k)^k} - 1 = \frac{P_{k+1}}{P_k} - 1 \tag{7.43}$$

Computing Multiyear Forward Rates from Spot Rates

$$i_{t,t+m} = {}_m f_t = \frac{(1+s_{t+m})^{\frac{t}{m}+1}}{(1+s_t)^{\frac{t}{m}}} - 1 \tag{7.44}$$

Computing Multiyear Forward Rates from Single Year Forward Rates

$$i_{t,t+m} = ((1+f_t)(1+f_{t+1})(1+f_{t+2})\cdots(1+f_{t+m-1}))^{\frac{1}{m}} - 1 \tag{7.45}$$

Computing Spot Rates from Forward Rates

$$s_{k+1} = ((1+f_0)(1+f_1)+\cdots(1+f_k))^{\frac{1}{k+1}} - 1 \tag{7.46}$$

Computing (implied) Forward Rates from Zero-Coupon Bond Prices

$$i_{n,n+1} = \frac{P_n}{P_{n+1}} - 1 \tag{7.47}$$

$$i_{n,m} = \left(\frac{P_n}{P_m}\right)^{\frac{1}{m-n}} - 1 \tag{7.48}$$

In some cases you will be asked to infer spot and forward rates from bond prices.

Example 7.34 The price of a two-year zero-coupon bond with par value $1,000 is $890. The price of a two-year par value bond with par value $1,000 and 5% annual coupons is $982.11. Compute $i_{1,2}$ = the one-year forward rate deferred one year.

Solution: From the first bond we have

$$\frac{1000}{890} = (1+s_2)^2 \text{ so } s_2 = 6.0\%$$

From the second bond we have

$$982.11 = \frac{50}{1+s_1} + \frac{1050}{1.06^2} \text{ so } s_2 = 5.01\%$$

Finally, we have

$$1.05(1+i_{1,2}) = 1.06^2 \text{ so } i_{1,2} = 7.0\%$$

7.9 Price of a Bond between Coupon Payments

Since bonds are traded on a daily basis, most transactions take place between coupon payments. We thus need a method to establish the price of a bound at any time prior to its call date. We assume that the interest conversion period matches the coupon payment period and seek to find the price of a bond at time $n + t$, where $0 < t < 1$. We compute P_n (the price at time n) using the earlier formulas and then accumulate this over the fractional time period t. The value obtained is referred to as the **price-plus-accrued** of the bond. As with other fractional time period problems, simple interest is sometimes used for this calculation. Here are the two formulas:

Bond Price-Plus-Accrued Using Simple Interest

$$P_{n+1} = P_n(1 + it) \tag{7.49}$$

Bond Price-Plus-Accrued Using Compound Interest

$$P_{n+1} = P_n(1 + i)^t \tag{7.50}$$

Example 7.35 A five-year bond with face and redemption values of $20,000 and a coupon rate of 8% is purchased to yield 4%. What is the price-plus-accrued value of the bond two and a half (or 30 months) years after it is purchased?

Solution: We compute the price of the bond using the TVM as shown in Table 7.46.

TABLE 7.46

N	I/Y	PV	PMT	FV
5	4	$CPT = 23,561.45$	$-1,600$	$-20,000$

After two years, there are three years remaining and the value at $N = 3$ is computed as shown in Table 7.47.

TABLE 7.47

N	I/Y	PV	PMT	FV
3		$CPT = 22,220.07$		

Using simple interest we compute the value at 2.5 as

$$P_{2.5} = P_2 \cdot (1 + .04 \cdot .5) = 22,664.47$$

Using compound interest we have

$$P_{2.5} = P_2 \cdot (1.04)^{.5} = 22,660.12$$

A second measure of the value is called simply the price. It is computed by subtracting the accrued value of the coupon payment (Frt) from the price-plus-accrued value. In this case Equation 7.50 is used for the price-plus-accrued giving us:

Price of a Bond between Coupons

$$\text{Price}_{n+t} = P_n(1+i)^t - Frt \qquad (7.51)$$

Example 7.36 Find the price of the bond in Example 7.23 at time 2.5

Solution Using Equation 7.51 we obtain

$$\text{Price}_{2.5} = 22,220.07 - 1,600 \cdot .5 = 21,420$$

7.10   Callable Bonds

A callable bond is a bond for which the seller reserves the right to pay off the bond at one or more dates prior to the end of its term. The redemption price at each **call date** is specified at the time of purchase. In many cases the redemption price is the same for all call dates. Callable bonds are a way for the seller to hedge against changes in interest rates and thus pose a greater risk to the investor. In order to evaluate the risk to the investor it is useful to be able to complete the yield rate for each of the call dates.

Suppose we have a bond with face with F, price P, and redemption value (for all call dates) of C. We assume a coupon rate of r and a period of n. We assume the bond is called just after coupon payment m. The **yield rate**, i_m is the interest rate at which the price of the bond is equal to the present value of the coupon payments (each at Fr) plus the present value of its redemption value (C). We have the following equation

$$P = Fra_{\overline{m}|,i_m} + Cv^m \qquad (7.52)$$

If $m = n$ this is just the price function. However, we are now solving for i_m rather than for P. The result of this calculation is referred to as the **yield to call** for the bond.

Example 7.37 We are given a callable four-year bond with par value of $5,000 and semiannual coupon rate of 6%. The price is $5,400 and the bond is redeemable at two, three, and four years at a redemption value of $6,000. What is the yield rate at each call date?

Solution: We are solving Equation 7.52 for i_m with $m = 4$, 6, 8 since we are counting in six-month intervals.

These are all computed using the TVM keys and solving for I/Y. At the end of year 2, $m = 4$ and we have the situation displayed in Table 7.48.

TABLE 7.48

N	I/Y	PV	PMT	FV
4	$CPT = 5.343$	5,400	-150	$-6,000$

At three years we have $m = 6$ and need only change set $N = 6$ and compute I/Y. See Table 7.49.

TABLE 7.49

N	I/Y	PV	PMT	FV
6	$CPT = 4.435$			

At four years we have $m = 8$ so $N = 8$. See Table 7.50.

TABLE 7.50

N	I/Y	PV	PMT	FV
8	$CPT = 3.984$			

Note that these are all semiannual rates. For this bond, the yield to call declines as the call date increases. The lowest yield rate corresponds to holding the bond to maturity. This is always the case for a bond which is purchased at a discount $(P < C)$. Because of this, callable bonds often offer a higher redemption rate for call dates prior to maturity. The amount of the additional payment is called the **call premium**. Conversely, for a bond purchased at a premium, the maximal yield to call occurs at the latest possible call date. There are two values you might be asked to compute: The maximum yield to a call date and the minimum yield to a call date. The minimum yield represents the yield rate an investor can be certain of, regardless of the call date.

Example 7.38 A six-year bond with par value of $4,000 has an annaul coupon rate of 7%. It was purchased for $4,400 and may be called at $m = 3$, 4, 5, or 6. It's redemption value is $4,000. The call premiums are $250 in year 3, $240 in year 4, $230 in year 5 and $0 in year 6. What is the yield to the investor in each case? Note that this bond was purchased at a premium $(4,400 > 4,000)$.

Solution: Here are the TVM calculations for $m = 3, 4, 5, 6$ (Table 7.51):

TABLE 7.51

N	I/Y	PV	PMT	FV
3	$CPT = 5.285$	4,400	-280	$-4,250$

N	I/Y	PV	PMT	FV
4	$CPT = 5.527$			$-4,240$

N	I/Y	PV	PMT	FV
5	$CPT = 5.674$			$-4,230$

N	I/Y	PV	PMT	FV
6	$CPT = 5.028$			$-4,000$

Figure 7.3 displays yield to call values based in whether a bond is purchased at a premium or a discount.

Maximizing Yield To Call

Bond Purchased at a Discount	Redeem at earliest allowable date
Bond Purchased at a Premium	Redeem at latest allowable date

Computing the Minimum Yield to Call

Bond Purchased at a Discount	Redeem at latest allowable date
Bond Purchased at a Premium	Redeem at earliest allowable date

FIGURE 7.3

In this case, the investor is ensured of a yield of 5.028% (if the bond is called at the earliest allowable date) and could earn as much as 5.674%. We can calculate the call premium required at each call date by using a constant yield rate of 5.028% and varying the term of the bond. For example, if the bond is called at the end of year 4 we can compute the required premium using the TVM keys.

TABLE 7.52

N	I/Y	PV	PMT	FV
6	5.028	4,400	-280	$CPT = -4,146.61$

The required call premium is $141.61. See Table 7.52.

Example 7.39. A seven-year bond with semiannual coupons and an annual coupon rate of 8.6% has a face value of $50,000. The bond is redeemable for $45,000 at end of term and is callable at this same amount in years 4 through 7. If the price is $56,973, what is the maximal yield to the investor? What is the minimal yield?

Solution: Since the bond was purchased at a premium the maximal yield to the investor is at year 7 or $m = 14$. See Table 7.53.

TABLE 7.53

N	I/Y	PV	PMT	FV
14	$CPT = 2.50$	56,983	$-2,150$	$-45,000$

The minimal yield is obtained at the end of year 4, or $n = 8$. See Table 7.54.

TABLE 7.54

N	I/Y	PV	PMT	FV
8	$CPT = 1.25$			

7.11   Bonds with Varying Payments

In some cases, a bond will make varying payments. These problems can be dealt with using the methods we developed for annuities with varying payments. Here are few examples.

Example 7.40 A bond has a redemption value of $50,000 and provides semi-annual coupon payments for ten years. The first payment is $1,000 and each subsequent payment is increased by 4%. The yield rate for this bond is 10%. Find the price of this bond.

Solution. The price of a bond is computed as the present value of its coupon payments plus the present value of its redemption value. In this case the payments are: 1000, 1000· (1.04), $1000 \cdot (1.04)^2$, ..., and we have

$$P = 1000v + 1000(1.04)v^2 + \cdots + 1000(1.09)^{19}v^{20} + 50000v^{20}$$

The first twenty terms (the payments) form a geometric series with $a = 1000v$, $r = 1.04 \cdot v$. We then have

$$P = 1000v \cdot \frac{1 - [1.04 \cdot v]^{20}}{1 - 1.04 \cdot v} + 50000v^{20} = \$18,670.57$$

Example 7.41 The present value of the redemption value of a ten-year bond with semiannual coupons is \$4,000. The initial coupon payment is \$50 and each subsequent coupon payment is 6% more than the previous payment. The bond is priced to yield a nominal annual 12%. Find the price of this bond.

Solution: This looks much like Example 7.40, but when we implement the formula we get a surprise. In this case $i = 6\%$, so that $v = \frac{1}{1.06}$. Proceeding as in Example 7.35 we obtain

$$P = 50v \cdot \frac{1 - [1.06 \cdot v]^{20}}{1 - 1.06 \cdot v} + 4000$$

Since $v = \frac{1}{1.06}$, the denominator of the fraction is 0!. That means we can't apply our formula. What's happening here is that the present value of each payment remains \$50. Each payment increases by 6%, but its present value is reduced by the same factor. Hence the present value of the payments is just $20 \cdot 50 = 1000$ and the price of this bond is $1000 + 4000 = \$5,000$.

Example 7.42 A ten-year bond offers annual premiums which start at \$40 and increase by \$30 with each subsequent payment. The bond is priced to yield 5% and has a redemption value of \$30,000. Find the price of this bond.

Solution In this case the bond premiums form a $P - Q$ annuity so we add the present value of this annuity to the present value of the redemption value to find the price of the bond.

$$P = Pa_{\overline{n}|,i} + Q\frac{a_{\overline{n}|,i} - nv^n}{i} + 30000v^n$$

We have $P = 40$, $Q = 30$, $i = .05$, $n = 10$. The result is $P = \$19,675.82$.

Example 7.43 A ten-year bond makes two annual payments of \$40 followed by two annual payments of \$60 and so on with the last two payments being \$120 each. The present value of the redemption value of this bond at an annual rate of interest of 5% is 2,000. What is the price of this bond if it is priced to yield 5% effective annual interest?

Solution The easiest way to do this problem is to accumulate the value of each pair of payments and view them as single payments taking at times, 2, 4, 6, 8, and 10. The accumulated value of the first two payments at $t = 2$ is $40(1.05) + 40 = 82$. Proceeding in a similar fashion with the other payments we obtain the sequence 82, 123, 164, 205, 246. We thus have a $P - Q$ sequence of payments being made once every two years with $P = 82$, $Q = 41$. To compute the present value of this series of payments we need the effective rate of interest per two years which is $(1.05)^2 - 1 = .1025$. Now we use the $P - Q$ formula with $n = 5$, $i = .1025$ to obtain 587.72 as the present value of the payments. Adding this to the present value of the redemption value results in a bond price of \$2,587.72.

Exercises for Chapter 7

1. Construct the bond amortization table for a five-year bond with par value of $10,000 and a redemption value of $12,000 if the annual coupon rate is 4.5% and the bond is purchased at a price which

   a) yields 7% annually

   b) yields 3.5% annually

2. A bond with par value of $5,000 with coupon rate of 6% and providing ten years of semiannual payments is redeemable at $5,500. If the bond is purchased to yield 7%, what is the purchase price?

3. A five-year bond with a par value of $700 and 10% bond semiannual coupons is purchased for $670.60. The present value of the redemption value (C) is $372.05. What is the redemption value?

4. A five-year par value bond with a par value of $100 and 10% semiannual coupons is purchased for $110 to yield a nominal rate of 4% converted semiannually. A five-year bond with a 3% coupon rate and semiannual coupons is purchased at $P to provide the same yield rate. Find P.

5. A ten-year with a par value of $100 bearing a coupon rate of 10% payable semiannually and redeemable at $105 is purchased to yield 8% convertible semiannually. What is the price?

6. Two $1,000 bonds redeemable at par at the end of the same period are purchased to yield 4% convertible semiannually. One bond costs $1,136.78 and has a coupon rate of 5% convertible semiannually. The other bond has a coupon rate of 2.5% convertible semiannually. What is the price of the second bond?

7. You are given two n-year par value ($C = F = 1,000$) bonds. Bond X has 14% semiannual coupons and price of $1,407.7 to yield i, compounded semiannually. Bond Y has the same yield rate, semiannual coupons of 12% and a price of $1,271.80. Find the price of bond X to yield $i - 1\%(i - .01)$.

8. An insurance company owns a $1,00 par value 10% bond with semiannual coupons. The bond will mature at par at the end of ten years. Current yield rates are 7% compounded semiannually. The company sells this bond and uses the proceeds to purchase a 6% bond with semiannual coupons which matures at par at the end of eight years. What is the par value of the second bond?

9. Marge has a five-year $1,000,000 face value bond with 6% coupons convertible semiannually. Fiona buys a ten-year bond with face amount

$X, 6\%$ coupons convertible semiannually. Both bonds are redeemable at par. Marge and Fiona buy their bonds priced to yield 4% compounded semiannually. They then sell them to an investor at a price which will yield the investor 2% compounded semiannually. Fiona earns the same amount of profit on the transaction as does Marge. Find X.

10. A ten-year bond with par value of $\$10,000$ and 8% semiannual coupons is purchased to yield 6% convertible quarterly. The redemption value of the bond at the end of ten years is $\$10,500$. Calculate the purchase price of this bond.

11. A $\$1,000$ par value ten-year bond with coupons at 8% convertible semiannually will be redeemed for $\$R$. The purchase price is $\$800$ and the present value of the redemption value is $\$301.51$. Find R.

12. A ten-year $\$1,000$ par value bond with 8% semiannual coupons is purchased for $\$1,204.15$ to yield 6% convertible semiannually. What would the price of this bond be if it was priced to yield 10% semiannually?

13. A $\$1,000$ par value ten-year bond with coupons at 5% convertible semiannually is selling for $\$1,081.78$. Calculate the exact value of the yield rate convertible semiannually.

14. An n-year zero coupon bond with a par value of $\$1,000$ was purchased for $\$600$. A second-year $\$1,000$ par value bond with annual coupons of $\$x$ was purchased for $\$850$. A 3n-year $\$1,000$ par value bond with annual coupons of $\$x$ was purchased for $\$P$. All three bonds have the same yield rate. Compute P.

15. A $\$1,000$ par value bond has a term to maturity of 4 years and is redeemable for $\$1,200$ and has an annual coupon rate of 8%. Compute the price of this bond given the spot rate shown in Table 7.55:

TABLE 7.55

Term	Annual Spot Interest Rate
1	.06
2	.08
3	.035
4	.07

16. For the bond in problem 15) calculate its effective annual yield rate if it is sold at a price equal to its value.

17. Suppose $i_k = .07 + .002 \cdot k - .001k^2$. Find the one-year forward rate for year 5 that is implied by this yield curve (this is $i_{5,6}$)

18. A three-year annuity-immediate will be issued one year from now with annual payments of $4,500. Given the spot rates shown in Table 7.56 below, compute the present value of this annuity as evaluated one year from now.

TABLE 7.56

Term in Years	Spot Interest Rate
1	5.2%
2	4.8%
3	6.2%
4	8.0%

19. For the same spot rates as in problem 18) compoute the at-par yield for an annual coupon bond with a term to maturity of three years.

20. Compute the one-year forward rates (f_t) implied by the following yield curve shown in Table 7.57.

TABLE 7.57

Term	Spot Rate
1	.04
2	.03
3	.05
4	.06

21. Find the price and yield to maturity of a three-year bond with 8% annual coupons and $F = C = \$10,000$. Assume the same spot rate structure as for problem 18)

22. Find the spot rates which would account for the following prices for bonds with $F = C = 800$ and 10% annual coupons. See Table 7.58.

TABLE 7.58

Term	Price
1 year	850
2 years	900
3 years	875

23. For the yield curve of Exercise 20, compute the two-year forward rates $(_2 f_t)$ which are implied by the yield curve using both methods.

24. An annuity provides payments of $500, $600, $700, and $800 at the end of each year.

    a) Compute the NPV of this annuity using the spot rate table of Exercise 20

    b) Find a level payment annuity-immediate with the same NPV. What is the IRR of this annuity?

    c) What is the required payment for a level-payment annuity-due with the same NPV as the varying annuity-immediate?

25. A ten-year bond with par value and redemption value of $50,000 has an annual coupon rate of 8%. If the bond is purchased to yield a nominal annual interest rate of 10% compounded semiannually what is the price-plus accrued and the price of the bond 54 months after it is purchased?

26. A eight-year bond is purchased for $50,000 and has a redemption rate of $47,000. It is callable at $47,000 at the end of years 5 through 8. At which call date does the investor realize the maximal yield to call?

27. Same as 26) but the purchase price is $56,000.

28. A nine-year bond with semiannual coupons at annual rate 6.4% has a face and redemption value of $60,000. The bond is purchased for $55,000 and is callable at the end of years 7, 8, and 9. What is the maximal yield to call in this situation?

Excel Projects for Chapter 7

1. Create an Excel Spreadsheet which computes the price of a bond given a spot rate table.

2. Create an Excel Spreadsheet which computes the price of a bond given the values of C, F, n, r and i.

3. Create an Excel Spreadsheet which creates the bond amortization table. Your spreadsheet should allow the user to enter the values of C, F, R, n, and i. Compute P (use part 2) if you like) and then use a macro to set up the table.

Chapter 8

Exact Asset Matching and Swaps

8.1   Exact Asset Matching

This section deals with two simple techniques used by companies to plan for anticipated future expenditures. The techniques are referred to as exact asset matching. A more complete (and complicated!) treatment of the asset matching problem will be considered in Chapter 9.

Introduction: Insurance companies collect premiums from policy holders and invest those premiums in anticipation of required payments to policy holders. In order to minimize risk and costs, it is usually desirable to invest in different types of instruments. The most direct method of doing this is called **exact asset matching**. In this case the company invests in securities which will provide exactly what is needed at the time it is needed. This method is 100% certain so long as the bonds do not default. The process of determining the bonds to be purchased and the cost is sometimes referred to as a project.

Example 8.1 A company has liabilities of $4,000 and $3,500 coming due at the end of years 1 and 2 respectively. They wish to use the zero-coupon bonds to "match" these assets. That is, they want the two bonds to produce $4,000 at the end of year 1 and $3,500 at the end of year 2 shown in Table 8.1

TABLE 8.1

Maturity	Effective Annual Yield - Spot Rates	$F = C$
1	5%	1,000
2	6%	1,000

What is the cost of this exact asset matching project?

Total cost is $6,924.51. See Table 8.2. Notice that this is a savings of $575.49 versus just putting $7,500 in a safe to be paid out as needed.

If coupon bonds are used, the situation becomes a bit more complex.

TABLE 8.2

One-year Bonds	$\frac{4000}{1.05} = 3,809.52$
Two-year Bonds	$\frac{3500}{1.06^2} = 3,114.99$

Example 8.2 A company has a liability of \$5,000 coming due in six months and a second liability of \$10,000 coming due in a year. They want to exactly match these liabilities using the following two bonds:

1. A six-month par value bond with face amount of \$1,000, a 4% nominal annual coupon rate convertible semiannually, and a 5% nominal annual yield rate.

2. A one-year par value bond with face amount of \$1,000, an 8% nominal annual coupon rate convertible semiannually, and a 7% nominal annual yield rate.

How much of each bond should they buy? What will be the cost of this project?

Solution: We assume that we can buy fractional portions of our two bonds. We will match the \$5,000 liability with a combination of the two bonds. The 10,000 liability must be met entirely by the one-year bond. Hence we begin with that calculation. The one-year bond will pay $1000 \cdot 1.04 = 1,040$ at time $t = 1$ for each \$1,000 purchased. Hence we have the equation

$$\frac{10000}{1040} = 9.61538 \tag{8.1}$$

The company needs to buy 9.6153 units or \$9,615.38 worth of the one-year bonds.

At six months the 9.61538 units of one-year bonds will provide $9,615.38 \times .04 = 384.62$ in coupon payments. The remainder of the \$5,000 obligation must be met with six-month bonds. Each \$1,000 in six-month bonds provides $1000 \cdot 1.02 = 1,020$ at trade-in. We thus have

$$\frac{5000 - 384.62}{1020} = 4.52489 \tag{8.2}$$

The total assets available at six months are $4.52489 \cdot 1020 + 384.62 = 5000$. To find the cost of matching these liabilities we need to price each bond. To do that we use the TVM keys and the stated yield rates for the bonds.

Six-month bond: See Table 8.3

TABLE 8.3

N	I/Y	PV	PMT	FV
1	2.5	$CPT = 4,502.82$	$-4,524.89(.02) = -90.5$	$-4,524.89$

One-year bond: Table 8.4

TABLE 8.4

N	I/Y	PV	PMT	FV
2	3.5	$CPT = 9,706.72$	$-9,615 \cdot .04 = -384.61$	$-9,615.38$

The total cost for this asset matching project is $14,209.51.

Example 8.3 What is the effective annual yield rate for the company in Example 8.2?

Solution: The company paid $14,209.51 in return for payments of $5,000 at six months and $10,000 at one year. We can use the Cash Flow Worksheet counting in units of six months to compute the effective rate of interest for six months. We then convert that to an annual rate of interest. See Table 8.5.

TABLE 8.5

Time	Symbol	Amount	Frequency
0	$CF0$	$14,209.54$	
1	$C01$	$-5,000$	1
2	$C01$	$-10,000$	1

We press IRR and CPT to obtain 3.3089% as the six-month interest rate. This would be a nominal 6.618 annual rate. We use the ICONV worksheet to compute the effective annual rate as shown in Table 8.6:

TABLE 8.6

NOM	6.618
EFF	$CPT = 6.73$
C/Y	2

The effective annual yield rate is 6.73%.

Example 8.4

A company expects liabilities of $5,000 in one year, $10,000 in two years, $25,000 in three years, and $40,000 in four years. They want to fund these liabilities using exact matching with a combination of the bonds in Table 8.7

TABLE 8.7

Bond	Yield Rate	Coupon Rate	Par Value	Maturity Term
A	4%	0%	1,000	1
B	5%	6%	1,000	2
C	7%	7%	1,000	3
D	3%	6%	1,000	4

a) How much of each bond should they buy?
b) What is the total cost of this asset matching strategy?
c) What is the yield rate for the company for this project?

Solution: Table 8.8 is a table with all the needed information in one place:

TABLE 8.8

Bond	Term	Coupon Rate	Yield Rate	Liability
A	1	0.000%	4.000%	$5,000.00
B	2	6.000%	5.000%	$10,000.00
C	3	7.000%	7.000%	$25,000.00
D	4	6.000%	3.000%	$40,000.00

The $5,000 liability will be covered by a combination of all four bonds, The $10,000 liability will be covered by a combination of Bonds B, C, and D. The $25,000 liability will be covered by bonds C and D. The $40,000 liability must be covered entirely by bond D. Hence we start with bond D and work backwards.

a) Bond D pays 1060 per 1000 at year 4. Hence we have:

$$\frac{40000}{1060} = 37.73585 \tag{8.3}$$

We need to buy $37,735.85 worth of bond D.

Bond C pays 1070 per 1000 at the end of year 3. In addition we will receive $60 * 37.7358 = 2264.15$ from bond D. Together these must account for the $25,000 obligation in year 3:

$$\frac{25000 - 2264.15}{1070} = 21.24846 \tag{8.4}$$

Hence, we must purchase $21,248.46 worth of bond C.

Bond B pays 1060 per 1000 at the end of year 2. In addition we will have 2264.15 from bond D and $21.248 * 70 = 1487.39$ from bond C. Altogether this must provide $10,000 so we have:

$$\frac{10000 - 2264.15 - 1487.39}{1060} = 5.89477 \tag{8.5}$$

We need to buy $5,894.77 worth of bond B.

Finally, bond A pays 1000 per 1000 at the end of year 1. In addition we will have 2264.15 from bond D, 1487.39 from bond C and $5.895 * 60 = 353.69$ from bond B. We have:

$$\frac{5000 - 2264.15 - 1487.39 - 353.69}{1000} = .89477 \qquad (8.6)$$

We need to buy \$894.77 worth of bond A. Note that the FM Exam sometimes asks for the fractional part of a bond (21.248, 5.895, and .8948 in this example). The question may be phrased as "how many of each bond should be purchased?"

b) We now price each bond based on its stated yield rate and add these prices to obtain the total cost.

Bond A (Table 8.9)

TABLE 8.9

N	I/Y	PV	PMT	FV
1	4	$CPT = 860.36$	0	-894.77

Bond B (Table 8.10)

TABLE 8.10

N	I/Y	PV	PMT	FV
2	5	$CPT = 6,004.38$	-353.69	$-5,894.77$

Bond C (Table 8.11)

TABLE 8.11

N	I/Y	PV	PMT	FV
3	7	$CPT = 21,248.46$	$-1,487.39$	$-21,248.46$

Bond D (Table 8.12)

TABLE 8.12

N	I/Y	PV	PMT	FV
4	6	$CPT = 41,943.88$	$-2,264.15$	$-41,943.88$

The total cost is $860.36 + 6004.38 + 21248.46 + 41943.88 = \$70,057.08$.

c) Yield rate:

The company paid $70,057 in year 0 in return for payments for $5,000, $10,000, $25,000 and $40,000 in years 1, 2, 3, and 4. We use the Cash Flow work sheet to compute the effective rate of interest displayed as Table 8.13.

TABLE 8.13

Time	Symbol	Amount	Frequency
0	$CF0$	70,057.08	
1	$C01$	$-5,000$	1
2	$C02$	$-10,000$	1
3	$C03$	$-25,000$	1
4	$C04$	$-40,000$	1

We compute the IRR as 4.19%.

8.2   Swap Rates

Definition: A **swap** is a contract for an exchange of multiple payments or agreed-upon interest rates over time.

Commodity Swaps

Suppose company A wishes to buy 100,000 units of a product at the end of one and two years. Suppose further that the predicted prices for the product and the current spot rates (LIBOR[1] rates typically) are as shown in Table 8.14:

TABLE 8.14

Term/Purchase Date	Price Per Unit	Spot Rate
1	110	6%
2	111	6.5%

Since the price varies, Company A might choose to enter into a contract with a second party (called the counter party) in order to ensure equal payments each year. The basic principal is that the net present value of the

---

[1]LIBOR = London Interbank Offer Rate. This rate is typically, but not always, used to determine spot rates (recall that $s_n = i_{0,n}$).

transaction must be the same for both parties at the time the swap is made. As we will see, if the spot rates change the value of the swap will also change.

While it might seem reasonable to use the average price of 110.5, this results in the following situation from the point of view of Company A: We pay .5 too much in year 1 and .5 too little in year two. We can think of this as a loan to Company B of .5 which is repaid in a year with a payment of .5. That's only fair if the interest rate is 0, which it is not!

We can calculate an appropriate payment (or payments in several ways). In each case the present value of the payments must equal the present value of the items purchased. In this case the present value per unit purchased is

$$\frac{110}{1.06} + \frac{111}{1.065^2} = 201.638$$

The simplest solution is for Company A to pay the counter party $210.38 per unit at the time of the agreement (one year prior to the first purchase) in return for an agreement to provide 100,000 units at time $t = 1$ and 10,000 units at time $t = 2$. This is known as a **pre-paid swap**. For Company A there is a substantial default risk, so this is not the usual technique.

A second method, known as a **physical settlement**, is for Company A to make equal payments at times 1 and 2. The present value of those payments must equal the present value of the product to be delivered and we have

$$\frac{x}{1.06} + \frac{x}{1.065^2} = 201.638$$

Solving yields the two-year swap physical settlement price of 110.483. Notice that this is slightly less than the average price. The length of a swap is also referred to as the **term** or **tenor** of the swap. And the dollar amount involved is called the **notional amount**. In this case we have a swap with a term or tenor of 2 and a notional amount of 201.638 per unit.

Finally, and most commonly, the counter party agrees to ensure a constant payment for Company A. This payment (known as the swap payment) would be $110.483 in our example. If the spot price exceeds the swap rate the counter party pays Company A the difference. If the swap rate exceeds the spot price, Company A pays the counter party the difference. The net result is that Company A is liable for the spot price at each purchase time. The situation from the point of view of Company A is

$$-\text{price} + (\text{spot price} - \text{swap price}) = -\text{swap price} \qquad 8.1$$

At the calculated swap rate Company A pays .483086 above the spot price at time 1 and .516914 below the spot price at period 2. This is the same as a loan of .483086 to be repaid in one year with .516914.

The interest rate is

$$\frac{.516914}{.483086} - 1 = 7.00024\%$$

We can compare this to the implied forward rate $i_{1,2}$ which computed using the spot rates

$$1.06 \cdot (1 + i_{1,2}) = 1.065^2 \Rightarrow i_{1,2} = 7.0024\% \tag{8.7}$$

In the general case, suppose we have a series of forward prices $F(0, t_i)$ for a product and a set of spot rates represented by $P_{t_i} = P(0, t_i) = \frac{1}{(1+s_{t_i})^{t_i}}$. We compute using the below formulas

Present Value of the Sequence of Purchases = Pre-Paid Swap Price

$$PV = \sum F(0, t_i)P(0, t_i) \tag{8.8}$$

Physical Settlement Swap Payment Per Period

$$R = \frac{\sum_i F(0, t_i)P(0, t_i)}{\sum_i P(0, t_i)} \tag{8.9}$$

We can rewrite this as

$$R = \sum \left(\frac{P(0, t_i)}{\sum_i P(0, t_i)} \right) \cdot F(0, t_i) \tag{8.10}$$

In this form, we can see that the swap payment is the present-value-weighted average of the forward prices.

The above analysis assumes that the same quantity of product is purchased at each time period. We can modify this to include varying quantities. If we purchase a quantity Q_{t_i} at t_i Equations 8.8, 8.9 and 8.10 become:

Present Value of the Sequence of Purchases

$$PV = \sum Q_{t_i} F(0, t_i)P(0, t_i) \tag{8.11}$$

Swap Payment with Varying Quantities

$$R = \frac{\sum Q_{t_i} F(0, t_i)P(0, t_i)}{\sum P(0, t_i)} \tag{8.12}$$

$$R = \sum \left(\frac{P(0, t_i)}{\sum P(0, t_i)} \right) \cdot Q_{t_i} F(0, t_i) \tag{8.13}$$

Example 8.7: Compute the swap price per unit based on Table 8.15 below.

TABLE 8.15

Term	Spot Price	Spot Rate
1	120	5.6%
2	115	6.0%
3	130	7.2%

Solution: We compute $P(0,1)$, $P(0,2)$, $P(0,3)$, and then use formula 8.6

$$P(0,1) = \frac{1}{1.056} = .94697$$

$$P(0,2) = \frac{1}{1.06^2} = .8900$$

$$P(0,3) = \frac{1}{1.072^3} = .81174$$

The swap price is then

$$R = \frac{120 * .947 + 115 * .890 + 130 * .811}{.947 + .890 + .811} = 121.38$$

Example 8.8 Compute the pre-paid swap price and swap payment for example 8.7 if we buy ten units at time 1, fifteen units at time 2, and twelve units at time 3.

Solution: We already have all the terms we need and merely subsitute into Equation 8.11 and 8.12

Pre-Paid Swap Price

$$10 \cdot 120 \cdot .947 + 15 \cdot 115 \cdot .890 + 12 \cdot 130 \cdot .870 = \$3,937.92$$

Physical Settlement Price

$$R = \frac{10 \cdot 120 \cdot .947 + 15 \cdot 115 \cdot .890 + 12 \cdot 130 \cdot .870}{.947 + .890 + .870} = \$1,486.73$$

8.3 Interest Rate Swaps

Interest rate swaps are used to convert a floating (variable) set of interest rates to a fixed rate. Floating rate loans are often made "at the LIBOR" – meaning that the interest paid each period is the implied forward rate for the current LIBOR spot rates. To begin we assume the loan amounts (notional value) are the same in each period. The easiest formula for the interest swap rate is given in terms of the values $P(0, t_i) = P_{t_i}$:

Swap Rate for an Interest Rate Swap

$$r = \frac{1 - P_{t_N}}{\sum_{i=1}^{N} P_{t_i}} = \frac{1 - \text{Last Price}}{\text{Sum of the Prices}} \qquad (8.14)$$

This is exactly the same formula used to compute at-par yield rate – the coupon rate for a bond with $F = P = C$.

This formula can also be written in terms of forward rates $i_{t-1,t}$

$$r = \frac{\sum_{i=1}^{n} P_{t_i} \cdot i_{t-1,t}}{\sum_{i=1}^{n} P_{t_i}} \tag{8.15}$$

Example 8.9 Calculate the swap rate for the spot rates in Example 8.7.

Solution: We substitute directly into Equation 8.13

$$r = \frac{1 - .812}{.947 + .890 + .812} = .071 = 7.1\%$$

8.4 Deferred Interest Rate Swaps

In some cases, you will be asked to compute the rate for a deferred interest rate swap. The standard (sometimes called "plain vanilla") swap starts in period one. A deferred swap starts at some later period. If the swap starts in period k we obtain

$$\frac{\sum_{i=k}^{n} P_{t_i} \cdot i_{t-1,t}}{\sum_{i=k}^{n} Pt_i} \tag{8.16}$$

The alternate formula is

$$\frac{P_{t_{k-1}} - P_{t_n}}{\sum_{i=1}^{n} P_{t_i}} \tag{8.17}$$

It's easier to remember this using the following

$$\frac{\text{Price just prior to start of swap} - \text{Last price during the swap}}{\text{Sum of all prices during the swap}} \tag{8.18}$$

This is consistent with Equation 8.14 as the price of a bond at time 0 is 1.

You might be asked to work with a LIBOR table such as we saw in Chapter 7.

Example 8.10 Fill in the remainder of the Table 8.16. Then compute the three-year swap rate and the two-year rate which starts in period 2 (deferred one year).

TABLE 8.16

Years to Maturity	0-Coupon Yield	0-Coupon Price	Implied Forward Rate	Par Coupon
1	6.0%			
2	6.5%			
3	7.0%			

Solution: We compute the 0-coupon prices using formula 7.37

$$P(0,1) = \frac{1}{1.060} = .9434$$

$$P(0,2) = \frac{1}{1.065^2} = .88166$$

$$P(0,3) = \frac{1}{1.07^3} = .81630$$

We compute the forward rates using the formula from Chapter 7

$$i_{0,1} = 6.0\%$$

$$i_{1,2} = \frac{P(0,1)}{P(0,2)} - 1 = 7.0$$

$$i_{2,3} = \frac{P(0,2)}{P(0,3)} - 1 = 8.0\%$$

The at-par coupon rates are the same as the interest rate swap rates and are computed using Equation 8.18

$$t = 1 \Rightarrow r = 6\%$$

$$t = 2 \Rightarrow r = \frac{1 - .88166}{.9434 + .88166} = 6.484\%$$

$$t = 3 \Rightarrow r = \frac{1 - .8163}{.9434 + .88166 + .8163} = 6.955\%$$

So the three-year swap rate is 6.95%.

Finally, the rate for a two-year swap deferred one year is computed as

$$\frac{.9434 - .8163}{.88166 + .81630} = 7.485\%$$

Table 8.17 displays the results of these calculations.

TABLE 8.17

Years to Maturity	0-Coupon Bond Yield	0-Coupon Bond Price	One Year Implied Forward	Par Coupon
1	6.000%	0.94340	6.0000%	6.000%
2	6.500%	0.88166	7.0024%	6.484%
3	7.000%	0.81630	8.0071%	6.955%
Two-year swap			6.4842%	
Three-Year Swap			6.9548%	
Two-Year Swap Delayed one year			7.4854%	

8.5 Varying Notional Amounts

The previous analysis of interest rate swaps assumed a constant notional amount (i.e., loan amount). If the notional amount varies we need a slight modification of our formula. If we suppose a notional amount Q_{t_i} at time t_i we obtain

$$R = \frac{\sum_{i=1} Q_{t_i} \cdot P(0,t_i) \cdot i_{t_{i-1},t_i}}{\sum_{i=1}^{n} Q_{t_i} \cdot P(0,t_i)} \tag{8.19}$$

In many cases, payments are made each period so the $t_i = i$. In this case we can greatly simplify this formula by using

$$i_{i-1,i} = \frac{P(0,i-1)}{P(0,i)} - 1 = \frac{P(0,i-1) - P(0,i)}{P(0,i)}$$

In this case Equation 8.19 becomes

$$R = \frac{\sum_{i=1}^{n} Q_i P(0,i) \left(\frac{P(0,i-1)}{P(0,i)} - 1 \right)}{\sum_{i=1}^{n} Q_i P(0,i)} = \frac{\sum_{i=1}^{n} Q_i [P(0,i-1) - P(0,i)]}{\sum_{i=1}^{n} Q_i P(0,i)}$$

Swap Rate Formula with Notionals if Payments are Made Each Year

$$R = \frac{\sum_{i=1}^{n} Q_i [P(0,i-1) - P(0,i)]}{\sum_{i=1}^{n} Q_i P(0,i)} \tag{8.20}$$

Example 8.11 Use Table 8.18 below to compute the three-year swap rate:

TABLE 8.18

Years to Maturity	0-Coupon Bond	$P(0,n)$	$i_{n-1,n}$	Q
1	5%	.95238	5.000%	100
2	6%	.89000	7.010%	200
3	5.5%	.85161	4.507%	150

Using Equation 8.19 or Equation 8.20 we obtain $r = 5.735\%$.

Table 8.19 displays the results as well as the four-year swap rate with notionals. Note: the par coupon in year n is the same as the n - year swap rate compouted without notionals.

TABLE 8.19

Years to Maturity	0-Coupon Bond Yield	LIBOR	One Year Implied Forward	Par Coupon	Notionals
1	5.000%	0.95238	5.0000%	5.0000%	$100.00
2	6.000%	0.89000	7.0095%	5.9707%	$200.00
3	5.500%	0.85161	4.5071%	5.5080%	$150.00
4	7.500%	0.74880	13.7304%	7.2964%	$100.00
Two-year swap			5.97%		
Three-Year Swap			5.51%		
Four-year Swap			7.30%		
Three-Year Swap with Notionals			5.74%		
Two-Year Swap Delayed one year			5.79%		
Four-year Swap with Notionals			6.99%		

There are two special cases of swaps with varying notional amounts:

An accreting swap is a swap in which the notional amounts increase. Unless stated otherwise, it is assumed that they are 1, 2, 3, ...

Example 8.12 Table 8.20 below presents the results of a four-year accreting swap. Note that difference in the three-year rates when notionals are included.

TABLE 8.20

Years to Maturity	0-Coupon Bond Yield	LIBOR	One Year Implied Forward	Par Coupon	Notionals
1	4.000%	0.96154	4.0000%	4.0000%	$100.00
2	4.500%	0.91573	5.0024%	4.4890%	$200.00
3	4.200%	0.88389	3.6026%	4.2052%	$300.00
4	4.600%	0.83536	5.8092%	4.5778%	$400.00
Two-year swap			4.49%		
Three-Year Swap			4.21%		
Four-year Swap			4.58%		
Three-Year Swap with Notionals			4.14%		
Two-Year Swap Delayed one year			4.31%		
Four-year Swap with Notionals			4.78%		

An amortizing swap is one in which the notional amounts decrease.

Example 8.13 Since the value of Q decreases over time, Table 8.21 represents an amortizing swap:

TABLE 8.21

Years to Maturity	0-Coupon Bond Yield	LIBOR	One Year Implied Forward	Par Coupon	Notionals
1	4.000%	0.96154	4.0000%	4.0000%	$500.00
2	4.500%	0.91573	5.0024%	4.4890%	$400.00
3	4.200%	0.88389	3.6026%	4.2052%	$100.00
4	4.600%	0.83536	5.8092%	4.5778%	$ 50.00
Two-year swap			4.49%		
Three-Year Swap			4.21%		
Four-year Swap			4.58%		
Three-Year Swap with Notionals			4.35%		
Two-Year Swap Delayed one year			4.31%		
Four-year Swap with Notionals			4.42%		

8.6   The Market Value of an Interest Rate Swap

At the time a swap contract is entered into the value to either party is zero or very nearly zero. Over time, its value changes. This is true whether the LIBOR rates change or not. Valuations are typically made based on three-month LIBOR rates. Suppose a loan of 2.5 billion (2,500 million) is to be paid off at LIBOR. The borrower equalizes the loan payments by using an interest rate swap. Table 8.22 is a hypothetical set of LIBOR rates (zero-coupon bond yields) and the calculated interest rate swap for four periods of three months each:

TABLE 8.22

Months	0-Coupon Bond Yield	0-Coupon Bond Price Price	One Year Implied Forward
3	0.2330%	0.99942	0.00233
6	0.3250%	0.99838	0.00104
9	0.4510%	0.99663	0.00176
12	0.5780%	0.99424	0.00240
	One-Year Swap Rate	0.578%	

Based on a loan amount of 2.5 billion, the swap payment per three months will be

$$2500 * .00575/4 = 3.61 \text{ million}$$

Suppose that we want to compute the value of the swap after sixty days – one month prior to the first required payment. To do that, we need the LIBOR rates for one, four, seven, and ten months at this time. Suppose the rates (and their associated prices) are as shown in Table 8.23:

TABLE 8.23

Months	LIBOR	0-Coupon Bond Price
1	0.3330%	0.99972
4	0.4250%	0.99859
7	0.5510%	0.99680
10	0.6780%	0.99438

The fixed payments are like a bond paying 3.6 each period and then 2500 as a balloon payment. The value (based on the LIBOR rates above) is

$$3.61 \cdot (.99972 + .99859 + .99680 + .99438) + 2500 \cdot .99438 = 2,500.36 \text{ million}$$

The first varying rate payment[2] is $2,500 \cdot \frac{.00233}{4} = 1.46$. The value of the varying payment scenario is

$$(1.46 + 2500) \cdot .99972 = 2,500.76$$

The value of this swap two months after it is entered into is then $2,500.36 - 2,500.76 = -.40$ million. We only use one payment to value the variable payment bond as that is all we know for sure. The LIBOR rates are assumed to vary over time. Note that we used the original forward rate to compute the coupon payment.

Exercises for Chapter 8

1. A company must pay liabilities of $4,500 and $6,000 at the end of years 1 and 2. They wish to fund these liabilities with one- and two-year zero-coupon bonds each with face values of $1,00. How much of each bond should they buy? Given the yield to maturity data below compute the cost to the company for this asset matching strategy. What is the yield rate for this strategy? See Table 8.24.

---

[2]This is calculated using the first quarter forward rate from the original table.

TABLE 8.24

Maturity	Effective Annual Yield	Par Value
1	5.6%	$1,000
2	7.2%	$1,000

2. Compute the cost for the situation in problem 1 if the liabilities are $6,000 in year 1 and $8,000 in year 2. Compute the yield rate.

3. A company has liabilities of $4,000 due in six months, $5,000 due in one year, and $7,000 due in eighteen months. They wish to asset match using three types of bonds:

   (a) A six-month bond with face amount of $1,000, a 5% nominal annual coupon rate convertible semiannually and a 3% nominal annual yield rate convertible semiannually.

   (b) A one-year bond with face amount of $1,000, a 4% nominal annual coupon rate convertible semiannually and a 5% nominal annual yield rate convertible semiannually.

   (c) An eighteen-month bond with face amount of $1,000, a 3% nominal annual coupon rate convertible semiannually and a 8% nominal annual yield rate convertible semiannually.

   How much of each bond do they need to buy? What is the total cost of the asset matching project? What is their yield rate for the project?

4. Given the Table 8.25 below compute:

   a) The pre-paid swap price

   b) The physical settlement price

TABLE 8.25

Purchase Date	Price Per Unit	Spot Rate
1	120	5.5%
2	115	6.6%

5. Given the Table 8.26 below compute:

   a) The pre-paid swap price

   b) The physical settlement price

TABLE 8.26

Term	Spot Price	Spot Rate
1	125	6%
2	135	6.6%
3	130	7%

6. Using the table for problem 5) compute the pre-paid and physical settlement price if we want to buy 100 units at $t = 1$, 150 units at $t = 2$, and 300 units at $t = 3$.

7. Using the Table 8.27 below compute:

a) The value of R for a three-year interest rate swap

b) The value of R for a two-year interest rate swap which is deferred one year

TABLE 8.27

Years to Maturity	Zero-coupon Bond Yield Rate
1	5%
2	5.6%
3	7.2%

8. The three-year interest rate swap rate is 6.256%, $s_1 = 5\%$, $s_3 = 6.4\%$. Compute s_2.

9. The rate for a two-year interest rate swap which is deferred one year is 7.606%, $s_1 = 4.0\%$, $s_3 = 6.4\%$. Compute s_2.

10. An interest swap rate is set up for two years with swap rate 8.534%. The one-year spot rate is 7%. Find the second year forward rate $i_{1,2}$.

11. Calculate the four-year annual interest rate swap that will enable you to pay a fixed rate and receive 1-year LIBOR. Use Table 8.28:

TABLE 8.28

Years to Maturity	1	2	3	4
Spot Rate	2.5%	3%	4.2%	3.2%

12. A company must pay liabilities of $30,000 and 45,000 at the end of years 1 and 2. The only investments to be used are the annual coupon bonds listed below in Table 8.29. What is the cost to the company of this transaction?

TABLE 8.29

Bond	Maturity	Coupon	Yield Rate	Par Value
A	1	7%	5%	X
B	2	4.75%	4.7%	1000

13. Complete the entries in the Table 8.30 below

TABLE 8.30

Years to Maturity	0-Coupon Bond Yield	LIBOR	One Year Implied Forward	Par Coupon	Force of Interest	Notionals
1	6.000%					$1,000.00
2	6.000%					$ 500.00
3	5.500%					$ 100.00
4	3.200%					$ 300.00

14. Using the results from problem 13 complete Table 8.31 below

TABLE 8.31

Two-year swap
Three-Year Swap
Four-year Swap
Three-Year Swap with Notionals
Two-Year Swap Delayed one year
Four-year Swap with Notionals

15. Complete Table 8.32 below

TABLE 8.32

Years to Maturity	0-Coupon Yield	0-Coupon Bond Price	One Year Implied Forward	Par Coupon	Force of Interest	Q
1		0.95694				100
2		0.89845				200
3		0.84679				150

16. Using the results from problem 15 complete Table 8.33 below

TABLE 8.33

Two-year swap deferred one year
Three-Year Swap
One-Year Swap
Three-Year Swap with Notionals

17. Complete Table 8.34 below

TABLE 8.34

Years to Maturity	0-Coupon Bond Yield	LIBOR	One Year Implied Forward	Par Coupon	Force of Interest	Notionals
1	4.500%					$ 400.00
2	5.200%	0.90358	5.905%	5.182%	5.0693%	$ 25.00
3			8.535%	6.219%	6.1095%	$ 100.00
4	6.500%	0.77732	7.102%		6.2975%	$ 300.00

18. Use the results from problem 17 to complete Table 8.35 below

TABLE 8.35

Two-year swap
Three-Year Swap
Four-year Swap
Three-Year Swap with Notionals
Two-Year Swap Delayed one year
Four-year Swap with Notionals

19-20. Create an Excel worksheet which will compute all the entries in the table for problem 13 and also compute the table for problem 14.

21. Complete Table 8.36 below

TABLE 8.36

Years to Maturity	0-Coupon Bond Yield	LIBOR	One Year Implied Forward	Par Coupon	Force of Interest
1	2.500%				2.4693%
2		0.94260	3.502%		2.9559%
3	3.100%	0.91248	3.300%		3.0529%

22. Complete Table 8.37 below. Amount needed is the redemption value of the bond. Cost is the cost of the bond. Coupon payment is the coupon payment provided by that bond.

TABLE 8.37

Bond	Term	Coupon Rate	Yield Rate	Liability	Amount Needed	Cost	Coupon Payment
A	1	0.000%	4.000%	$ 5,000.00			
B	2	6.000%	5.000%	$10,000.00			
C	3	7.000%	7.000%	$25,000.00			
D	4	6.000%	3.000%	$40,000.00			
Yield Rate for this Project:					Total Cost of the Project		

23. Complete Table 8.38

TABLE 8.38

Bond	Term	Coupon Rate	Yield Rate	Liability	Amount Needed	Cost	Coupon Payment
A	1	0.000%	6.000%	$10,000.00			
B	2	5.000%	4.500%	$40,000.00			
C	3	5.600%	5.000%	$25,000.00			
D	4	4.000%	5.600%	$25,000.00			
Yield Rate for this Project:					Total Cost of the Project		

24. Create an Excel worksheet to do problems like 22 and 23.

25. Complete Table 8.39 below

TABLE 8.39

Bond	Term	Coupon Rate	Yield Rate	Liability	Amount Needed	Cost	Coupon Payment
A	1			$ 50,000.00	$28,515.67	$27,977.64	$1,140.63
B	2	5.000%	7.000%	$45,000.00	$24,656.30	$23,764.72	$1,232.82
C	3	5.600%	5.000%	$25,000.00	$ 5,889.12	$ 5,985.34	$ 329.79
D	4	302.000%	4.000%	$25,000.00	$ 6,218.91	$73,489.35	$18,781.09
Yield Rate for this Project:					Total Cost of the Project	$131,217.05 $131,217.05	

Chapter 9

Interest Rate Sensitivity

9.1 Introduction

In this chapter we discuss a few more of the ways investors deal with the fact that interest rates change over time. As an example, suppose that you are obligated to pay $50,000 in five years. You have a choice of three-year, five-year or seven-year zero-coupon bonds[1] as a means of meeting this obligation. Let's do a brief analysis of these three possibilities to see what the issues are.

If you purchase five-year bonds which mature at $50,000 you are certain to meet your obligation. If the interest rate for five-year bonds is s_5 you would pay $\frac{50000}{(1+s_5)^5} = 50000 \cdot P_5$ to acquire the needed bonds.

However, it may be cheaper to purchase three-year bonds and reinvest the proceeds in two-year bonds at the end of three years. This decision would be made based on the current rates for three-year bonds and the projected rates for two-year bonds in three years.

Finally, you could purchase seven-year bonds with the intention of turning them in at the end of five years. In this case, your return will depend on the rates for two-year bonds at that time since the redemption value of your seven-year bonds (which have two years remaining) will depend on the then current rate for two-year bonds. If the rates for two-year bonds are higher than the rate your seven-year bonds carry, you will realize less than par on your bonds. Thus, if you buy shorter term bonds (three-year in this example) you will benefit if interest rates rise, while if you buy longer term bonds you will benefit if interest rates will fall.

Based on this analysis it might make sense to buy a mixture of bonds with the hope that the mixture would minimize your losses whatever future interest rates may be. Such a strategy is known as **immunization**. The most conservative strategy (buying five-year bonds) is known as **exact asset matching** or **a dedicated portfolio**. These techniques were discussed in Chapter 8. We now consider more sophisticated ways of dealing with anticipated future liabilities. To see the issues involved, we begin with an example. Its purpose is to illustrate how a varied portfolio can protect an investor in the face of fluctuating interest rates.

---

[1]Reminder - a zero-coupon bond pays interest only when it is redeemed. If the bond has a face value of P and offers a yield rate of i it pays $P(1+i)^n$ if it is redeemed in period n.

Example 9.1 You need to have $50,000 on hand in five years. Suppose you purchase two zero-coupon bonds.

1) A three-year bond which matures at $22,675 priced to yield 5% and

2) A seven-year bond which matures at $27,562 and is also priced to yield 5%.

Your plan is to turn in both bonds at the end of three years and put the assets realized into a two-year bond. Show that this strategy will produce the required $50,000 at interest rates of $i = 1\%$, 5%, and 20%.

Solution: We analyze each situation.

a) If $i = 1\%$, The seven-year bond will be worth $\frac{27,562}{(1.01)^4} = \$26,486$ at the end of three years. Together with the $22,675 realized from the three-year bond we have $49,161 to invest at 1% for two years. This will realize $49,161 \cdot (1.01)^2 = \$50,149$

b) If $i = 5\%$, the seven-year bond will realize $\frac{27,562}{(1.05)^4} = \$22,675$ at the end of three years. Along with the 22,675 from the three-year bond the investor will have a total of $45,351 to invest for two years at 5% yielding $45,351 \cdot (1.05)^2 = \$49,999$. Just $1 under our goal!

c) If $i = 20\%$, the seven-year bond will realize $\frac{27,562}{(1.2)^4} = \$13,291$ in addition to the 22,675 from the three-year bond for a total of $35,967 to invest for two years at 20%. This will accumulate to $35,967.85 \cdot (1.2)^2 = \$51,792$.

This is obviously a contrived example, but it does illustrate the concept of immunization. The three-year bond provides a constant $22,675 while the seven-year bond provides less at higher interest rates and more at lower interest rates. The total available for investment at the end of three years fluctuates in present value in a direct relation to the seven-year bond and in inverse relation to the prevailing interest rates. However, it then accumulates in direct relation to the prevailing interest rates. This allows us, in this case, to have enough on hand in either case. We can compute the accumulated amount as a function of the interest rate , i at the three-year mark:

$$amt(i) = \left(22675 + \frac{27562}{(1+i)^4}\right) \cdot (1+i)^2$$

We can then graph the accumulated value of our investment as a function of i. As the graph below shows, there is a minimum at just around 5% which is very nearly 50,000. See Figure 9.1.

Of course at this point we have no idea how the two numbers in this example were arrived at. The remainder of this chapter will shed some light on that topic. We begin with a measure of the sensitivity[2] of a series of cash

---

[2]That is, we consider how a change in the prevailing interest rate affects the present value of the sequence of cash flows. Sequence of cash flows whose present value remain nearly constant with respect to changes in interest rates are said to have a low sensitivity – that's what most investers want.

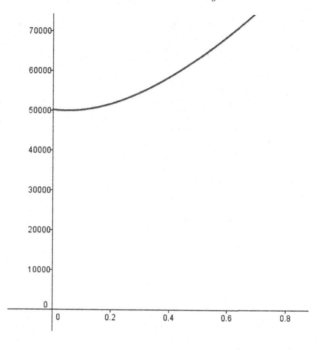

FIGURE 9.1

flows to fluctuations in the interest rates and the related concepts of **duration** and **convexity**.

9.2 The Price Curve: Approximations Using Tangent Lines and Taylor Polynomials

We consider a set of cash flows C_t made at time t where $t \geq 0$ and wish to compute the present value of this investment at time $t = 0$ for a given interest rate i. Since this also represents the price at which the purchase of this set of cash flows would produce a yield rate of i, we refer to the present value of the investment as $P(i)$. We can compute the value of $P(i)$ fairly easily using cash flow analysis. The present value of each cash flow C_t is $C_t(1 + i)^{-t} = C_t \cdot P_t$. To obtain $P(i)$, we sum the present the value of each individual payment

$$P(i) = \sum_{t \geq 0} C_t(1 + i)^{-t} \tag{9.1}$$

For our analysis, we will assume that there are only a finite number of non-zero transactions and that there is a rate i_0 such that $P(i_0) > 0$. We are not assuming that $C_t > 0$ for all t.

Example 9.2 Suppose that our cash flows consist of: $C_2 = 100$, $C_3 = 300$, and $C_5 = 400$.

Compute the price function and graph it. Note that there is no activity in this account at times 1 and 4.

Solution: Remember that the subscript indicates the time at which the amount is to be received/paid out. With this in mind, our price function is

$$P(i) = 100(1 + i)^{-2} + 300(1 + i)^{-3} + 400(1 + i)^{-5}$$

Its graph appears below as Figure 9.2

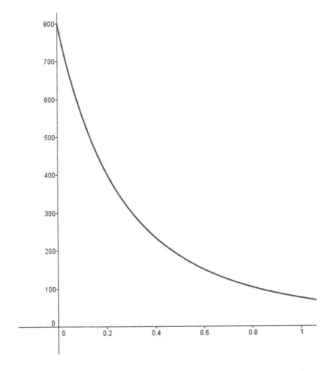

FIGURE 9.2

The price function in this example is a decreasing function of the interest rate. The graph is also concave up. From calculus we know that such a graph occurs when $P'(i) < 0$ and $P''(i) > 0$. The graph of price as a function of the interest rate is known as the **price curve**. As a general rule, the price curve looks much like Figure 9.2. That is, they are typically decreasing and concave up.

We now begin our analysis of the way changing interest rates affect the value of a portfolio of mixed-term bonds. The rate of change of the price function is its derivative with respect to the interest rate:

$$\frac{d}{di}P(i) = P'(i) \tag{9.2}$$

Increasing the interest rate decreases the value of future payments. This means that $P(i)$ is usually a decreasing function and so its derivative is typically negative.

We will also use the second derivative of the price function with respect to the interest rate:

$$\frac{d^2}{di^2}P(i) = P''(i) \tag{9.3}$$

This number tells us how fast $P'(i)$ is changing. It is almost always positive since the graph of $P(i)$ is almost always concave up.

Sensitivity analysis seeks to approximate the rate of change of the price function at a given value of i. As you may recall from calculus, the tangent line is a good approximation to a function near the point in question. As a first approximation we will use the tangent line to estimate values of $P(i)$ near a known point $(i_0, P(i_0))$ on the curve. The tangent line approximation to the graph of $P(i)$ at the point $(i_0, P(i_0))$ is given by

Tangent Line Approximation To The Price Curve

$$P_1(i, i_0) = P(i_0) + P'(i_0)(i - i_0) \tag{9.4}$$

The subscript of 1 in Equation 9.4 reminds us that the degree of this approximation is equal to 1. We will also compute a second degree (quadratic) approximation to the price function.

We are computing the derivative with respect to i, keeping all other variables constant. This derivative is fairly easy to compute:

$$P'(i) = \sum_{t \geq 0} C_t(-t)(1+i)^{-t-1} = -\sum_{t \geq 0} C_t \cdot t \cdot (1+i)^{-(t+1)} \tag{9.5}$$

An alternate expression for $P'(i)$ can be obtained by recalling that $v = (1+i)^{-1}$ which yields

$$P'(i) = -\sum_{t \geq 0} C_t \cdot t \cdot v^{t+1} \tag{9.6}$$

We then evaluate the derivative and the function at the point $i = i_0$ and use Equation 9.4 to compute the equation of the tangent line at this point.

Example 9.2 Find the tangent line approximation to the price function of Example 9.2 at $t_0 = .05$. Graph $P(i)$ and its tangent line on the same graph

Solution: Using Equation 9.5, we have

$$P'(i) = -\left(\frac{200}{(1+i)^3} + \frac{900}{(1+i)^4} + \frac{2000}{(1+i)^6}\right)$$

Direct calculation gives us $P'(.05) = -2405.63$ and $P(.05) = 663.26$. The equation of the tangent line at $i = .05$ is then

$$P_1(i) = 783.5 - 2405.63$$

The price function and its tanget line approximation are graphed below (Figure 9.3) for $0 \leq i \leq .2$. The tangent line seems to provide a fairly accurate approximation so long as i is close to $i_0 = .05$.

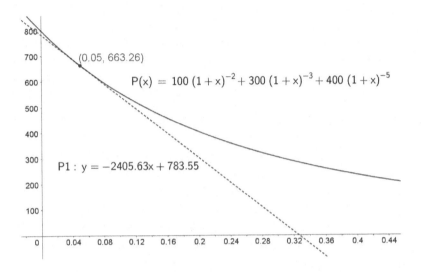

FIGURE 9.3

We can compute an even better approximation to the price curve by using the second degree Taylor polynomial centered at $(i_0, P(i_0))$ instead of the tangent line. The second degree Taylor Polynomial is a quadratic polynomial $P_2(i, i_0)$ with three properties:

1) $P_2(i_0, i_0) = P(i_0)$
2) $P_2'(i_0 i_0) = P'(i_0)$
3) $P_2''(i_0, i_0) = P''(i_0)$

If we let $P_2(i, i_0) = a \cdot (i - i_0)^2 + b \cdot (i - i_0) + c$ we have

$$P_2(i_0, i_0) = c = P(i_0)$$
$$P_2'(i_0, i_0) = b = P'(i_0)$$
$$P_2''(i_0, i_0) = 2a = P''(i_0) \tag{9.7}$$

Solving for a, b, c we obtain the following expression for the second degree Taylor Polynomial which you may recall from calculus:

$$P_2(i, i_0) = P(i_0) + P'(i_0)(i - i_0) + \frac{P''(i_0)}{2}(i - i_0)^2 \tag{9.8}$$

Using the price curve from Example 9.2 we have

$$P''(i) = \frac{600}{(1+i)^4} + \frac{3600}{(1+i)^5} + \frac{12000}{(1+i)^7}$$

$$P''(.05) = 11842.49$$

As are result, we obtain the following equation for the second order Taylor Polynomial:

$$P_2(i, .05) = 798.35 - 2997.76i + 5921.24i^2$$

Graphing $P_1(i)$, $P_2(i)$ and $P(i)$ together reveals that $P_2(i)$ appears to be a better approximation than the tangent line. See Figure 9.4. It is possible to create polynomials of any degree which approximate the price function. In practice, only the first and second degree Taylor polynomials are used in this context.

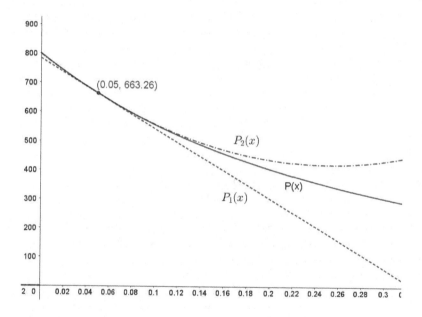

FIGURE 9.4

Example 9.3 Compare the tangent line and second degree Taylor polynomial approximations computed at $i = .05$ to the actual price in Example 9.2 at $i = .06$.

Solution. $P(.06) = 639.7887$. The tangent line approximation yields 639.208 and the second degree Taylor polynomial produces 639.8005.

9.3 Measuring Sensitivity to Interest Rate Fluctuation: Duration

As we have seen, the tanget line approximation provides a reasonable estimate for $P(i)$ if i is close to i_0. Using Equation 9.4, we have

$$P(i, i_0) \approx P(i_0) + P'(i_0)(i - i_0) \qquad (9.9)$$

Subtracting $P(i_0)$ from both sides and then dividing both sides by $P(i_0)$ gives us a measure of the relative price change as a function of P'.

$$\frac{P(i) - P(i_0)}{P(i_0)} \approx \frac{P'(i_0)}{P(i_0)} \cdot (i - i_0) \qquad (9.10)$$

Increases in interest rates are often described in terms of **basis points.** One hundred basis points is equal to one percent. Thus if the interest rate (or the yield rate if we are thinking in terms of cash flows) increases from 6% to 6.5% we can say it has increased by fifty basis points. If $i - i_0 = .01$ (or 100 basis points) we have

$$\frac{P(i) - P(i_0)}{P(i_0)} \approx \frac{P'(i_0)}{P(i_0)} \cdot .01 \qquad (9.11)$$

This can be written as

$$100 * \frac{P(i) - P(i_0)}{P(i_0)} \approx \frac{P'(i_0)}{P(i_0)} \qquad (9.12)$$

With this language we can say that the relative percentage change (left-hand side of Equation 9.11) in price for each 100 basis points change in i is $\frac{P'(i_0)}{P(i_0)}$. We can re-write Equation 9.10 in yet one more way:

$$\frac{P(i) - P(i_0)}{i - i_0} \frac{1}{P(i_0)} \approx P'(i_0) \frac{1}{P(i_0)} \qquad (9.13)$$

Disregarding the $\frac{1}{P(i_0)}$ factor on both sides, we see that Equation 9.13 is just another way of saying that the slope of the secant line (left-hand side) approximates the derivative (right-hand side) when the change in i ($i - i_0$) is small. It is common to use the notation $\Delta i = i - i_0$. Using this notation we can write $P(i) - P(i_0) \approx P'(i_0)\Delta i$.

The first derivative, $P'(i)$ is known as the **duration** of the portfolio. While this number can be useful, a more useful measure is the relative rate of change: $\frac{P'(i)}{P(i)}$. Finally, since $P'(i)$ is usually negative, a negative sign is attached to produce a (usually) positive measure of interest sensitivity known as the **modified duration.**

Definition: If $P(i)$ represents the yield (or price) of a sequence of cash flows with interest being converted once per period, the **modified duration** for a sequence of cash flows is denoted by $D(i, 1)$ and is defined as

$$D(i, 1) = -\frac{P'(i)}{P(i)} \tag{9.14}$$

The notation $D(i, 1)$ is used to remind us that the calculation is based on interest being compounded once per period.

Since $P'(i) < 0$ in most cases, it is also the case that $D(i, 1) > 0$ in most cases. As a result, larger values of $D(i, 1)$ correspond to smaller (more negative) values of $\frac{P'(i)}{P(i)}$ and hence to greater sensitivity to interest rate fluctuations.

If interest is converted m times per year with a nominal interest rate of $i^{(m)}$ we have (from Chapter 2):

$$i = \left(1 + \frac{i^{(m)}}{m}\right)^m - 1 \tag{9.15}$$

We define $D(i, m)$ by analogy with $D(i, 1)$ as

$$D(i, m) = -\frac{\frac{dP}{di^{(m)}}}{P(i)} \tag{9.16}$$

We now use the chain rule to obtain an explicit expression for $D(i, m)$. We first compute:

$$\frac{dP}{di^{(m)}} = \frac{dP}{di}\frac{di}{di^{(m)}}$$

$$= P'(i)\left(1 + \frac{i^{(m)}m}{m}\right)^{m-1}$$

$$= P'(i)\frac{\left(1 + \frac{i^{(m)}}{m}\right)^m}{1 + \frac{i^{(m)}}{m}}$$

$$= P'(i)\frac{1 + i}{1 + \frac{i^{(m)}}{m}} \tag{9.17}$$

Using Equation 9.16, we then have

$$D(i, m) = -\frac{\frac{dP}{di^{(m)}}}{P(i)}$$

$$= -\frac{P'(i)\frac{1+i}{1+\frac{i^{(m)}}{m}}}{P(i)}$$

$$= D(i, 1)\frac{1 + i}{1 + \frac{i^{(m)}}{m}} \tag{9.18}$$

If we cross multiply, we obtain

$$D(i, m)\left(1 + \frac{i^{(m)}}{m}\right) = D(i, 1)(1 + i) \tag{9.19}$$

Hence the left-hand side of Equation 9.19 is independent of the value of m.

9.4   Macaulay Duration $(D(i, \infty)) = D_{\mathrm{mac}}(i)$

Recall from Chapter 2, that we defined the force of interest as $\delta_t = \frac{a'(t)}{a(t)}$. In the case of compound interest we found that δ_t is a constant and that $\delta_t = \delta = \ln(1 + i)$. In the case of continuous compounding we define $D(i, \infty)$ as follows

$$
\begin{aligned}
D(i, \infty) &= -\frac{\frac{dP}{d\delta}}{P(i)} \\
&= -\frac{1}{P(i)}\frac{dP}{di}\frac{di}{d\delta} \\
&= D(i, 1)(1 + i) \tag{9.20}
\end{aligned}
$$

Using Equation 9.19 we obtain

$$D(i, \infty) = D(i, m)\left(1 + \frac{i^{(m)}}{m}\right) \tag{9.21}$$

The function $D(i, \infty)$ is known as the Macaulay Duration and is often denoted by $D_{\mathrm{mac}}(i)$. The Macaulay Duration is most often expressed using $P(i)$ and $P'(i)$ as

$$D_{\mathrm{mac}}(i) = -\frac{P'(i)}{P(i)}(1 + i) = D(i, 1)(1 + i) \tag{9.22}$$

Another useful formula relates the value of $P'(i)$ to the Macaulay Duration:

$$P'(i) = -\frac{D_{\mathrm{mac}}(i) \cdot P(i)}{(1 + i)} \tag{9.23}$$

Example 9.4: A cash flow has a price of $50,000 and Macaulay Duration of 5.6 at $i = 4.5\%$. Compute the approximate change in price if the interest rate changes to 4.6%.

Solution: We use $P(i) - P(i_0) \approx P'(i_0)\Delta i$. In this case $\Delta i = .001$. We compute $P'(.045)$ using Equation 9.23:

$$P'(.045) = -\frac{5.6 \cdot 50000}{1.045} = -297,942.58$$

Hence $P(i) - P(i_0) \approx P'(i_0)\Delta i = -296972.57 \cdot .001 = -\$267.94$.

$D(i, \infty)$ measures the sensitivity of a sequence of cash flows to changes in interest in the case that interest is converted continuously. Recall that in the case that interest is compounded continuously, we have $a(t) = e^{\delta t}$ and hence that the present value of payment C_t is given by $C_t e^{-\delta t}$. In this case we obtain

$$P(\delta) = \sum_{t \geq 0} C_t e^{-\delta t} \tag{9.24}$$

Thus

$$
\begin{aligned}
D(i, \infty) &= -\frac{\frac{dP}{d\delta}}{P(i)} = -\frac{\sum_{t \geq 0} C_t(-t) e^{-\delta t}}{\sum_{t \geq 0} C_t e^{-\delta t}} \\
&= \frac{\sum_{t \geq 0} C_t \cdot t \cdot e^{-\delta t}}{\sum_{t \geq 0} C_t e^{-\delta t}}
\end{aligned}
\tag{9.25}
$$

We can expand the denominator of this last fraction and simplify to obtain

$$
\begin{aligned}
D(i, \infty) &= \sum_{t \geq 0} \frac{t C_t e^{-\delta t}}{\sum_{t \geq 0} C_t e^{-\delta t}} \\
&= \sum_{t \geq 0} \left(\frac{C_t e^{-\delta t}}{\sum_{t \geq 0} C_t e^{-\delta t}} \right) \cdot t \\
&= \sum_{t \geq 0} \left(\frac{C_t e^{-\delta t} \cdot t}{P(i)} \right)
\end{aligned}
\tag{9.26}
$$

We used the fact that $\delta = \ln(1 + i)$, so that $e^{-\delta t} = e^{-\ln(1+i)t} = \frac{1}{(1+i)^t} = (1 + i)^{-t}$. The notation $D_{\text{mac}}(i)$ is also used for the Macaulay Duration.

This can be written in a slightly easier form using $v^t = (1 + i)^{-t}$

$$D_{\text{mac}} = \frac{\sum t \cdot C_t v^t}{P(i)} \tag{9.27}$$

Since we are using the time of payment to evaluate D_{mac} we must convert nominal interest rates per year to effective rates at the frequency of bond payments to compute both the coupon payments and the price of the bond. Note that for bonds this differs from our previous practice of computing coupon and yield rates as nominal rates.

Example 9.5 Compute the Macaulay Duration of a three-year bond with annual coupons. The coupon rate is 6%, the face value is $12,000, the redemption value is $15,000. Evaluate at $i = 7\%$.

Solution: To compute the duration, we first need the price. We use the cash flow model to compute the price. We have $n = 3$, $C = 15,000$, $F = 12,000$, and $r = .06$. See Table 9.1.

TABLE 9.1

N	I/Y	PV	PMT	FV
3	7	$CPT = 14,134$	-720	15,000

The price is the denominator of our duration formula. We now compute the numerator noting that $C_t = Fr = 720$.

$$\sum t \cdot C_t v^t = 1 \cdot 720 \cdot v + 2 \cdot 720 \cdot v^2 + 3 \cdot 15720 \cdot v^3 = 40,427$$

We then have

$$D_{\text{mac}}(.07) = \frac{40427}{14134} = 2.86$$

Example 9.6 Compute the Macaulay Duration of a two-year bond with semiannual coupons if the coupon rate is a effective 5% per year, the face value is \$15,000, the redemption value is \$14,000, and the yield rate is an effective annual rate of 4%.

Solution: We convert to the effective semiannual interest rate for both coupon payments and yield rate. For the coupon rate we can use ICONV; See Table 9.2

TABLE 9.2

Key	
C/Y	2
EFF	5
NOM	$CPT = 4.939$

The coupon payment is then $15000 \cdot \frac{4.939}{2} = 370.43$.

To price the bond we need the semiannual yield rate. Since we will enter this into the TVM keys, its easier to use the TVM keys to compute the interest rate as well: See Table 9.3

TABLE 9.3

N	I/Y	PV	PMT	FV
2	$CPT = 1.98039$	1	0	-1.04

We can now compute the price of the bond: See Table 9.4

TABLE 9.4

N	I/Y	PV	PMT	FV
4		$CPT = 14,354.96$	-370.43	$-14,000$

We use Equation 9.27 to compute the Macaulay Duration. We are counting in half-year periods. The denominator is just the price: \$14,354.96. The numerator is

$$\sum t \cdot C_t v^t = \sum_{k=1}^{4} \left(\frac{k}{2}\right)(370.43)v^{\frac{k}{2}} + 14354.96v^2 = 27,634.22$$

Recall that v in this case is based on the effective annual interest rate i. That is $v = \frac{1}{1+i}$. Hence $v^{\frac{k}{2}} = (1+i)^{-\frac{k}{2}}$. We can use this expression or note that $(1+i)^{\frac{1}{2}} = 1.0198039$. Finally, we compute the Macaulay Duration: $\frac{27,634.22}{14354.96} = 1.93$.

If we look at the expression $\sum t \cdot Fr \cdot v^t = \sum (t \cdot Fr)v^t$ in the case of a coupon bond we see that we can interpret this as the present value of a $P - Q$ annuity with $P = Q = Fr$. We can then use the formula for a P-Q annuity to obtain the following formula for the duration of a coupon bond:

$$\frac{Fr\left(a_{\overline{n}|,i}\right) + \frac{a_{\overline{n}|,i}-nv^n}{i}}{P(i)} \tag{9.28}$$

9.4.1 Duration of a Coupon Bond for which $F = P = C$

We will derive formulas for $D(i, m)$ and $D(i, \infty)$ for a par value bond with a life of n years with m coupons per year purchased at its redemption value. In the notation of Chapter 7, this means the $F = P = C$.

If the bond earns an effective annual interst rate of i, the interest per coupon period is $\frac{i^{(m)}}{m} = (1+i)^{\frac{1}{m}} - 1$. Since $F = P = C$, $i = r$ and so the coupon payment is $P \cdot \frac{i^{(m)}}{m}$. If we let $N = n \cdot m$, the bond provides a total of N coupon payments at times $\{\frac{1}{m}, \frac{2}{m}, \frac{3}{m}, \ldots \frac{N}{m} = n\}$ plus its redemption value at time n. We have $P(i) = P$ and now use the same technique as in Example 9.4 to compute the Macaulay Duration

$$D(i, \infty) = \frac{\sum_{k=1}^{N} P \cdot \left(\frac{i^{(m)}}{m}\right)(1+i)^{-\frac{k}{m}} \cdot \left(\frac{k}{m}\right) + P \cdot (1+i)^{-n} \cdot n}{P}$$

$$= \left(\sum_{k=1}^{N} \left(\frac{i^{(m)}}{m}\right) \cdot (1+i)^{-\frac{k}{m}} \cdot \left(\frac{k}{m}\right)\right) + (1+i)^{-n} \cdot n$$

$$= \left(i^{(m)} \sum_{k=1}^{N}(1+i)^{-\frac{k}{m}} \cdot \left(\frac{k}{m^2}\right)\right) + (1+i)^{-n} \cdot n \tag{9.29}$$

Now we engage in a bit of fancy foot work. We note that $(1+i) = \left(1 + \frac{i^{(m)}}{m}\right)^m$ so that $(1+i)^{-\frac{k}{m}} = (1 + \frac{i^{(m)}}{m})^{-k}$. We now rewrite Equation 9.29 so as to be able to use some earlier formula for increasing payment annuities.

$$D(i, \infty) = \left(\frac{i^{(m)}}{m^2} \sum_{k=1}^{nm} \left(1 + \frac{i^{(m)}}{m} \right)^{-k} \cdot k \right) + (1+i)^{-n} \cdot n$$

$$= \frac{i^{(m)}}{m^2} (Ia_{\overline{nm}|, \frac{i^{(m)}}{m}}) + (1+i)^{-n} n$$

$$= \frac{i^{(m)}}{m^2} \cdot \frac{\ddot{a}_{\overline{nm}|, \frac{i^{(m)}}{m}} - nm \left(\frac{1}{1 + \frac{i^{(m)}}{m}} \right)^{nm}}{\frac{i^{(m)}}{m}} + (1+i)^{-n} n \qquad (9.30)$$

One more tedious round of algebra yields the following formula

$$D(i, \infty) = \ddot{a}_{\overline{n}|,i}^{(m)} = \left(1 + \frac{i^{(m)}}{m} \right) \cdot \frac{i}{i^{(m)}} \cdot a_{\overline{n}|,i} \qquad (9.31)$$

For an annual coupon bond we have $m = 1$ and equation 9.31 simplifies to $(1+i)a_{\overline{n}|,i}$. Dividing by $(1+i)$ yields the following formula for an annual coupon bond with $F = P = C$ (and hence $r = i$)

$$D_{\mathrm{mac}}(i) = D(i, \infty) = a_{\overline{n}|,i} \qquad (9.32)$$

Example 9.7 Find the Macaulay Duration for a 20 year $F = C = P$ bond with quarterly coupons with an annual effective yield of 7.5%

Solution:

$$\left(1 + \frac{i^{(4)}}{4} \right)^4 = 1.075 \Rightarrow \left(1 + \frac{i^{(4)}}{4} \right) = 1.01824 \Rightarrow i^{(4)} = .072978.$$

We compute $a_{\overline{n}|,i}$ (Table 9.5)

TABLE 9.5

N	I/Y	PV		PMT	FV
20	7.5	$CPT = 10.194$		−1	0

$$D(.075, \infty) = \left(1 + \frac{i^{(4)}}{4} \right) \left(\frac{i}{i^{(m)}} \right) a_{\overline{n}|,i} = 1.075 \cdot \frac{.075}{.0729} \cdot 10.19 = 11.26$$

You might be wondering why the term duration is used. To see why, we compute the duration of a zero-coupon bond. Such a bond has only one payment made at the end of its life Equation 9.25 reduces to a single term and we find that the Macaulay Duration of a zero-coupon bond is the term of the bond

$$D_{\mathrm{mac}}(i) = D(i, \infty) = \frac{C \cdot (1+i)^{-n} \cdot n}{C \cdot (1+i)^{-n}} = n$$

In Example 9.5 we found that duration of a twenty-year bond with quarterly payments was 11.26 years. That means it has roughly the same risk (sensitivity to changes in interest rates) as a 11.26 year zero-coupon bond.

9.4.2 Duration of an Amortized Loan

We now compute the duration of an amortized loan. For simplicity, we assume that the loan has no early repayment option. We consider a loan being amortized over n years with payments of R each mth of a year. We let $N = n \cdot m$ be the total number of payments. The numerator of Equation 9.25 is the present value of the loan. If the effective annual interest rate is i, then

$$PV = \sum_{k=1}^{N} R \cdot (1+i)^{-\frac{k}{m}} \tag{9.33}$$

The denominator of Equation 9.25 is the interest-weighted sum of the payments, which are made (as in the previous example at times $\{\frac{1}{m}, \frac{2}{m}, \frac{3}{m}, \dots \frac{N}{m}\}$):

$$\sum_{k=1}^{N} \frac{k}{m} R(1+i)^{-\frac{k}{m}} \tag{9.34}$$

We proceed as follows

$$
\begin{aligned}
D(i,\infty) &= \frac{\sum_{k=1}^{N} \frac{k}{m} R(1+i)^{-\frac{k}{m}}}{\sum_{k=1}^{N} R(1+i)^{-\frac{k}{m}}} \\
&= \frac{\sum_{k=1}^{N} \frac{k}{m^2}(1+i)^{-\frac{k}{m}}}{\sum_{k=1}^{N} \frac{1}{m}(1+i)^{-\frac{k}{m}}} \\
&= \frac{\frac{\ddot{a}_{\overline{n}|}^{(m)} - nv^n}{i^{(m)}}}{a_{\overline{n}|}^{(m)}} \\
&= \frac{1}{d^{(m)}} - \frac{n}{(1+i)^n - 1}
\end{aligned}
\tag{9.35}
$$

Duration of an Amortized Loan

$$D_{\text{mac}}(i) = \frac{1}{d^{(m)}} - \frac{n}{(1+i)^n - 1} \tag{9.36}$$

Example 9.8 What is the Macaulay Duration of a fifteen-year mortgage with monthly payments if the effective annual interest rate is 6%? Compute the duration for a thirty-year mortgage at the same interest rate.

Solution: We need to find $d^{(12)} = 12(1 - (1+i)^{-\frac{1}{12}}) = .058127667$, or on the TVM keys (remember we use negative values for both N and I in the case of discount calculations). See Table 9.6.

TABLE 9.6

N	I/Y	PV	PMT	FV
−12	$CPT = -.48439$	1	0	−1.06

$$\times 12 = 5.81276\% = .05812766$$

For $n = 15$, we have

$$D(i, \infty) = \frac{1}{d^{(m)}} - \frac{n}{(1+i)^n - 1}$$

$$= \frac{1}{.0581276} - \frac{15}{(1.06)^{15} - 1}$$

$$= 17.20351157 - 10.74069099 = 6.46$$

For $n = 30$

$$D(i, \infty) = 17.20351157 - \frac{30}{(1.06)^{30} - 1} = 10.88$$

We can interpret these numbers using our result for zero-coupon bonds. The fifteen-year mortgage carries roughly the same risk as a 6.5 year zero-coupon bond. The thirty-year mortgage carries roughly the same risk as a 10.9 year zero coupon bond. This is part of the reason that fifteen-year mortgages typically carry a lower interest rate than do thirty-year mortgages.

9.5   Duration of a Portfolio

We assume an effective annual interest rate of i and a set of s assets $A^k\{k = 1, 2, \ldots, s\}$. A set of assets is referred to as a portfolio. We further assume that each asset A^k consists of a set of cash flows $\{C_t^k\}$. That is $A^k = \{C_1^k, C_2^k, C_3^k, \ldots, C_{nk}^k\}$. If we denote the present value of asset A^k by $P^k(i)$ we have

$$P^k(i) = \sum_{t=1}^{n_k} C_t^k (1+i)^{-t} \tag{9.37}$$

We can then compute the Macaulay durations of asset A_k:

$$D^k(i, \infty) = \sum_t \frac{C_t^k \cdot t \cdot (1+i)^{-t}}{P^k(i)} \tag{9.38}$$

Finally, if $P(i)$ represents the price for the entire portfolio, we have $P(i) = \sum_{k=1}^{s} P^k(i)$. We can then compute the Macaulay Duration of the portfolio by adding up all the weighted present values of all future cash flows and dividing the total by $P(i)$

$$D^{\text{portfolio}}(i, \infty) = \frac{\sum_{k=1}^{s} \left(\sum_{t=1}^{n_k} C_t^k (1+i)^{-t} \cdot t \right)}{P(i)}$$

$$= \left(\sum_{k=1}^{s} \sum_{t=1}^{n_k} \frac{C_t^k (1+i)^{-t} \cdot t}{P^k(i)} \right) \cdot \frac{P^k(i)}{P(i)}$$

$$= \left(\sum_{k=1}^{s} D^k(i, \infty) \right) \cdot \frac{P^k(i)}{P(i)} \tag{9.39}$$

In practice this is usually computed using the very first line of Equation 9.39. We compute the total price by adding the prices of each asset. We then compute the term $\sum_t C_t^k (1+i)^{-t} \cdot t$ for each asset and add these. Finally, we divide.

Example 9.9 Compute $D_{\text{mac}}^{\text{portfolio}}(.04, \infty)$ for a portfolio which consists of the following bonds:

Bond A: A 3-year \$5,000 par value $F = C$ bond with a 5% coupon rate and semiannual coupons.

Bond B: A 4-year \$8,000 zero-coupon bond.

Bond C: A 6-year \$10,000 par value $F = C$ bond with an 8% coupon rate and semiannual coupons.

Solution: We compute the prices of each bond so as to yield an effective annual rate of .04. We will designate the prices using P^A for the price of bond A, P^B for the price of bond B, etc.

Bond A: The coupon payment is $5000((1.05)^{\frac{1}{2}} - 1) = \$123.48$. The yield rate per term is $(1.04)^{\frac{1}{2}} - 1 = 1.9804$. We compute the price using the TVM keys as shown in Table 9.7

TABLE 9.7

N	I/Y	PV	PMT	FV
6	1.9804	$CPT = 5,137.10$	-123.48	-5000

Bond B: This one is easy, it's just the present value of \$8,000

$$P^B = 8000(1.04)^{-4} = \$6,838.43$$

Bond C: Proceeding as with Bond A, we obtain a coupon payment of 392.30 and a price of \$12,056.90

We now add these three numbers to obtain the numerator, $P(.04) = \$24,032.43$

Now we compute term $\sum_t C_t^k (1+i)^{-t} \cdot t$ for each bond:

Bond A: $\sum_t C_t^A (1+i)^{-t} \cdot t = 123.48 \cdot \sum_{k=1}^{6} (1.04)^{-\frac{k}{2}} \cdot \frac{k}{2} + 5000 \cdot 3 \cdot (1.04)^{-3} = \$14,526.37$

Bond B: $\sum_t C_t^B (1+i)^{-t} \cdot t = 8000(1.04)^{-4} \cdot 4 = 4 \cdot P^B = 27,353.734$

Bond C:

$$\sum_t C_t^C (1+i)^{-t} \cdot t = 392.30 \cdot \sum_{k=1}^{12} (1.04)^{-\frac{k}{2}} + 10000 \cdot \sum_{k=1}^{6} \cdot (1.04)^{-6} = 60,433.53$$

The total of $\sum_t C_t (1+i)^{-t} \cdot t$ for all three bonds is 102,313.75. Thus

$$D_{\text{mac}}(.04) = D(.04, \infty) = \frac{102,313.75}{24,0032.43} = 4.256$$

9.5.1   Duration of a Portfolio of Instruments of Known Duration

If we already know the durations and prices of a collection of instruments, we can compute the duration of the portfolio as the dollar-weighted average of the individual instruments. In particular if we have n instruments with prices P_i and durations D_i the duration of the portfolio is given by:

$$D(i, \infty) = \frac{\sum_{i=1}^{n} P_i \cdot D_i}{\sum_{i=1}^{n} P_i} \tag{9.40}$$

9.6   Convexity

Convexity is derived from the second degree Taylor polynomial for the price curve and is closely related to the concavity of the graph. Recall that the second degree Taylor Polynomial is given by

$$P_2(i, i_0) = P(i_0) + P'(i_0)(i - i_0) + \frac{P''(i_0)}{2}(i - i_0)^2 \tag{9.41}$$

The idea is that $P(i) \approx P_2(i)$ so we have

$$P(i) \approx P_2(i) = P(i_0) + P'(i_0)(i - i_0) + \frac{P''(i_0)}{2}(i - i_0)^2$$

$$P(i) - P(i_0) \approx P'(i_0)(i - i_0) + \frac{P''(i_0)}{2}(i - i_0)^2$$

$$\frac{P(i) - P(i_0)}{P(i_0)} \approx \frac{P'(i_0)}{P(i_0)}(i - i_0) + \frac{P''(i_0)}{2 \cdot P(i_0)}(i - i_0)^2 \tag{9.42}$$

Recall that we defined the modified duration by $D(i_0, 1) = -\frac{P'(i_0)}{P(i_0)}$. We define the **modified convexity**[3] as:

$$C(i_0, 1) = \frac{P''(i_0)}{P(i_0)} \tag{9.43}$$

With these definitions in hand we can re-write the last of Equations 9.42 as

$$\frac{P(i) - P(i_0)}{P(i_0)} \approx -D(i_0, 1)(i - i_0) + C(i_0, 1)\frac{(i - i_0)^2}{2} \tag{9.44}$$

For most price curves, convexity is positive. This is true because the price is positive, and the price curve is concave up – hence both terms in Equation 9.44 are positive. Figure 9.5 is a plot of the convexity and duration for the price function of Example 9.2

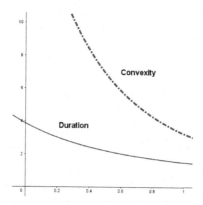

FIGURE 9.5

In a manner analogous to the Macaulay Duration, the Macaulay Convexitity for a sequence of cash flows C_t at times t and interest rate i is defined by

$$C_{\text{max}}(i) = C(i, \infty) - \frac{\sum_t (C_t(1 + i)^{-t} \cdot t^2)}{P(i)} \tag{9.45}$$

We can write this in terms of v as

$$C(i, \infty) = \frac{\sum t^2 \cdot C_t \cdot v^t}{P(i)} \tag{9.46}$$

The duration and convexity are similar to the notion of first and second moments you may have encountered in statistics. We find the convexity of

---

[3]The convexity is just $P''(i_0)$, the second derivative of the price function. Recall that if $P'' > 0$, the graph is concave up, while if $P'' < 0$, the graph is concave down.

a portfolio in a manner similar to that for the duration.

$$C_{\text{mac}}^{\text{portfolio}}(i, \infty) = \frac{\sum_{k=1}^{s} \sum_{t}(C_t^k(1+i)^{-t} \cdot t^2)}{\sum_{k=1}^{s} P^k(i)} \tag{9.47}$$

Example 9.10 Find the convexity of the portfolio in Example 9.7

Solution: The numerator remains the same. We just need to compute each bond's contribution to the denominator.

Bond A: $\sum_t C_t^A(1+i)^{-t} \cdot t^2 = 123.48 \cdot \sum_{k=1}^{6}(1.04)^{-\frac{k}{2}} \cdot (\frac{k}{2})^2 + 5000 \cdot 3^2 \cdot (1.04)^{-3} = 42,560$

Bond B: $\sum_t C_t^B(1+i)^{-t} \cdot t^2 = 8000(1.04)^{-4} \cdot 16 = 109,415$

Bond C: $\sum_t C_t^C(1+i)^{-t} \cdot t^2 = 392.30 \cdot \sum_{k=1}^{12}(1.04)^{-\frac{k}{2}} \cdot (\frac{k}{2})^2 + 10000 \cdot 36 \cdot (1.04)^{-6} = 337,632$

Thus $C^{\text{portfolio}}(i, \infty) = \frac{489,607}{24,032.43} = 20.37$

As was the case for duration, **zero-coupon bonds** have a particularly simple convexity function

The Convexity of a Zero-Coupon Bond

$$C(i, \infty) = n^2 \tag{9.48}$$

9.6.1 Convexity of a Portfolio of Instruments with Known Convexities

We can compute the convexity of a portfolio in the same we as we computed its duration – it is the weighted average of the convexities of the instruments in the portfolio. If we have a set of instruments with prices P_i and convexities C_i the convexity of the portfolio is given by

$$C(i, \infty) = \frac{\sum_{i=1}^{n} C_i \cdot P_i}{\sum_{i=1}^{n} P_i} \tag{9.49}$$

9.7 Immunization

We now address the problem which motivated the ideas of duration and convexity: How can a company with a set of known future obligations create a portfolio of investments which will satisfy that obligation even if interest rates fluctuate? In Example 9.1 we studied a portfolio that was capable of meeting a known future obligation for interest rates ranging from 1% to 20%. The process of creating a portfolio whose predicted future value is relatively stable with respect to changes in interest rates is known as **immunization**.

Assume that we have a set of cash flows which consist of assets and of liabilities which become available or fall due, respectively at various times. In particular, suppose there is a set of assets A_t which become available at time t, and a set of liabilities L_t which fall due at time t. We can compute the **net value at time** t by the expression $A_t - L_t$. Note that we might well have either $A_t = 0$ or $L_t = 0$ for some values of t. Remember also that t is not just a subscript – it records the time at which an asset becomes available or a liability comes due. The present value of the sequence of differences $A_t - L_t$ computed for a particular interest rate, i, is called the **surplus** and is denoted by $S(i)$. This is easy to compute – we just compute the present value of each term $A_t - L_t$ and add:

$$S(i) = \sum_t (A_t - L_t)(1 + i)^{-t} = \sum_t A_t(1 + i)^{-t} - \sum_t L_t(1 + i)^{-t} \quad (9.50)$$

We note a few facts.

1. $S(i) > 0$ indicates that the present value of the assets exceeds the present value of the liabilities. The account is said to show a surplus of $S(i)$.

2. $S(i) < 0$ indicates that the present value of the liabilities exceeds the present value of the assets. This is referred to as a negative surplus!

3. $S(i) = 0$ indicates that assets and liabilities have equal present value.

If you are putting together a portfolio of assets to pay off a known future obligation (such as a life insurance policy) you want to ensure that $S(i_0) \geq 0$, but don't want to tie up any more money than is needed. The same would apply for an escrow account created to ensure funds for a future tax obligation. The goal, then, is to keep $S(i_0) \geq 0$, but as small as possible.

Since we can't be sure of future interest rates, we also want to create a portfolio with a positive surplus over a range of interest rates. We thus want to minimize the surplus subject to the additional condition that this minimum be positive. We are looking for a non-negative minimum value of $S(i_0)$ for a particular i_0. From calculus, we know that a minimum occurs for a continuous differentiable function when $S'(i_0) = 0$ and $S''(i_0) > 0$. As you may recall, the second condition is known as the second derivative test for a minimum.

Redington Immunization is a technique for putting together a portfolio which makes use of this simple fact from calculus. We construct our portfolio so that

$$S(i_0) = 0$$
$$S'(i_0) = 0$$
$$S''(i_0) = 0 \quad (9.51)$$

A function which satisfies these three conditions at a point i_0 has a local minimum since its derivative is 0 and the graph is concave up.

We can translate each condition in Equation 9.37 into the language of cash flows as follows

Redington Immunization Conditions

$S(i_0) = 0 \Rightarrow$ Present Value of the Assets = Present Value of the Liabilities

$S'(i_0) = 0 \Rightarrow$ The Macaulay Duration of the Assets = Duration of the Liabilities

$S''(i_0) \geq 0 \Rightarrow$ Convexity of the Assets $\geq$ Convexity of the Liabilities

$$(9.52)$$

    The first step $(S(i_0) = 0)$ is sometimes called asset matching. The second step is then called duration matching. The third requirement is called the convexity condition. As we shall see, we won't have enough information to satisfy all three every time. In practice we find an i_0 which matches assets and durations and then check to see if it also satisfies the convexity condition. If it does, then we have found a solution to the Redington Immunization problem. We will always use the Macaulay Duration in this context.

    We will consider the special case where we wish to match one liability with two assets. All three items will usually be zero-coupon bonds so the duration will match the period. See Table 9.8.

TABLE 9.8

Item	Value When Due	Time Due	Duration
Liability	L	t	d
Asset 1	A_1	t_1	d_1
Asset 2	A_2	t_2	d_2

We will match assets at time $t = 0$. The asset matching equation is then

Asset Matching Equation

$$\frac{L}{(1+i)^t} = \frac{A_1}{(1+i)^{t_1}} + \frac{A_2}{(1+i)^{t_2}} \tag{9.53}$$

Note: the numbers L, A_1, A_2 represent that values of the assets at different times d, d_1, and d_2 respectively. Equation 9.53 equates these values at time $t = 0$.

    If we want the duration of the liability to equal the combined durations of the two assets we have (using Equation 9.33):

$$\frac{d\frac{L}{(1+i)^t}}{\frac{L}{(1+i)^t}} = \frac{d_1\frac{A_1}{(1+i)^{t_1}} + d_2\frac{A_2}{(1+i)^{t_2}}}{\frac{A_1}{(1+i)^{t_1}} + \frac{A_2}{(1+i)^{t_2}}} \tag{9.54}$$

If we have already matched assets the denominators of these fractions are equal and we have the following equation for **duration matching**

Duration Matching Equation

$$d\frac{L}{(1+i)^t} = d_1\frac{A_1}{(1+i)^{t_1}} + d_2\frac{A_2}{(1+i)^{t_2}} \qquad (9.55)$$

The convexity condition results in the following requirement:

Convexity Condition

$$d^2\frac{L}{(1+i)^t} \le d_1^2\frac{A_1}{(1+i)^{t_1}} + d_2^2\frac{A_2}{(1+i)^{t_2}} \qquad (9.56)$$

We can simplify the calculations quite a bit by solving for the purchase price (present value) of each bond. We see set $B_1 = \frac{A_1}{(1+i)^{t_1}}$, $B_2 = \frac{A_2}{(1+i)^{t_2}}$ we obtain

Immunization Formulas

$$\frac{L}{(1+i)^t} = B_1 + B_2 \qquad (9.57)$$

$$d\frac{L}{(1+i)^t} = d_1 B_1 + d_2 B_2 \qquad (9.58)$$

$$d^2\frac{L}{(1+i)^t} = d_1^2 B_1 + d_2^2 B_2 \qquad (9.59)$$

Example 9.11 A company is obligated to pay $250,000 in four years. They will purchase a combination of two-year and five-year zero-coupon bonds each of which will be sold to yield 4.5%. Show that is possible to do so in such a way as to satisfy the Redington conditions and investigate the stability of the surplus of this portfolio with respect to fluctuation in interest rates.

Solution: Assume x is the redemption value of the two-year bonds and y is the redemption value of the five-year bonds. We have $i_0 = .045$. We start with $S(.045) = 0$.

$$\frac{x}{1.045^2} + \frac{y}{1.045^5} = \frac{250000}{1.045^4} = 209,640.34$$

The duration of the portfolio must be equal to the duration of the obligation, which is 4 (it's just like a zero-coupon bond). We compute the weighted average of the durations to obtain the duration of the portfolio. Hence we have the additional equation:

$$\frac{4\cdot 250000}{1.045^4} = \frac{2x}{1.045^2} + \frac{5y}{1.045^5}$$

Solving the system of two equations in two unknowns results in $x = 76,310.83$, $y = 174,166.67$. We need to purchase $\frac{76,310.83}{1.045^2} = 69,880.11$ worth of the two-year bonds and $\frac{174,166.67}{1.045^5} = 138,760.22$ worth of the five-year bonds.

Now we analyze the situation with respect to fluctuating interest rates. The two-year bonds will provide 76,310.82 at the end of two years. The 5-year bonds have a future value (at the end of five years) of 174,166.64. Suppose that the prevailing interest rate at that time is i. Then the present value (at time $t = 2$) of the five-year bonds is $174,166(1+i)^{-3}$ while the two-year bonds are worth 76,310.82. We thus have $76,310.82 + 174,166.64(1+i)^{-3}$ to accumulate at an interest rate of i over two years. After two years this will accumulate to

$$FV(i) = (76,310.82 + 174,166.64(1+i)^{-3}) \cdot (1+i)^2$$

A plot of this function reveals that the accumulated value does have a minimum at $i = .045$ and stays above \$250,000 near that value. See Figure 9.6.

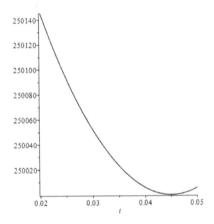

FIGURE 9.6

If we use the B_i formulation we solve for the present value using

$$x + y = 209640$$
$$2x + 5y = 4 \cdot 209640$$

This yields $x = 69,880$, $y = 139,760$ computing future values yields \$76,310 and \$174,166 which is consistent with our first method. The convexity check goes the same way regardless of how we find the values of our bonds.

9.8   Full Immunization

In certain special cases it's possible to ensure that the net present value of a portfolio is never negative. This is called full immunization. Here's the theorem we need:

Theorem. Suppose that the force of interest is δ_0. We assume a portfolio of a single liability of value L to be paid at time t. This is to be immunized with two assets X, Y to be paid at t with $0 < x < t$ and $t + y$ with $y > 0$. In this case the following two conditions ensure that $S(\delta) > 0$ for $\delta \neq \delta_0$.

$$S(\delta_0) = 0$$
$$S'(\delta_0) = 0 \tag{9.60}$$

A tedious algebra exercise yields the following system of two equations in four unknowns X, x and Y, y:

$$Xe^{\delta_0 x} + Ye^{-\delta_0 y} = L$$
$$xXe^{\delta_0 x} - yYe^{-\delta_0 y} = 0 \tag{9.61}$$

Since $\delta_0 = -\ln(v_0)$ we can convert these to an equivalent set of equations involving v_0:

$$Xv_0^{-x} + Yv_0^y = L$$
$$xXv_0^{-x} - yYv_0^y = 0 \tag{9.62}$$

The first of equations 9.62 computes the value of L, X and Y time t. Since L is due at t, its value at this time is just L. X is due x periods prior to time t, while Y is due y periods after time t.

Returning to Example 9.9, we can compute the two values in Equation 9.62. The first is just the asset matching equation. For the second we obtain a value of -1.17 which shows we very nearly obtained full immunization!

We can't solve the system in 9.62 unless we assume values for two of the four variables. Here's an example:

Example 9.12 A portfolio consists of a liability of \$50,000 due in five years. It is to be immunized with assets of X, Y due in two and seven years respectively. Write down and solve the immunization equations at $i_0 = .02$.

Solution: We have $v_0 = .98038$. The equations are

$$X.98038^{-3} + Y.98038^2 = 50,000$$
$$3X.98038^{-3} - 5Y.98038^2 = 0$$

Solving yields $X = -94232$, $Y = 1.56060$. So no full immunization exists in this case.

Exercises for Chapter 9

1. Suppose that $C_0 = 150$, $C_3 = 200$, and $C_5 = 100$.
   a) Find an equation for $P_1(i,j)$ at $j = .03$
   b) Find an equation for $P_2(i,j)$ at $j = .03$

c) Graph $P(i)$ for $i = 0...5$

d) Compare $P(i)$, $P_1(i, j)$, and $P_2(i, j)$ for $j = .03$ and $i = .01, .025, .04$. Explain the magnitudes of these quantities

e) Graph $P(i) - P_1(i)$ for $i = .02...04$

f) Graph $P(i) - P_2(i)$ for $i = .02...04$

2. Repeat the analysis of Example 9.1 for $i = .08$.

3. Compute the minimum value of the portfolio of Example 9.1 using calculus.

4. Compute and graph the price curve for the sequence of cash flows:
$$C_1 = 300, \ C_3 = 400, \ C_5 = -200, \ C_7 = 300$$

5. For problem 4):

a) Find the tangent line approximation at $j = .03$

b) Find the second order Taylor approximation at $j = .03$

c) Graph the function and its linear and quadratic approximations in the region around $j = .03$

6. Compute and graph the modified duration for the cash flow defined by
$$C_1 = 400, \ C_3 = 500, \ C_5 = 550 \text{ for } 0 \leq i \leq 1.$$

7. Compute the Macaulay Duration of a ten-year 8.8% bond with a par value of \$25,000 with semiannual coupons and a redemption value of \$22,000. The bond is priced to yield an effective annual interest rate of 6%.

8. Compute the Macaulay Duration of a fifteen-year $F = P = C$ bond with monthly coupons and an effective annual yield of 8.5%.

9. Compute the Macaulay Duration for a thirty-year mortgage with an effective annual interest rate of 6%

a) Payments are made monthly

b) Payments are made quartely

c) Explain the relative sizes of your answers to a) and b)

10. Compute $D^{\text{portfolio}}(.03, \infty)$ for a portfolio which consists of the following bonds:

Bond A: A three-year \$6,000 bond with a 5% coupon rate and semiannual coupons.

Bond B: A six-year \$12,000 zero-coupon bond.

Bond C: A five-year \$25,000 par value bond with an 8% coupon rate and semiannual coupons.

11. Compute the convexity of the portfolio in problem 10).

12. Consider a sequence of cash flows defined by:

$C_2 = 450$, $C_3 = 500$, $C_5 = 350$.

Find an expression for the convexity and modified duration as a function of i. Graph these two functions on the same coordinate axes.

13. Do the analysis as in Example 9.11 if the interest rate is changed from 4.5% to 6%. Include a graph such as that presented in the solution to Example 9.9.

14. Portfolio A consists of four-year bonds with 7% annual coupons and five-year zero-coupon bonds. Portfolio B consists of a single four-year zero-coupon bond with a maturity value of $10,000. All bonds yield 7% effective. The amount of each bond in Portfolio A is such that the two portfolios have the same present value and the same modified duration. How many dollars of portfolio A are invested in zero-coupon bonds?

15. A $1000 twenty-year bond with 7% annual coupons is purchased to yield $\delta = .07$. Find the modified duration of this bond.

16. Suppose a bond has a face amount of $F = P = C.n$ annual coupon payments based on rate r and is priced to yield j per year. Find a formula for $D(i, \infty)$ and show that it does not depend on F.

17. A three-year annual coupon bond has $F = C = P = 100\%$ and a 10% coupon rate. Compute the McCauley Duration at 4 $j = 11.8$.

18. Compute the McCauley Convexity of a ten-year 6% $1,000 bond having annual coupons and a redemption value of $1,200 if the yield to maturity is 8%.

19. An investor has a portfolio containing $30,00 worth of two-year bonds with duration 1.96, $20,000 worth of three-year bonds with duration 2.88, and $50,000 worth of five-year bonds with a duration of 4.59. What is the Macaulay Duration of the portfolio?

20. An investor wishes to immunize a liability of $500,000 due in 5 years by purchasing a portfolio consisting of zero-coupon bonds of durations two and eight years. Using an interest rate of 3.5% compute the required portfolio and verify the convexity condition.

21. Find the McCauley Duration and the convexity of the portfolio of zero-coupon bonds in Table 9.9:

TABLE 9.9

Term	Price
2	50,000
4	75,000
6	110,000

Chapter 10

Determinants of Interest Rates

10.1 Introduction

As we have seen, interest rates are involved in every calculation in actuarial science. This brief chapter provides an introductory look at the factors which determine interest rates. Since government issued securities (T-Bills) are among the chief determinants of interest rates, they are included here as well.

We begin with a brief discussion of how interest rates are determined. As we will see government issued securities set the lowest rate sellers must offer for bonds and other securities – the so call zero-risk rate. We will then consider the effect of default risk, inflation, and other factors on the value of interest rates. The total interest rate can often be decomposed into several component parts with each of these components viewed as compensation for a particular risk. A company (individual) with a poor credit rating would pay higher interest rates due to the default risk. You may be aware of your own credit rating which determines, in part, the rate you have to pay to purchase a loan.

10.2 Equilibrium Interest Rates

As we have seen earlier, interest is compensation for a deferred payment. If we think of a loan as a product with the interest rate charged as the price of that product we can use the concept of supply and demand to analyze this situation. The number of persons willing to take out a loan will decrease as the interest rate (price) increases. Likewise, the number of lenders willing to make (sell) a loan increases as the interest rate (price) increases. As you no doubt learned in microeconomics: at equilibrium, the demand for loans is exactly equal to the supply.

As you also learned in microeconomics, the demand curve for products varies from product to product. The same thing applies to financial instruments. In particular, demand for bonds varies with the length of the bond. The equilibirum rate for five-year bonds might be different than that for three-year bonds. We will discuss this situation in Section 10.5.

10.3   T-Bills

Both the United States and Canadian governments issue zero-coupon bonds which are often referred to as T-Bills. They are typically short-term instruments (four, thirteen, twenty-six, or fifty-two weeks). T-Bills are sold at a price P and offer a payment M (called the maturity value) after a specified period called the term of the bond. T-Bills are thus a special type of zero-coupon bond. The maturity value is also called the face amount – not to be confused with the face value of a standard bond. Since they are all short-term intruments the term of aT-Bill is referred to as days to maturity. The interest earned is the face amount minus the price. As the following example shows we can calculate the effective annual yield for T-bills just like we did for zero-coupon bonds. We assume that there are 365 days in a year when computing the effective rate of interest.

Example: 10.1 A 200-day T-bill sells for \$950 and matures at \$1,000. What is the effective annual yield rate?

Solution: Let r be the effective daily rate and i be the effective annual rate. We have

$$(1 + r)^{200} = \frac{1000}{950}$$

$$1 + r = \left(\frac{1000}{950}\right)^{\frac{1}{200}}$$

$$1 + i = (1 + r)^{365} = \left(\frac{1000}{950}\right)^{\frac{365}{200}} = 1.09813$$

Hence the effective annual rate of return on this T-bill is 9.813%.

Unfortunately, T-bill rates are not expressed as effective annual but rather as "quoted rates." For U.S. T-Bills this calculation assumes a 360-day year!

Quoted Rate for a U.S. Government T-Bill

$$\text{Quoted Rate} = \frac{360}{\text{Days to Maturity}} \times \frac{\text{Interest Earned}}{\text{Maturity Value}} \qquad (10.2)$$

For Example 10.1 this yields

$$\frac{360}{200} \times \frac{50}{1000} = 9\%$$

You might also be asked to compute the effective rate of discount which is calculated as

$$d = \frac{i}{1+i} = \frac{.09813}{1.09813} = 8.936\%$$

Example 10.2 A 250-day T-bill has a quoted reate of 2.75% and matures at $2,000.

    a) What is the price?
    b) What is the effective annual rate of interest?
    c) What is the effective annual rate of discount?

a) The Amount of Interest = Maturity − Price so we have

$$.0275 = \frac{360}{250} \times \frac{2000 - P}{2000}$$

Solving for P yields $P = \$1,961.81$.

b) We have

$$1 + i = \left(\frac{2000}{1961.81}\right)^{\frac{365}{250}} = 1.0285$$

so $i = 2.85\%$

c) $d = \frac{.0285}{1.0285} = 2.78\%$

Example 10.3 A 200-day T-bill sells for $5,000 and has a quoted rate of 2.5%

    a) What is the maturity value?
    b) What is the interest earned?
    c) What is the effective annual rate of interest?
    d) What is the effective annual rate of discount?

Solution:
    a) If we let M represent the maturity value we have

$$.025 = \frac{360}{200} \times \frac{M - 5000}{M}$$

$M = \$5,070.42$
    b) The interest earned is $5,070.42 - 5,000 = 70.42$.
    c) $i = 2.59\%$
    d) $d = 2.52\%$

The Canadian government also issues T-bills. They are also described in terms of a quoted rate computed as follows:

Canadian T-Bill Quoted Rate Formula

$$\text{Quoted Rate} = \frac{365}{\text{Days to Maturity}} \times \frac{\text{Amount of Interest}}{\text{Current Price}} \qquad (10.3)$$

Example 10.4 A 300-day T-Bill sells for $970 and matures at $1,000.
a) What is the U.S. quoted rate?
b) What is the Canadian quoted rate?
c) What is the effective annual rate of interest?
d) What is the effective rate of discount?

Solution

a) $\frac{360}{300} \times \frac{30}{1000} = .036$, 3.6%

b) $\frac{365}{300} \times \frac{30}{970} = .0371$, 3.76%

c) $1 + i = \left(\frac{1000}{970}\right)^{\frac{365}{300}} = 1.0377$, $i = 3.78\%$

d) $d = 3.64\%$

10.4   Effective and Continously Compounded Rates

As we have seen, there are two ways to state the effective rate for compound interest.

Effective Interest Rate
Effective Annual Interest Rate

$$P_t = P_0(1 + i)^t \tag{10.4}$$

Continuously Compounded Rate
Force of Interest

$$P_t = P_0 e^{rt} = P_0 e^{\delta t} \tag{10.5}$$

We can convert back and forth using the formulas

$$\delta = \ln(1 + i)$$
$$i = e^\delta - 1 \tag{10.6}$$

Finally, recall that APY or APR are typically nominal rates.

The advantage of using continuously compounded rates is that the average rate over several periods can be computed directly as arithmetic average of the rates for each period. If the rate for period i is r_i and r is the continuously compounded average we have

$$P_t = P_0 e^{r_1} e^{r_2} \ldots e^{r_n} = P_0 e^{\sum_{i=1}^n r_i} = P_0 e^{rn}$$

Hence $r = \frac{\sum_{i=1}^n r_i}{n}$, the arithmetic average. Note that this only holds when using continuously compounded interest rates.

10.5 Interest Rates Assuming No Inflation or Risk of Default

Suppose we price loans of $1,000 which are to be repaid in one, two, three, four, or five years with all interest paid at the end of the loan term. We first observe that the amount repaid must always increase as the term of the loan increases. If a two-year loan repays $1,100, a three-year loan must repay more than this else a lender would sell a two-year loan and let the money sit idle for a year.

Table 10.1 is a hypothetical schedule of equilibrium repayment and interest rates for a loan of $1,000

TABLE 10.1

Repayment Amounts for a Loan of $1,000

Term	1	2	3	4	5
Repayment Amount	1,025	1,065	1,110	1,190	1,290
Effective Per Annum Rate	2.50%	2.96%	3.54%	4.44%	5.22%
Continuous Per Annum Rate	2.47	2.91%	3.48%	4.35%	5.09%

There are several possible explanations for the varying effective rates.

Market Segmentation Theory: The pool of buyers and sellers for loans for different terms varies, hence the equilibrium rates will also vary.

Liquidity Preference Theory (also known as Opportunity Cost Theory): If the rates for a one-year and a two-year loan are the same a lender would prefer the one-year loan as the money is available for new investments sooner. Hence, a premium must be paid for the longer term.

A plot of interest rates as a function of length of the loan is known as a yield curve. As a general rule interest rates increase with the term of the loan so the slope of the yield curve is usually positive. If rates go down as the term increases the yield curve is said to be inverted. There are four fundamental types of curves:

Normal: Interest rates increase with time.

Inverted: Interest rates decrease with time.

Flat: Interest rates are independent of time.

Humped: Interest rates increase for a while and then decrease.

10.6   Interest Rates Assuming a Default Risk but No Inflation

Corporations as well as local and state governments often issue zero-coupon bonds as a means of generating cash. These bonds are essentially loans to the issuer. These bonds are issued ratings which represent the default risk (if any) of the issuer. U.S. Government T-Bills are assumed to have a default risk of 0 and hence define the lowest rate any other entity would have to pay. If the effective rate for a U.S. T-Bill is 3.2%, a corporation would have to offer at least that much to attract investors. Suppose a 240-day T-Bill sells at $980 and matures at $1,000. This is the zero-risk price. Hence a corporation or government could not charge any more than $980 and most likely would have to charge less. Bond issuers are rated based on their payment history. Lower ratings result in higher interest rates (and resulting lower price for bonds with equal maturity values). Yield rates are quoted in terms of the investor. The T-Bill rate is the lowest yield rate which will attract investors.

In some cases the earnings from government-issued bonds are tax-exempt. This results in the bonds being issued at a lower yield rate. For example, if an investor has a marginal tax rate of 15% a municipal bond with a price of $800 which matures at $1,000 would provide the entire $1,000 as income while a corporate bond would only provide (after taxes) $850. In order to provide the same after-tax yield rate the corporate bond would have to be priced lower than $800.

Bonds with the highest risk (and which must thus pay the highest interest rate) are referred to as **junk bonds**. In this section, we discuss the relation between an expected default rate (the probability that the loan will not be paid off in full) and the interest rate which must be paid. In terms of corporate and government bonds, the lender is the purchaser of the bond and the corporation or government is the borrower.

We now consider how interest rates change with a default risk. Suppose we compute the interest to be charged on a loan of L for which there is probability of p that the borrower defaults. Let the required single payment be represented by P. Suppose further that in the case of a default the payment is estimated to be $K < P$. A complete default would correspond to $K = 0$. A risk-free loan would have $p = 0$. If the lender charges a rate i, the yield rate will be less than i because of the risk of default. We will refer to the yield rate as s. In this section interest rates will be assumed to be continuously compounded rates unless stated otherwise.

We use the concept of expected value to compute the value of s. As shown in Table 10.2 we have the following situation.

TABLE 10.2

Outcome	Value	Probability
Full Payment	P	p
Default Payment	K	$1-p$

We compute the expected value using the standard formula from probability theory:

Expected Value of the Payment

$$Pp + K(1-p) \qquad (10.7)$$

Example 10.5: Suppose we charge 5% continously compounded interest on a two-year loan with an expected default rate of 2%. Suppose further that the expected default payment is 65% of the amount due. What is the effective yield rate for the lender? What rate must be charged in order for the yield rate to be 5%?

Solution: Suppose the loan amount is L. At a rate of 5%, the required payment would be $Le^{.05t}$. If the loan defaults the payment will be $.65Le^{.05t}$. Using Equation 10.7, the expected value of the loan repayment is

$$.98Le^{.05t} + .02 \cdot .65 \cdot L \cdot e^{.05t}$$

To compute the yield rate we solve the equation below for s.

$$.98Le^{.05t} + .02 \cdot .65 \cdot L \cdot e^{.05t} = Le^{st}$$

Simplifying we obtain

$$(.98 + .013)e^{.1} = 1.0974 = e^{2s}$$

which yields $s = 4.65\%$. As would be expected, the yield rate is reduced by the default risk.

If a lender wants to earn 5% effective on this loan, a rate higher than 5% will need to be charged. To calculate this we suppose that the rate charged is i and that we want $s = .05$. At a rate of i, the required payment will be Le^{2i} and we obtain

$$.98Le^{2i} + .02 \cdot .65 \cdot Le^{2i} = Le^{.1} = L \cdot 1.105$$

Simplifying we obtain
$$(.98 + .013)e^{2i} = 1.105$$

Solving for $i = 5.35\%$.

Note that the change in rates does not depend on the amount of the loan. The key variables are the default rate and the fraction of the required amount which is assumed to be paid in the event of a default. Suppose that p represents probability of full payment and that a is the fractional amount to be paid in the case of a default. Our relation between i and s becomes

$$pLe^{it} + (1-p)aLe^{it} = Le^{st} \tag{10.8}$$

We can eliminate the common factor of L and obtain an equation which relates the charged rate i to the effective rate s:

$$(p + (1-p)a)e^{it} = e^{st} \tag{10.9}$$

We can solve Equation 10.9 for either i in terms of s or s in terms of i. Table 10.3 collects all the variables in one place.

TABLE 10.3

Summary

Symbol	Meaning
i	Rate charged by the lender
s	Effective rated earned by the lender
p	Probability of full payment
a	Fraction of amount due paid in the case of a default

The value for effective rate is the continously compounded rate which is computed as below.

Continuous Effective Rate Given Rate Charged

$$s = \frac{\ln(e^{ti}(p + (1-p)a))}{t} \tag{10.10}$$

Required Rate to Produce a Desired Continuous Effective Rate

$$i = \frac{\ln\left(\frac{e^{ts}}{p+(1-p)a}\right)}{t} \tag{10.11}$$

Example 10.6 A lender requires a continuously compounded rate of 6.3% for a six-year loan of $3,500 to be paid in full with interest at the end of the loan period. The predicted default rate is .04 and it is predicted that a default will result in a payment of .75 of the required payment.
  a) What is the yield rate if the lender charges 6.3%?
  b) What rate must be charged if the yield rate is to be 6.3%

Solution:
  a) We use Equation 10.10 to obtain $s = 6.13\%$
  b) We use Equation 10.11 to obtain $i = 6.47\%$
  The difference between the rate charged and the desired yield rate is known as the **compensation for default risk**. In Example 10.6 this value is $6.47 - 6.3 = .17\%$.

10.7   Inflation

Inflation results in a reduction in the value of the loan payment. As a result a lender wishing to earn a given effective rate must charge a premium over that rate even after the default risk has been accounted for. We will assume that the lender requires a yield of r and computes the compensation for default risk at s. The rate absent inflation would then be $R = r + s$. If an inflation rate of i is expect the lender must charge $R + i = r + s + i$.

In many cases inflation is dealt with at the time of repayment. If inflation (using an agreed upon index) has increased prices by a factor of f then the default adjusted payment will be multiplied by this factor. Loans of this type are referred to as inflation protected loans.

The Consumer Price Index (CPI) reports the price of a selected "market basket" of items as a multiple of the price of that same collection at some past time. The original price is usually normalized to $100. Table 10.4 is a sample set of CPI values

TABLE 10.4

Year	Price
2016	100
2017	101
2018	104
2019	105

Suppose we take out a loan in 2017 to be repaid in 2019. The $f-$ factor in this case would be

$$f = \frac{105}{101} = 1.0396$$

Example 10.7   A loan of $5,000 originates in 2017 and is to repaid in 2019. The lender requires a yield rate of 5.2% and the required compensation for default risk is .03%. What is the required payment?

What is the effective rate of this loan for the borrower?

Solution: The rate charged is $.052 + .0003 = .0523$ so the pre-inflation payment is $5000e^{.0523 \cdot 2} = 5551.33$. The $f-$ factor for inflation is $\frac{105}{101} = 1.0396$. So the adjusted final payment is $5551.33 \cdot 1.0396 = 5771.19$. To compute the effective rate for the lender in the event of full payment we use the TVM keys (Table 10.5):

TABLE 10.5

N	I/Y	PV	PMT	FV
2	$CPT = 7.44$	5,000	0	−5771.19

This number is the effective annual rate, not the continuously compounded rate. To obtain the final answer we compute $\ln(1.0744) = 7.18\%$

Inflation protection is a benefit to the lender and will result in a reduction in the rate that can be charged. The rate reduction symbol is c. The continuously compounded interest rate for a loan with inflation protection and 0 default risk is given by

$$R_1 = r - c \tag{10.12}$$

The actual rate earned if the inflation rate is i_a is[1]

$$R_1^{(a)} = r - c + i_a \tag{10.13}$$

We can further break down inflation into the expected (i_e) and unexpected (i_u) resulting in

$$R_2 = r + i_e + i_u \tag{10.14}$$

In many cases we have $R_2 - R_2 = i_e + i_u + c \approx i_e$. This is valid when i_u is small. This is measured by the volatility in interest rates. Volatility refers to uncertainty in expected interest rates (i.e., a large estimated i_u). One final bit of interest rate jargon: the "spread" refers to the sum of all adjustments to the interest rate due to inflation and default risk: $i_e + i_u + s$. This results in

$$R^* = r + s + i_e + i_u \tag{10.15}$$

While this looks very "official" it is very difficult in practice to tease out these individual terms. Equation 10.15 is thus a theoretical rather than a practical equation.

Exercises for Chapter 10

1. Define liquidity preference theory and its effect on the structure of the yield curve.

2. Define market segmentation theory and its effect on the structure of the yield curve.

3. A loan of $6,000 is to paid off in a single lump sum at the end of five years. The lender estimates the default rate at 2% and that a default will result in a payment of .8 of the required payment.

---

[1] The a here is short for "actual." i_a is the actual inflation, usually measured by the CPI or other index.

(a) What is the yield rate to the lender if the lender charges 3.2%?

    i. Continuously compounded

    ii. Effective annual interest rate

(b) What rate must the lender charge in order to have a yield rate of 4%?

    i. Continuously compounded

    ii. Effective annual interest rate

4. A loan of $10,000 to be repaid in two years carries a rate of 3.2% plus a .3% default protection. The loan requires inflation protection. Given the CPI table below (Table 10.6) calculate a) The required repayment and b) The effective rate for the borrower for the two scenarios

i) The loan originates in 2017 and is to be repaid in 2019

ii) The loan originates in 2018 and is to be repaid in 2020

TABLE 10.6

Year	Price
2017	100
2018	103.3
2019	104.5
2020	101.2

5. A lender is willing to make loan of $50,000 to be repaid in two years so long as the effective yield is 3.3%. The lender estimates the default rate at 2% with an expected default payment of $44,000. Using the table in problem 4) compute the required payment and effective rate for the borrower if the loan originates in 2017 and is repaid in 2019.

6. A 240-day U.S. T-Bill has a quoted rate of 2.4% and matures at $4,500

a) What is the price?

b) What is the effective annual rate of interest?

c) What is the effective annual rate of discount?

7. A 240-day Canadian T-Bill has a quoted rate of 2.4% and matures at $4,500

a) What is the price?

b) What is the effective annual rate of interest?

c) What is the effective annual rate of discount?

8. What can you say about the relation between the quoted prices of U.S. and Canadian T-bills which have the same term, maturity value, and price?

9. What can you say about the effective rate of two T-Bills with the same term, price and maturity value?

Final Thoughts

As stated in the introduction the purpose of this text is to prepare you to begin studying for the SOA Exam FM/2. While it includes all the the information you need to succeed, it does not include enough practice problems to prepare you to succeed on this very difficult multiple choice exam.

Your next step should be to visit the SOA website (beanactuary.org) and begin research the resources available to you as you prepare for Exam FM. These include both review manuals and online instruction. Here are a list of a few of them.

Coaching Actuaries
https://www.coachingactuaries.com/

Coaching Actuaries provides an online text as well as practice exams. They will keep track of your progress and let you know when your skill level seems sufficient to pass the exam.

Actex
https://www.actexmadriver.com/

Actex provides textbooks, study guides and online materials. They update their texts each year.

Actuarial Study Materials
https://www.studymanuals.com/

ASM has the most challenging review manual – lots of very difficult problems.

Actuarial Bookstore
https://www.actuarialbookstore.com/

Similar to ASM and Coaching Actuaries

Studying with one or more other people is probably a really good idea. You can share the cost of materials and encourage one another.

Good Luck.

Appendix: Basic Setup for the TI BA II Plus

1) Formatting output (page 4 of the TI BA II Plus owner's manual)

There are several formatting settings you should change from default. These settings only need to be changed once so long as you don't use the 2nd RESET key sequence which clears all memories and restores all default settings.

a) Enter 2ND FORMAT to access the format menu.

b) You will see DEC. Enter the number of decimals places to show (I recommend 8).

c) Use the down arrow key (top row of buttons) to scroll down to the units for angles. It comes set to DEG (degrees) to change to RAD (radians) key in 2ND SET and it will switch to RAD each time you key in 2ND SET toggles the setting.

d) Use the down arrow key to scroll down to the calculation method which is either Chn(chain) or AOS (algebraic). We want AOS, so use the 2ND SET sequence to toggle it so AOS shows up.

e) Key in CE/C until you see 0.00000000.

Answers To Odd-Numbered Exercises

Chapter 1

1)
a) 5.999
b) 8
c) 3.846405533*10^7
d) 274.086

3)
a) $\frac{1}{2}\ln(x^2+3)|_1^3 = \frac{1}{2}\ln\left(\frac{12}{4}\right) = .549$

b) $\ln(t+1) - \ln(2) = \ln\left(\frac{t+1}{2}\right)$

5)
a) $\ln(1+x) = 0 + x - \frac{x^2}{2} + \frac{x^3}{3}$
$T_1(x) = x$, $T_2(x) = x - \frac{x^2}{2}$
Graphed as Figure A.1

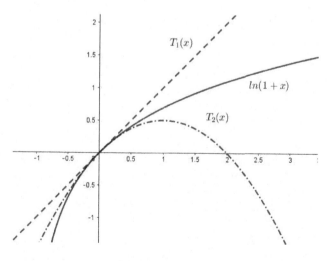

FIGURE A.1

b)
$$\sin(x) = 1 + 0(x - \pi/2) + \frac{-1}{2!}(x - \pi/2)^2$$
$$T_1(x) = 1, \ T_2(x) = 1 - \frac{(x-\pi/2)^2}{2}$$
Graphed as Figure A.2

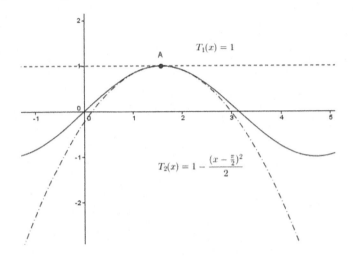

FIGURE A.2

Chapter 2

1) $500 \cdot (1 + i \cdot 12) = 1250$
$\Rightarrow 1 + 12i = \frac{1250}{500} = 2.5 \Rightarrow 12i = 1.5$
so $i = .125$

3)
$127.50

5)
$n = z - x - y$

7)
a) $450(1 + .056t) = 450 + 25.2t$
b) A line with slope 25.2 and intercept 450.
c) $\frac{.056}{1+.056(t-1)}$
d)
$n = 1 \Rightarrow .056 = 5.6\%$
$n = 2 \Rightarrow i_n = 5.30\%$
$n = 3 \Rightarrow i_n = 5.04\%$

e) See Figure A.3

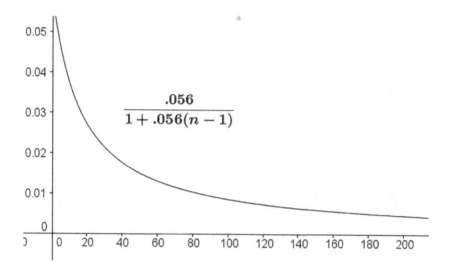

$$\frac{.056}{1 + .056(n-1)}$$

FIGURE A.3

9)

a) $(1 + i \cdot t) - (1 + i)^t$ $\underline{i = .05}$

See Figure A.4

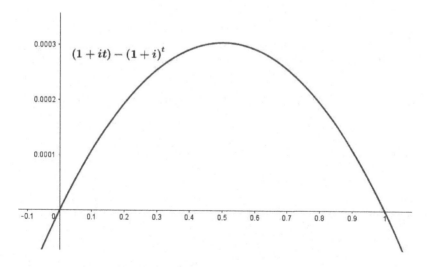

$(1 + it) - (1 + i)^t$

FIGURE A.4

About $t = .5$

b) $t = \frac{\ln\left(\frac{i}{\ln(1+i)}\right)}{\ln(1+i)}$

c) For $i = .05$, $t = 0.502$

11)

a)

$789.00

b)

$816.52

13) 19.56%

17)

$(1 + \frac{i}{12})^{12} = (1 - \frac{.056}{4})^{-4} = 1.058016255$

$i = .05653 = 5.65\%$

19)

a) $e^{\int_0^t \frac{t}{1+t^2}\,dt} = \sqrt{1 + t^2}$

b) $A(2) = A(4)\frac{a(2)}{a(4)} = 30\frac{\sqrt{5}}{\sqrt{17}} \approx 16.27$

21)

a)

$152.997 \approx 153$

b) 4.34%

c) $i = 0.0529974 = 5.3\%$

23)

a) $i = 0.06314 = 6.31\%$

b) $i = 0.0716 = 7.16\%$

c) $d = 0.0594$

25) $933,395.81 will result in a zero balance after the last payment of $250,000.

27) $i = .125 = 12.5\%$, $x = 4,500$ The equations are $ix = 562.5$, $\frac{i}{1+i} \cdot x = 500$.

29)

$15,000

31)

a)

$$
\begin{aligned}
A(n) &= A(n-1) + I_n \\
&= I_n + A(n-2) + I_{n-1} \\
&= I_n + I_{n-1} + A(n-3) + I_{n-1} \\
&= \ldots \\
&= I_n + I_{n-1} + \ldots I_1 + A(0)
\end{aligned}
$$

b) The change in the principal is the sum of all interest earned.

c) No. $i_n = \frac{I_n}{A(n-1)}$ so the collapsing sum technique fails.

33)

a) $i_4 = \frac{3}{59}$

b) $i_8 = \frac{3}{71}$

35) $840 Hint: Solve for it.

37) Simple Interest: $1,532.5

Compound Interest: $1,557.36

39)

a)

Simple Interest:

$n = 4.237$ years.

Compound Interest:

$n = 3.635$ years

b)

Simple Interest:

$i = .1 = 10\%$

Compound Interest:

$i = .0845 = 8.45\%$

41) $10,982.55

43) $i = .0082071 = .821\%$

45)

a) $16,045.01, b) 3.289 years

Chapter 3

1) Exact Method: There are 759 days with 365 days per year: $768.38

Ordinary Simple Interest: There are 747 days with 360 days per year: $766.73

3) See Table A.1

TABLE A.1

t	Simple	Compound	Frac.
1	532	532	532
5	660	681.83	681.83
$\frac{1}{2}$	516	515.75	516
$5\frac{1}{4}$	668	692.49	692.74
$12\frac{3}{4}$	908	1102.75	1103.14

5)
The simple interest approximation (as shown in 4) constructs the secant line. Since the graph of PV $(1+i)^t$ is concave up the secant line is above the graph.

Maximum difference occurs when $q = \dfrac{\ln\left(\frac{i}{\ln(1+i)}\right)}{\ln(1+i)}$

7) $i = .02305 = 2.305\%$

9)
a) $\dfrac{\ln(2)}{\ln(1.06)} = 11.896$
b) $1 + .06 \cdot t = 2 \Rightarrow t = 16.667$
c) 11.8929
d) $e^{.06t} = 2 \Rightarrow t = 11.552$

11)
$914.84

13)
$i = \frac{.04}{12} = .0033 = 3.33\%$ Count in months!

$$x = \frac{10000}{(1+i)^{36} + 2(1+i)^{24}} \approx 3036.23$$

15) $i = .06138 = 6.138\%$

17) $k = .01386$

19) $i = .015 = 1.5\%$ per quarter. Count in quarters.
$1831.37

21) 14.18%

23) $1000(1.01)^{36}\sqrt{26} \approx 7{,}295.85$

25) $i^{(12)} = 17.819\%$ nominal annual interest rate.

27) Compound interest rate $= 14.870\%$
Continuous interest rate $= 13.863\%$

29)
Use CF worksheet: 6,159.45

31) $4,117.40

33) 3.824%

35) 5.9657%

Chapter 4

1) $2,332.0931

3)
a) $3,355.04
b) $3,623.44

5) Algebra Problem

7) The value of the 114 remaining payments is $5000a_{\overline{114}|,.025} =$ $188,018.5511

9) $\frac{1}{.49} = 2.040816327\%$

11) $3,037.6807 per year.

15) $277,238.63

17) 122.63 months.

19) $15,031.19

21)
a) 5.00%
b) $215,252.4202 Take the lump sum, it pays more!
c) $228,167.55654

23) Mork should look for the BGN symbol – he computed the value as an annuity-due!

25) The interest rate per month is .0070337438 = .703%
The exact number of months is 75.98828761
After 75 months the remaining amount is 835.867

27) 25,085.92

29) $140,339.57

31) 0.0849572937

33)
a) $k = 3$. 10 years is 60 two-month periods, so $n = 60$. $i = \frac{.08}{6} = .0133 = 1.33\%$
$PV = \$74,603.32536$
b) six month interest rate is $(1.0133333)^3 - 1 = .040535601$
$PQani(600, 600, 20, .040535601) =$
$PV = \$74,603.33$

35) The difference in present values is 123.21. The difference in future values is 198.64

Chapter 5

1)
a) $1,154.87
b) $774.37
c) See Table A.2.

TABLE A.2

Year	Payment	Interest Due	Amount to Principal	Remaining Balance
0		$ -	$ -	$5,000.00
1	$1,154.87	$250.00	$904.87	$4,095.13
2	$1,154.87	$204.76	$950.12	$3,145.01
3	$1,154.87	$157.25	$997.62	$2,147.38
4	$1,154.87	$107.37	$1,047.50	$1,099.88
5	$1,154.87	$ 54.99	$1099.88	$ 0.00

3) There will be 120 payments the interest rate per period is $\frac{.12}{12} = .01$ per period.
a) $717.35
b) $36,082.57

5) For the first ten years the effective annual interest rate is
$(1 + \frac{.06}{6})^6 - 1 = 0.061520156$
For the refinanced loan the monthly interest rate is $(1 + \frac{.06}{4})^{\frac{1}{3}} - 1 = 0.004975206$.
a) $4,413.1772
b) $32,248.71
c) $26,380.48
d) $207.19
e) $56,289.60

7) 48 payments
The Monthly Interest Rate is .80170558%
The APR is 9.62%
The effective annual interest rate is 10.0562%

60 payments
The Monthly Interest Rate is .95282858%
The APR is 11.43394299%
The effective annual interest rate is 12.053%
Longer payment plan almost always increases the interest rate. Time value of money.

9) Graphs for $n = 100$ and $n = 50$ appear as Figures A.5 and A.6.

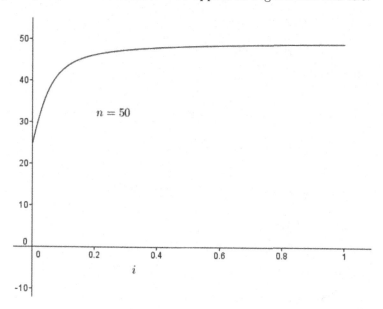

FIGURE A.5

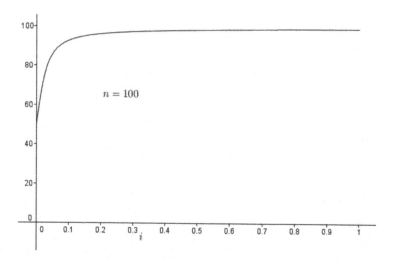

FIGURE A.6

11)

$$1200 = \frac{\frac{L}{2}}{ani(18,.08)} + .075 \cdot \frac{L}{2} + \frac{\frac{L}{2}}{sni(18,.03)}$$

$$= \text{amortized pmt.} + \text{interest} + \text{sinking fund pmt.}$$

$L = \$10,694.673$

13)
The amount financed is \$382,500. Interest due at signing is

$$\frac{19}{365} \cdot 382500 * .075 = \$1,493.321918$$

The Six-Step Program

1) Payment: \$2,674.49

2) $L^* = 382500 - 8150 = \$374,350$

3) $K = \$589,961.70$

4) ..64345031\% per month

5) The APR is 7.721403699\%

6) The effective annual interest rate is $I = 8.00\%$

15) \$1,030.22

17) \$1,178.41

19) \$258.29 not including the interest due on the loan.

21) \$36.52

23)
a) \$375.75
b) \$528.26

25) \$36.30

27) \$1,867.99

29) \$2262.68

31) 14.479\% down payment. Remember that APR is $12 \cdot$ monthly rate.

33) There are 27 payments if a balloon payment is used. The value of the balloon payment is \$2,388.27. There are 28 payments if a drop payment is used. The value of the drop payment is \$941.13.

Chapter 6: Yield Rates

1) 4.65%. The account balance is always positive at this interest rate, so the yield rate is unique.

3) .03593834302 = 3.59%

5) 6.16%
Hint: A single deposit of $10,000 should accumulate to the same amount as the sequence of $1,000 payments.

7) Use Equation 6.12 to compute the accumulated value if payments were made for the entire ten years and then subtract the accumulated value of the payments for the final five years (which were note made!)

$$1000\left((10+.05\cdot\frac{s_{\overline{11}|,.04}-11}{.04})-(5+.05\cdot\frac{s_{\overline{6}|,.04}-6}{.04})\right)$$

Now simplify this expression.

9)
a) 6.5%
b)
$10210(1+i)+4000(1+i)^{\frac{10}{12}}-3000(1+i)^{\frac{7}{12}}+1000(1+i)^{\frac{1}{12}}=12,982$
c) 6.4985% Hint: Use CF worksheet to compute monthly interest rate and convert to annual effective rate.

11)
$W=\$24$, $x=\$104.4$

13)
The unknown deposit is $870.8133971. The effective annual interest rate for the first six months is 13.28%.
Hint: Find the unknown deposit (x) using the value of the dollar-weighted yield. Then use the time-weighted yield to find the effective annual interest rate.

15) Bert has $5,984.71, Ernie has $6,083.251. Difference is $98.54.

17) $184,465.83

19) 9.500%

21) 6.129%

23)
a) $1,420.90, 5.79%
b) $1,439.25, 6.26%
c) $1,486.51, 7.40%

Chapter 7

1)

a) The price is $10,400.923

Table A.3 displays the amortization table for this loan.

TABLE A.3

Year	Coupon Payment	Interest at Yield Rate	Change in Price	Balance
0			$10,400.92	
1	$450.00	$728.06	$278.06	$10,678.99
2	$450.00	$747.53	$297.53	$10,976.52
3	$450.00	$768.36	$318.36	$11,294.87
4	$450.00	$790.64	$340.64	$11,635.51
5	$450.00	$814.49	$364.49	$12,000.00

b) The price is: $12,135.45

Table A.4 displays the amortization table for this loan.

TABLE A.4

Year	Coupon Payment	Interest at Yield Rate	Change in Price	Balance
0			$12,135.45	
1	$450.00	$424.74	$(25.26)	$12,110.19
2	$450.00	$426.86	$(26.14)	$12,804.08=5
3	$450.00	$422.94	$(27.06)	$12,056.99
4	$450.00	$421.99	$(28.01)	$12,028.99
5	$450.00	$421.01	$(28.99)	$12,000.00

3) $500.03

5) $115.87

7) $1,497.42

9) $504,568.86

11) $799.89

13) Nominal 3.89% Effective 4.039%

15) $1,192.72

17) 10.117%

19) 7.723%

21)
Price $= \$10,095.08$ Yield to Maturity $= 7.715\%$

23) $2f_1 = f(1,3) = .05504 = 5.504\%$
$2f_2 = f(2,4) = .0909 = 9.09\%$

25)
Price-plus accrued using simple interest $= 48,383$
Price-plus-accrued using compound interest $= 48,331$
Price $= 47,183$

27) The bond was purchased at a discount so the maximal yield is achieved at the end of year 5.

Chapter 8

1) Buy 4.5 units of one-year bonds and six units of two-year bonds. Cost is $12,963.19 Yield rate is 6.89%

3) Buy 6.896 units of eighteen-month bonds, 4.777 units of one-year bonds and 3.777 units of six-month bonds. Cost is 14,825.62 and the nominal annual interest rate is 7.13%

5)
a) 342.84
b) 129.88

7)
a) 7.075%
b) 8.23%

9) 5.65%

11) 3.213%

13) - 14) See Table A.5.

TABLE A.5

Years to Maturity	0-Coupon Bond Yield	LIBOR	One Year Implied Forward	Par Coupon	Force of Interest	Notionals
1	6.000%	0.94340	6.000%	6.000%	5.8269%	$1,000.00
2	6.000%	0.89000	6.000%	6.000%	5.8269%	$ 500.00
3	5.500%	0.85161	4.507%	5.526%	5.3541%	$ 100.00
4	3.200%	0.88162	-3.403%	3.319%	3.1499%	$ 300.00
Two-year swap			6.000%			
Three-year swap			5.5265%			
Four-year swap			3.3191%			
Three-year swap with Notionals			5.9137%			
Two-year swap Delayed one year			5.2700%			
Four-year swap with Notionals			4.4959%			

15) 16) See Table A.6.

TABLE A.6

Years to Maturity	0-Coupon Bond Yield	LIBOR	One Year Implied Forward	Par Coupon	Force of Interest	Notionals
1	4.500%	0.95964	4.500%	4.500%	4.4017%	$ 100.00
2	5.500%	0.89845	6.510%	5.473%	5.3541%	$ 200.00
3	5.700%	0.84679	6.101%	5.670%	5.5435%	$ 150.00
4	3.200%	0.88162	-3.951%	3.303%	3.1499%	$ 300.00
Two-year swap			5.4731%			
Three-year swap			5.6699%			
Four-year swap			3.3032%			
Three-year swap with Notionals			5.9028%			
Two-year swap Delayed one year			6.3114%			
Four-year swap with Notionals			1.9949%			

17) 18) See Table A.7.

TABLE A.7

Years to Maturity	0-Coupon Bond Yield	LIBOR	One Year Implied Forward	Par Coupon	Force of Interest	Notionals
1	4.500%	0.95964	4.500%	4.500%	4.4017%	$ 40.00
2	5.200%	0.90358	5.905%	5.182%	5.0693%	$ 25.00
3	6.200%	0.83488	8.229%	6.126%	6.0154%	$ 100.00
4	6.500%	0.77732	7.405%	6.412%	6.2975%	$ 300.00
Two-year swap			5.1822%			
Three-year swap			6.1258%			
Four-year swap			6.4122%			
Three-year swap with Notionals			6.8763%			
Two-year swap Delayed one year			7.0207%			
Four-year swap with Notionals			7.2029%			

21) See Table A.8.

TABLE A.8

Years to Maturity	0-Coupon Bond Yield	LIBOR	One Year Implied Forward	Par Coupon	Force of Interest	Notionals
1	2.500%	0.97561	2.500%	2.500%	2.4693%	$ 40.00
2	3.000%	0.9426	3.502%	2.993%	2.9559%	$ 25.00
3	3.100%	0.91248	3.300%	3.092%	3.0529%	$ 100.00

23) See Table A.9.

TABLE A.9

Bond	Term	Coupon Rate	Yield Rate	Liability	Amount Needed	Cost	Coupon Payment
A	1	0.000%	6.000%	$10,000.00	$5,965.42	$5,627.76	$ -
B	2	5.000%	4.500%	$40,000.00	$35,965.42	$36,302.18	$1,798.27
C	3	-5.600%	5.000%	$25,000.00	$22,763.69	$23,135.64	$1,274.77
D	4	4.000%	5.600%	$25,000.00	$24,038.46	$22,693.43	$ 961.54
Yield Rate for this Project:	5.096%				Total Cost of the Project	$87,759.01	

25) See Table A.10.

TABLE A.10

Bond	Term	Coupon Rate	Yield Rate	Liability	Amount Needed	Cost	Coupon Payment
A	1	0.000%	6.000%	$50,000.00	$45,623.60	$43,041.13	$ -
B	2	5.000%	7.000%	$45,000.00	$40,623.60	$39,154.63	$2,031.18
C	3	5.600%	5.000%	$25,000.00	$22,654.78	$23,024.94	$1,268.67
D	4	4.500%	4.000%	$25,000.00	$23,923.44	$24,357.64	$1,076.56
Yield Rate for this Project:	5.390%				Total Cost of the Project	$129,578.35	

Chapter 9

1)

$$P(i) = 150 + 200 \cdot (1+i)^{-3} + 100 \cdot (1+i)^{-5}$$
$$P'(i) = -600 \cdot (1+i)^{-4} - 500 \cdot (1+i)^{-6}$$
$$P''(i) = 2400(1+i)^{-5} + 3000 \cdot (1+i)^{-7}$$

a)
$P_1(i, .03) = 447.84 - 951.83i$
b)
$P_2(i, .03) = 449.87 - 1087.12i + 2254.77i^2$
c) See Figure A.7.

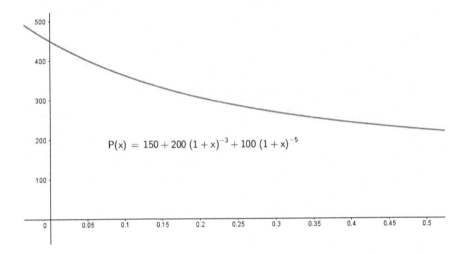

$$P(x) = 150 + 200\,(1 + x)^{-3} + 100\,(1 + x)^{-5}$$

FIGURE A.7

d) See Table A.11.

TABLE A.11

i	P	P_1	P_2
.01	439.26	438.33	439.23
.025	424.11	424.05	424.10
.04	409.991	409.77	409.996

$P_1(i)$ is the tangent line approximation. From the graph, the function is concave up, hence the tangent line is always below the function.

e) See Figure A.8.

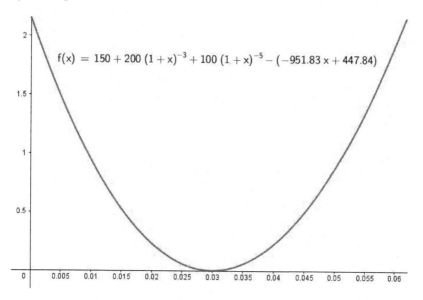

$$f(x) = 150 + 200\,(1+x)^{-3} + 100\,(1+x)^{-5} - (-951.83\,x + 447.84)$$

FIGURE A.8

f) See Figure A.9.

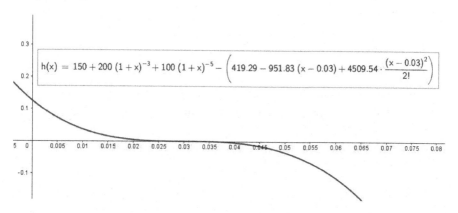

$$h(x) = 150 + 200\,(1+x)^{-3} + 100\,(1+x)^{-5} - \left(419.29 - 951.83\,(x - 0.03) + 4509.54 \cdot \frac{(x - 0.03)^2}{2!}\right)$$

FIGURE A.9

3) The minimum occurs at $i = 0.05$. Price at this point is $49,999.76

5)
a) $P_1(i, .03) = 793.8 - 2169.24i$
b) $P_2(i, .03) = 799.51 - 2549.84i + 6343.43i^2$

c) See Figure A.10

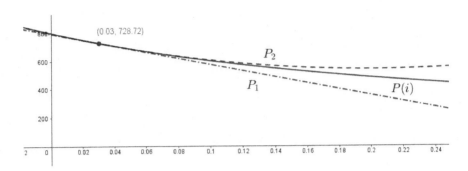

FIGURE A.10

7)

$$M(.0609, \infty) = 7.01$$

9) Formula is $\frac{1}{d^{(m)}} - \frac{n}{(1+i)^n - 1}$

a) $m = 12$, $n = 30$

$d^{(12)} = 12 * (1 - 1.06^{-\frac{1}{12}}) = .058127667$

$\frac{1}{d^{(m)}} - \frac{n}{(1+i)^n - 1} = 10.88$

b) $m = 4$, $n = 30$

$d^{(4)} = 4 * (1 - 1.06^{-\frac{1}{4}}) = .057846553$

$\frac{1}{d^{(m)}} - \frac{n}{(1+i)^n - 1} = 10.96$

11) 22

13) Let x = amount spent on the two-year bonds and y = amount spent on the five-year bonds.

$x + y = 250000(1.06)^{-4} = 198,023.4158$. As in Example 9.11, we have

$$\frac{2x + 5y}{x + y} = 4$$

So $y = 2x$ and

$x + y = 3x = 198,023.4158$.

$x = 66,007.81$, $y = 132,015.61$.

The convexity is

$$\frac{x \cdot 2^2 + y \cdot 5^2}{198023.4158} = 18$$

See Figure A.11.

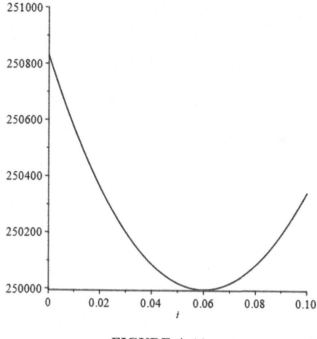

FIGURE A.11

15) Convert to effective annual interest. $e^\delta = 1 + i \Rightarrow i = .072508181$. Use the BOND worksheet to obtain
DUR $= 10.463$

17) Use the BOND worksheet: DUR $= 2.441$

19) 3.46

21) MacD $= 4.51$ MacC $= 22.81$

Chapter 10

1) Liquidity theory posits that investors value liquidity (access to funds), and therefore that longer term instruments must offer a higher yield to attract investors. This results in the "normal" yield curve.

3)
a) 3.2%
b) 4.16%

5)
a) 55,948.96
b) 5.62%

7)
a) 4,430.09
b) 2.41%
c) 2.35%

9) The effective rate of interest is the same.

Index

Note: Page numbers followed by "n" indicate footnotes.

Account balance, 217
Accumulated amount, 69
Accumulated value, 19, 113
    of annuities under varying
        interest rates, 126–128
    of annuity, 179
    of investment, 219–220, 235–236
    of payments of annuity, 100
Accumulation
    bonds, *see* Zero-coupon bonds
    function, 19–21, 23–24, 27–28,
        32–36
Actual monthly interest rate, 43–44
Allocation of loan payments between
        principal and interest,
        161–165
Alternate bond pricing formula, 252
Alternative solution, 58
Amortization
    annualized amortization tables
        using Excel, 176–178
    Excel projects, 204–205
    exercises and answers, 198–203,
        360–362
    of loan, 156–157
    method, 153–156
    with non-standard payments,
        186–188
    table, 161–165
    TI BA II Plus to create
        amortization tables,
        170–176
    worksheet, 168

Amount functions, 19–20
    calculator, 33
    for compound interest, 28
    for simple interest, 25
Amount to interest, formulas for,
        165–170
Amount to principal at any
        time, formulas for,
        165–170
Annual interest rate, 225, 342
Annualized amortization tables
        using Excel, 176–178
Annual percentage rate (APR), 40,
        46, 189, 342
Annuities-due, 98, 108, 137, 144;
        see also Perpetuity-due
    accumulated value of, 109
    fixed term, 108–114
    PV of, 109
Annuities-immediate, 98, 100, 102,
        109–111, 114
    Accumulated value (FV), 100
    Value at inception (PV), 100
Annuities-payable, 138
Annuity/annuities, 6, 8n3,
        97, 104
    alternative method, 131–135
    annuity at any time between
        first and last payments,
        value of, 117–118
    continuous, 146–148
    exercises and answers, 148–152,
        359

Annuity/annuities (*Continued*)
    fixed term annuities-immediate
        with constant payments,
        98–108
    non integer time periods,
        122–123
    paid more frequently than
        interest is converted,
        135–138
    payable at different frequencies
        than interest is convertible,
        129–131
    payments period, 129
    perpetuities, 119–122
    perpetuities paid more
        frequently than interest is
        convertible, 138–139
    present value of, 98–100
    price of, 103–104
    types of, 98
    unknown rate of interest,
        124–125
    unknown time, 123–124
    value after final payment is
        made, 116–117
    value at any date, 114
    value of annuity at any date,
        114–118
    value prior to inception, 114–116
    varying annuities paid less
        frequently than interest is
        convertible, 146
    with varying payments, 139–142
    varying perpetuities, 142–145
    varying rates of interest,
        125–128
Annuity-certain, 97
Annuity-immediate, 98, 137, 139
Annuity-immediate mode, 125
Approximation techniques, 9
    approximations using Taylor
        series, 11–14
    Newton's method, 10–11
Approximation to IRR, 236
APR, *see* Annual percentage rate

APY, 342
Arithmetic series, 7–9
Asset matching equation, 332
Average rate of return, 60

Balance, formulas for, 165–170
Balloon payments, 195
Base amount, 246
    valuation formula, 248–249
Block payments, 142
Bond(s), 241
    amortization schedule, 253
    amortization tables, 255–260
    bonds with varying payments,
        283–285
    callable bonds, 280–283
    determination of yield rates,
        260–262
    Excel projects, 288
    exercises and answers, 285–288,
        364–365
    forward rates, 271–273
    premium and discount, 252–260
    price adjustment formulas, 254
    price of bond between coupon
        payments, 279–280
    pricing, 247–252
    terminology, 244–247
    term structure of interest,
        262–270
    types of, 241–244
    with varying payments, 283–285
Book value, 253

Calculators, 1
    exercises and answers, 16–17,
        353–354
    sequences and series, 6–9
    Texas Instrument calculator
        models, 1–2
    TI-30XS MultiView, 2–5
    TI BA II Plus Professional
        Edition, 2–5
Callable bonds, 280–283
Call premium, 281

Cash flow (CF), 207, 222, 225
    analysis method, 249
    keys, 246–247
    valuation formula, 247–248
CF, *see* Cash flow
Commodity swaps, 294
Comparison date, 70
Compensation for default risk, 346
Compound discount, 50
Compounded monthly, 52
Compounded quarterly, 39, 52, 218
Compound interest, 23, 38, 50, 54,
    279–280
Computers, 1
    Excel, 6
    sequences and series, 6–9
Consumer Credit Protection Act
    (1968), 189
Consumer price index (CPI), 347
Continuous annuities, 146–148
Continuous compounding, formulas
    for, 55
Convexity, 328–330
    condition, 333
    portfolio of instruments
        with known convexities,
        330
Counter party, 294
Coupon
    bonds, 241–243
    payment, 322
    rate, 245
CPI, *see* Consumer price index
CPT, 211
Credit cards, 189

Decreasing annuity of term, 144–145
Dedicated portfolio, 311
Deferred interest rate swaps, 298–300
Deposits, 86, 207, 211–212, 221,
    224, 226
Determinants of interest rates;
    see also Interest rate
    effective and continuously
        compounded rates, 342

equilibrium interest rates,
    339–340
exercises and answers, 348–351,
    371–372
inflation, 347–348
interest rates assuming default
    risk no inflation, 344–346
interest rates assuming no
    inflation or risk of
    default, 343
T-bills, 340–342
Discount, 252–260
    bonds, 245–246
    factor, 36
    forces of, 54
Discounted cash flow analysis,
    207–213
    investment scheme, 209–211
    notation, 207–208
Discount rates, 48
    nominal rate of, 51–54
Discount valuation formula,
    248–249, 260
Dollar amount, 21
Dollar-weighted estimate,
    223–226, 236
Doubling time, 79
    interest rate required, 81
    rule of 72, 82
Drop payments, 195
Duration
    and convexity, 313
    matching equation, 333

Earned interest, 20–21
Effective annual rate of interest, 22,
    40–41
Effective rate of discount, 49
Effective rate of interest, 22, 25,
    40, 42
    accumulation function, 23
    calculator cautions, 23
    notation, 42
    TVM Keys, 44
Equated time method, 79

Equation of value at any time, 70–72

Equilibrium interest rates, 339–340

Exact asset matching, 289–294, 311
  exercises and answers, 303–309,
      365–367

Exact method, 68

Excel, 6, 75
  annualized amortization tables
      using, 176–178
  Excel projects in bonds, 288
  Excel projects in sinking fund,
      204–205
  financial functions, 9
  sinking fund schedule in,
      183–186

Exponential function, 28

Exponents, 14–16

Exposure, 224

Face amount, 244

Face value, 242

Finance charge, 189–190

Financial Mathematics (FM), 1

Financial transaction, 19, 108

First order Taylor approximation, 13

Fixed term
  annuities-due, 108–114
  annuities-immediate with
      constant payments, 98–108
  level payment annuity, 97

FM, *see* Financial Mathematics

Forward rates method, 125, 271,
      277–278
  alternative notation and
      calculations for, 274–279
  converting from, 273–279
  multi-year, 272–273
  spot rates, 271–272
  from yield curve, 271

Fractional Time Periods, 69–70

Future value (FV), 6, 20, 36, 102, 111

Geometric average, 60

Geometric increase in payments,
      perpetuities with, 143–144

Geometric series, 6–7, 138

Goal Seek, 210–211

Home loans, 191–194

ICONV worksheet, 41, 131, 244

Immunization, 330–334
  formulas, 333–334

Implied forward rates, 275

Inception, 48
  date as comparison data, 71

Increasing annuity of term, 144

Inflation, 347–348

Initial transaction as positive
      number, 84

Integration, 15–16

Interest, 19, 67; *see also*
      Determinants of interest
      rates
  accumulation and amount
      functions, 19–20
  accumulation function, 20–21
  allocation of loan payments
      between principal and,
      161–165
  compound interest, 23–35
  conversion periods, 20, 131
  deposits and withdrawals, 86–92
  discount rates, 48–54
  doubling time, 79
  effective rate of, 22–23
  equations of value at any time,
      70–74
  exercises and answers, 61–65,
      354–357
  exercises and answers, 92–96,
      357–358
  finding rate of, 82–86
  forces of interest and discount,
      54–60
  fractional time periods, 69–70
  measuring time periods, 67–69
  method of equated time, 79
  mixture of, 108
  nominal and effective rates of
      interest, 40–48

portion of payment, 166
present and future value, 36–40
sensitivity, 318
simple interest, 24–26
theory, 1
unknown time, 74
Interest conversion period, 20
Interest forces, 54
varying rates of, 57–60
Interest rate, 342
fluctuation, 318–320
implied by zero coupon bond,
243
Interest rate sensitivity; *see also*
Determinants of interest
rates
convexity, 328–330
duration of portfolio, 326–328
exercises and answers, 335–337,
367–371
full immunization, 334–335
immunization, 330–334
Macaulay Duration, 320–326
measuring sensitivity to interest
rate fluctuation, 318–320
price curve, 313–317
Interest rate swaps, 297–298
deferred, 298–300
market value of, 302–303
varying notional amounts,
300–302
Internal rate of return (IRR), 78, 86,
207–208, 211, 216, 221–225,
246, 260, 270
approximation of, 223–226
dollar-weighted estimate for,
223–226
TI BA II plus cash flow
worksheet, 211–213
time-weighted estimate for,
226–230
Investment rate of return, *see*
Internal rate of return
(IRR)
Investment scheme, 86, 207

Investment year, 233–236
IRR, *see* Internal rate of return

Junk bonds, 344

LIBOR, *see* London Inter Bank
Offered Rate
Linear function, 24, 28
Liquidity preference theory, 343
Loan
amortization of, 156–157
origination fee, 189
payments, 233
problems involving, 196–198
Loan balance, 157, 166
methods for computing, 157–161
prospective method, 157–161
retrospective method, 157–161
Logarithms, 14–16
London Inter Bank Offered Rate
(LIBOR), 274, 294n1

Macaulay Duration, 320–326
duration of amortized loan,
325–326
duration of coupon bond,
323–324
Makeham valuation formula, 248–249
MAPLE software, 9, 10n4, 75
Market value of interest rate swap,
302–303
Maturity date, 241
Maturity value, *see* Redemption
value
Mean Value Theorem, 210
Measurement of interest, 1
Microeconomics, 339
Modified convexity, 329
Modified coupon rate, 246
Modified duration, 318–319
Monthly interest rate, 43

Natural logarithm, *see* Logarithms
NCC, *see* North Central College
Negative values, 53
Net future value (NFV), 92, 207

Net interest in period, 180
Net present value (NPV), 78, 85, 89,
        209–211, 270
New money rate, 233–234
Newton-Raphson method, *see*
        Newton's method
Newton's method, 9–11
NFV, *see* Net future value
Nominal rate of interest, 40–41
        notation, 42
        TVM Keys, 44
Non-integer time periods, 70,
        122–123
Non-standard accumulation
        functions, 32
Non-standard payments,
        amortization with, 186–188
North Central College (NCC), 121
Notional amount, 295
NPV, *see* Net present value

One-year forward rates, 58, 271–272,
        275, 278
Opportunity cost theory, *see*
        Liquidity preference theory
Ordinary simple interest method, 68
Outstanding balance, *see* Loan
        balance
Outstanding principal, *see* Loan
        balance

Par coupon, 276–277
Par value, 242–245
Payment (PMT), 67, 232, 245
        key, 99
        perpetuities with arithmetic
                increase in, 142–143
        perpetuities with geometric
                increase in, 143–144
        varying according to arithmetic
                progression, 139–141
        varying in geometric
                progression, 141–142
Period, *see* Interest conversion
        period
Perpetuity-due, 138

Perpetuity-immediate, 138, 143;
        see also Annuity-immediate
Perpetuity/perpetuities, 6, 97,
        119–122
        with arithmetic increase in
                payments, 142–143
        with geometric increase in
                payments, 143–144
Per quarter compounded quarterly, 39
Physical settlement method, 295
PMT, *see* Payment
Polynomial, 213
Portfolio, 326–328
        instruments of known
                duration, 328
        methods, 233–236
        rate, 234
Premium, 252–260
        bonds, 245–246
        valuation formula, 248–249, 260
Pre-paid swap, 295, 297
Present value (PV), 6, 20, 36, 99,
        102, 106, 109, 247
        of annuities with varying
                interest rates, 126
        of purchase sequence, 296
Price curve, 313–317
Price-plus-accrued, 279
Pricing bond, 247–252
Principal, 19, 108
        allocation of loan payments
                between interest and,
                161–165
        mixture of, 108
        portion of tth payment, 166
Prospective method, 157–161
Purchase
        at discount, 242
        price, 244
Purchased at premium, 242
PV, *see* Present value

Quadratic equations, 230
Quadratic formula, 213
Quoted rates, 340–342

Rate of return (ROR), *see* Internal rate of return (IRR)

Real estate loans, 191–194

Reasonable value, 213

Redemption dates, 242–243

Redemption value, 241, 244, 340

Redington immunization, 331–332

Reinvestment, 218–220, 235
    notation, 218
    single investment with proceeds, 218–219

Retrospective method, 157–161

Rule of 72, 82

2ND Date, 68

2ND SET, 68

Second order Taylor approximation, 13

Segmentation theory, 343

Selling loans, 231–233

Selling price, 242–243

Series
    arithmetic series, 7–9
    geometric series, 6–7
    sequences and, 6

Simple interest, 23, 27, 38, 69, 279–280
    effective rate interest in case of, 25

Sinking fund
    Excel projects, 204–205
    exercises and answers, 198–203, 360–362
    method, 153
    payments, 179–183
    schedule in Excel, 183–186

Six-step program, 193, 199, 205, 362

Spot rates, 125

Spot rates always, 263, 265

Spot rates method, 271–272, 277–278
    alternative notation and calculations for, 274–279
    converting from, 273–279

Stripped bond, 264

Surplus, 331

Swap, 294
    payment, 295
    tenor of, 295
    term of, 295

Swap rates, 294–297
    deferred interest rate swaps, 298–300
    exercises and answers, 303–309, 365–367
    interest rate swaps, 297–298

Tangent line approximation, 315

Taylor polynomials, 317, 328

Taylor series, 11–14

T-bills, 340–342, 344

Tenor of swap, 295

Term of swap, 295

Texas Instrument calculator models, 1–2

TI-30XS, 15, 72–73, 76, 82, 101
    interest, 23, 30, 35, 37–38, 59–60
    MultiView, 2–5, 7

TI BA II Plus, 15, 23, 30–31, 35, 37–40, 43, 46–47, 49, 72, 74–77, 91, 101–104, 113, 116–117, 123, 132, 144
    AMORT worksheet, 242
    calculator, 29
    cash flow worksheet, 211–213, 216, 236
    to creating amortization tables, 170–176
    FV, 102
    keystrokes, 213–214
    notation, 208, 212, 219
    Professional Edition, 2–5
    professional edition provides NFV, 87
    PV, 102
    solution, 29, 34, 41, 48, 56, 59–60, 71, 74, 80–83, 87–88
    TVM keys, 67
    TVM worksheet, 53, 250, 261–262, 266

TI calculator internal calculation, 9

Time lines, 31

Time periods measurement, 67
    exact method, 68
    ordinary simple interest
        method, 68

Time value of money (TVM), 1, 107
    keys, 44, 50, 99, 232, 244–245,
        247, 322, 327

Time-weighted estimate for,
    226–230, 236

Total accumulated value, 218

Truth in Lending, 189–191

TVM, *see* Time value of money

Unpaid balance, *see* Loan balance

Value at any time, equations of, 70

Value of annuity, 269–270

Varying perpetuities, 142
    decreasing annuity of term,
        144–145
    increasing annuity of term, 144
    perpetuities with arithmetic
        increase in payments,
        142–143
    perpetuities with geometric
        increase in payments,
        143–144

Varying rates of interest, 57, 125
    accumulated value of annuities
        under varying interest
        rates, 126–128
    present value of annuities
        with varying interest rates,
        126

Withdrawals, 86–92, 207, 211–212,
    221, 224, 226

Yield curve, 262–263, 265

Yield rates, 246, 260–262, 266,
    271
    discounted cash flow analysis,
        207–213
    exercises and answers, 236–240,
        363
    interest measurement of fund,
        220–230
    investment year and portfolio
        methods, 233–236
    reinvestment, 218–220
    selling loans, 231–233
    uniqueness of, 213–218

Yield to maturity, 246

Zero-coupon bonds, 241, 243, 330

Printed in the United States
by Baker & Taylor Publisher Services